Basic Composition Skills

ERING

S

OLOGISTS

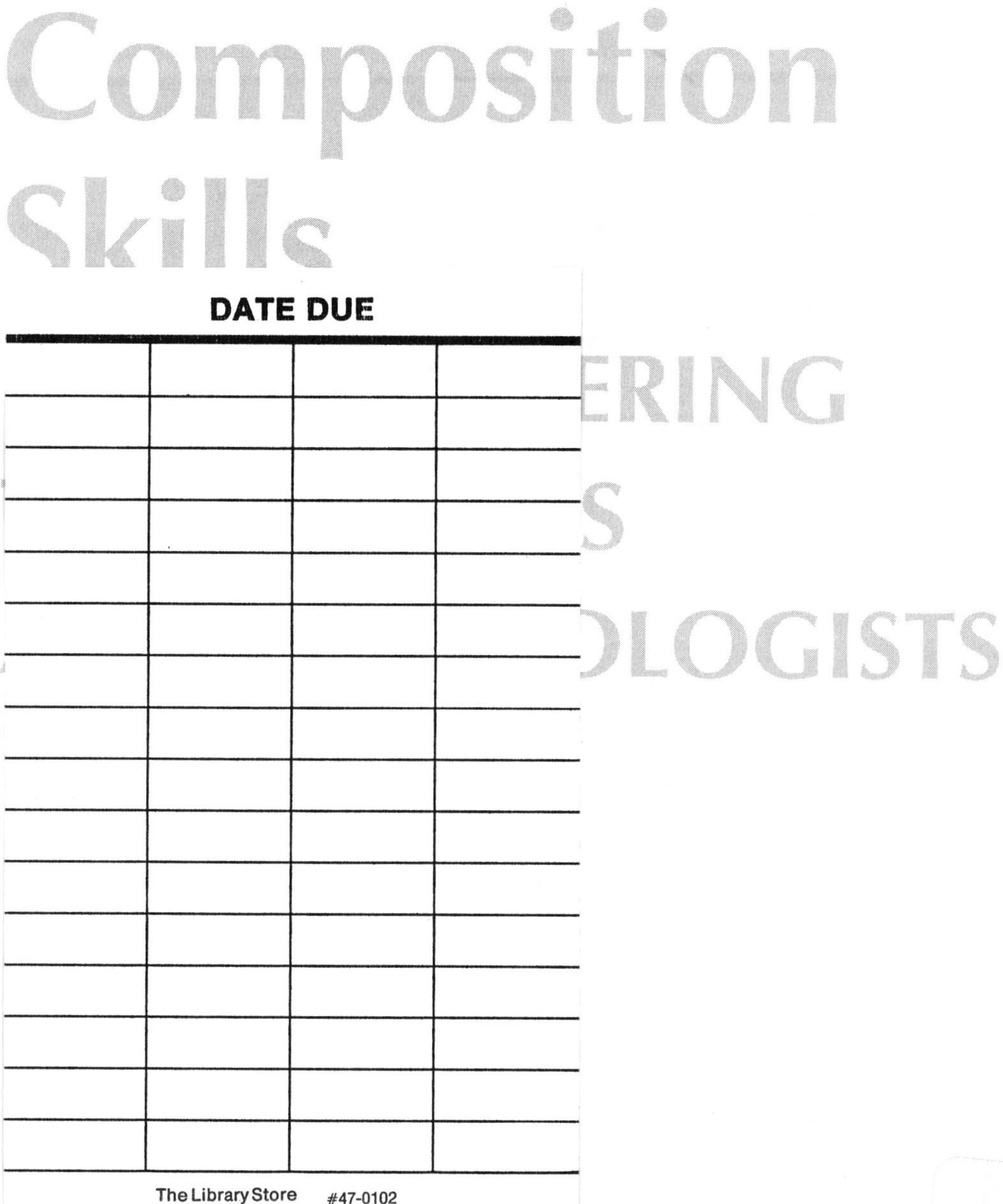

Basic Composition Skills

FOR ENGINEERING TECHNICIANS AND TECHNOLOGISTS

DAVID W. RIGBY

North Seattle Community College

The Wordworks™ Series

Prentice Hall

Upper Saddle River, New Jersey
Columbus, Ohio

Library of Congress Cataloging in Publication Data

Rigby, David W.
Basic composition skills for engineering technicians and technologists / David W. Rigby.
p. cm. -- (Wordworks)
Includes index.
ISBN 0-13-490111-8
1. English language--Technical English. 2. English language--Rhetoric. 3. English language--Grammar. 4. Technical writing. I. Title.

PE1475 .R55 2001
808'.042--dc21

00-034667

Vice President and Publisher: Dave Garza
Editor in Chief: Stephen Helba
Executive Editor: Debbie Yarnell
Associate Editor: Michelle Churma
Production Editor: Louise N. Sette
Production Supervision: Clarinda Publication Services
Design Coordinator: Robin G. Chukes
Text Designer: Ceri Fitzgerald
Cover Designer: Ceri Fitzgerald
Production Manager: Brian Fox
Marketing Manager: Jimmy Stephens

This book was set in Optima by The Clarinda Company. It was printed and bound by Victor Graphics, Inc. The cover was printed by Victor Graphics, Inc.

10 9 8 7 6 5 4 3 2 1
ISBN 0-13-490111-8

Welcome to Wordworks™

Wordworks™ is a series of four communication skills manuals. The manuals consist of three writers' guides for engineering and technical applications and an additional guide to in-service spoken communication. The manuals are designed to provide in-demand information in a readable fashion. They are matter-of-fact and use-oriented.

Each manual focuses on specific exit skills that are necessary for job performance. In this respect the texts are somewhat unique. They were inspired by the carefully tailored goal orientation of corporate seminar manuals. This strategy is at the heart of the streamlined manuals of corporations where specific skill outcomes are the narrow focus of company class time. For skills manuals this strategy can accelerate and focus the learning process, and the approach is particularly useful in college programs where English components are never more than a course or two of the total learning experience.

The *Wordworks™* manuals are conversational, visual, and practical so that the learning experience is accessible. Chapter-by-chapter discussions encourage a learning curve of understanding. Important concepts and practical applications are identified and explored. Models and other illustrations are also features of the texts.

The texts rely on an extensive use of graphics to conceptualize ideas, and models are provided to draw attention to desirable skills. This approach is intended to help students build a strong understanding of logical design features that they can use to construct any writing project.

Three of the texts in the series deal with the basics of the craft of writing, and the series uses a writer-to-writer strategy to explore and explain this craft. The manuals are intended to be learning tools, but because they focus on a craft, they are not intended to be overly academic. To the extent to which streamlining can be achieved in a college environment, the *Wordworks™* series provides a practical approach to skills training that is compatible with the limited time available for communications offerings in technical programs. The manuals encourage curiosity and provide a learning-oriented climate, but they also simplify the path to practical knowledge and skills.

Each title in the *Wordworks™* series is intended to complement the other titles:

Basic Composition Skills for Engineering Technicians and Technologists. This first text of the series is intended to help upgrade fundamental skills in writing. It is a thorough discussion of the problems that are encountered by writers and the solutions to those problems. This book is uniquely designed to build upon existing skills that are part of every work day.

Writer's Handbook for Engineering Technicians and Technologists. The second title in the series is a writer's handbook of the rules and practices of writing. Part style guide, part grammar book, part technical writing reference, the *Writer's Handbook* is designed to be a bridge that can support both *Basic Composition Skills* and *Technical Document Basics.* The *Writer's Handbook* is specifically intended for engineering and technical students.

Technical Document Basics for Engineering Technicians and Technologists. The third title in the *Wordworks*™ series develops a concentrated focus on basic technical writing know-how. The text is designed to identify and explore the documentation standards that are used to develop and produce technical projects. The basic skills are condensed into a short, readable text.

Workplace Communications for Engineering Technicians and Technologists. An additional text that supports the *Wordworks*™ concept with an exploration of spoken communication. Studies reveal that 80% or more of our work-related communcation in trade and industrial settings is handled in conversation. The absence of training tools in this area is an invitation for communication problems if only because spoken messages outnumber our memos by four to one. *Workplace Communications* helps identify and improve in-service communication skills.

About Basic Composition Skills

Basic Composition Skills is designed to be used in an introductory composition class for two-year programs in vocational technical environments, and in engineering technologies in particular. The intent of *Basic Composition Skills* is to discuss effective writing habits, the strategies of efficient and orderly methods of writing, and practical procedures for practical writing. The focus concerns composition methods and work-related writing for occupational applications. The book was primarily designed to address an audience of readers who need more than "vocational English" but who also need a more applications-oriented course than "English 101."

The audience for *Basic Composition Skills* includes a broad spectrum of potential engineering transfer students who may be destined for university programs, and an equally diverse group of engineering technology students who are completing two-year degrees for trade and industry applications. To meet the needs of both groups, *Basic Composition Skills* explores two avenues of writing strategies. The traditional standards of college transfer texts—in particular, the emphasis on formal compositions—are part of the focus of this book. Mutual concern for technical interests *and* the workplace setting are also central to the text.

For pre-engineering students, the focus on composition found in *Basic Composition Skills* should be helpful in a host of academic writing chores. However, many technical students have not completed college prep programs, many attended vocational high schools or military programs, many are older people who are retraining, and some have GEDs. For this audience, the conversational approach of *Basic Composition Skills,* and its practical focus on writing procedures, should be particularly helpful. The business communication standards that are explored in the later chapters of *Basic Composition Skills* will serve the purposes of all readers.

The text is divided into three sections: the creative process, the writing activity, and the written product. The text develops a detailed account of the writing process, a subject often addressed in college transfer composition texts. This important element of the writer's craft is also much needed as a skill for today's engineering technicians. The text is a balance between theory and practice, and it attempts to keep the workplace setting in focus through applications-oriented discussions of the thought process and the writing process. Grammar is not included in the text, which assumes that instructors will use either the *Writer's Handbook* (another title in the *Wordworks*™ series) or a conventional grammar text.

The writing samples that appear in *Basic Composition Skills* are student products. Many of the samples were designed for both college settings and industrial applications in that the documents were college projects that were concurrently developed for a number of companies. This bridge between the needs of engineering technical students at college and at work is the unique focus of the *Workworks* series, and it is important to place student writing in *both* academic and occupational contexts.

Drawing on study skills, standard writing practices, the practices of students and professionals, psychological considerations, and the like, *Basic Composition Skills* encourages positive writing skills. Written in a series of clear, uncomplicated discussions, *Basic Composition Skills* should be easy and interesting reading that will help engineering and engineering technology students build practical skills for college and for work.

NOTE: Readers of this text do not have to be familiar with computers, even though the text frequently comments on the use of computer technology. The text assumes that most college-level readers have some word processing knowledge. Students who have no computer skills, cannot afford the equipment, or do not have access to a computer lab are still welcome to learn about the craft of writing.

This book is designed as a stand-alone text, but readers are encouraged to also use a desktop reference guide to rules for writers, preferably the *Writer's Handbook for Engineering Technicians and Technologists* or a similar guide intended for technicians and engineers.

Contents

Basic Composition Skills

FOR ENGINEERING TECHNICIANS AND TECHNOLOGISTS

Work In Progress

The Public World

At the beginning of each of the following chapters you will have the opportunity to read a page from the working notes of a writer in the workplace. Like yourself, the author is an engineering technician. This recent graduate has an associate's degree and works for a local company. She will take you through the paces of a typical writing project that evolved out of work-related demands within her company.

Hello, my name is Laura. I work for Pacific Aero Tech, Inc., a small company based in Kent, Washington, that specializes in repairs, overhauls and rebuilds of aircraft windows and avionic equipment. I have worked at Pacific Aero Tech for four years while I worked on my degree at a local community college. I have now completed my degree in electronics technology. At work, I was recently assigned to research, document and implement a work safety program for Pacific Aero Tech (PAT). Upon accepting this project I encountered interesting challenges that I would like to share with you.

Pacific Aero Tech is a small company. Although a young business, after only seven years it has two partners, one president, two working supervisors, one accountant, one receptionist, and thirteen specialized, technical employees trained in their specific craft. PAT has established a strong client-base for the airline repair industry and supports the aviation window and avionics industries locally, throughout the nation, and internationally.

Precisely because of the growth of the business, the work started to back up, so Karen Stevenson, the President, had to hire more people. The company size doubled in two years, which increased the number of employees. The president was aware that the Occupational Safety and Health Administration (OSHA) requires added regulatory compliance with certain safety guidelines when a company has more than eleven employees.

My job is a technical one that involves a number of skills. I am an avionics technician. My job includes testing, repairing and overhauling specific avionics equipment for various airline customers. But as companies grow, employees find themselves involved in new and sometimes unexpected areas of development. Recently my job description was extended to include research and development of a written safety program especially adapted to the needs and goals of PAT. This task was a direct result of the success of Pacific Aero Tech. This assignment presented some unique challenges for me.

So, with the newly acquired title of Safety Officer and a newly acquired role as committee leader, I marched straight into the project. I realized that sorting and sifting through the mire of safety information could quickly overwhelm me, which could slow down the program's developmental process. I had to organize a plan.

The next chapters will discuss my "plan" and the challenges that I faced in order to write and implement this program.

L.C.L.

Writing at Work

In corporate or business environments, many companies have staff writers to handle the tasks involved in creating reports, guides, manuals, data sheets, and an endless variety of documents that must be generated with professional skill. These professional writers are a very small part of the workforce of a company. Considerable paperwork is generated as a routine activity by the rest of the employees—and you may soon be counted among them. Writing is, to a lesser or greater extent, part of the workday for millions of employees of companies large and small.

As an employee of a company, you will be part of a team, whether you are in a lab or office or in the field. Your professional skills will function as a type of technological support for a company's services. Your daily activities will focus on the application of your technical knowledge to the tasks involved in your job.

Documentation will be part of your workday because you will be expected to provide written records or explanations of your technical activities. The demand for writing can vary from infrequent status reports to daily correspondence by e-mail.

Writing at work may not be an easy undertaking, especially if you do not write very much or very often. If you are a data processor, a civil engineer, a digital electronics technician, a mechanical engineer, or an aviation technician, you are not a writer, except incidentally. You will, however, share in generating the paperwork to some degree. If you find you are an employee who must write from time to time, you are what I call a "working writer."

In engineering and industrial settings, most technical writing must be done in the field by people who know what they are talking about. You can see this distinction between yourself and someone in the front office who may be thoroughly trained for writing. You know *what* has to be said. The office staff or the technical writing department may know *how* to present attractive-looking results, but the raw material is in your hands. Every item that you write will in some way contribute to company output because all the technical documentation is part of the record of the company's progress. The elegant brochures, the manuals on glossy paper, and the legal proceedings submitted to the patent office are all end results of the routine writing of employees.

There is a catch, though. Writing is often a barrier for working writers who must do occasional writing. This book was written to help you in this task. The text focuses on writing strategies for work, and it assumes that you have done only a small amount of writing.

Work demands speed and efficiency from employees. Writers in the workplace are often frustrated in their writing attempts because they demand speed and efficiency from their writing as well, often without training. This text will show you how to generate writing with speed *and* efficiency. There is a good chance that any misgivings you may have about writing are related to the time it takes and the quality of the outcomes. Perhaps you can change all that by learning about some paths of least resistance that you can take.

I would like to be able to simply jump into the tactics of writing that will improve your productivity, but I must define the problems before I define the solutions. You need to see the causes of the problems so that the usefulness of the strategies make sense. In addition, if you understand the causes of the challenges you face when writing, you can also seek out your own solutions.

You know how to speak English better than anyone beyond our borders. You could even teach the language if you had to. Writing the language seems to be a different matter. This confusing situation is one reason why you want to look carefully at the uniqueness of writing. Apart from having what seem to be *language* problems (grammar, spelling, and so on), many writers seem to have trouble getting their *thinking* on paper. Somehow the thoughts don't show up exactly as authors imagined them: parts of the reports are vague; other sections ramble; chunks of the writing might not always be coherent.

Your background is also an important consideration. It might have included good courses at excellent schools, or your training might have been minimal if you were a high school dropout or spent years with the armed services.

In sum, all writers have occasional difficulty with some aspect of writing, whether it be language fundamentals, organization of ideas, or skill levels. Most working writers could use a little encouragement.

Let me give you two incentives before I go on. First, you no doubt can speak energetically and with confidence. Remember that writing is related to speaking, and *speaking* can play a major role in work-related writing. Turn this skill to your advantage as often as you can. Second, you probably desire to be successful at what you do. Remember that in business, as responsibility increases, your need to produce written documents is likely to increase also. Increased responsibility usually goes hand-in-hand with increased income and status. Thus, writing skills have a bearing on your potential to succeed.

Look carefully, however, and you will see that you have probably had very little coaching for writing, and you have had very little practice. The basic problem is that part-time skills rarely get full-time attention. Many people simply lack experience. If you are out of practice or in need of basic writing skills, you need both coaching and practice. You also need to learn the tricks of the trade to become a successful writer.

> *Ten years ago I completed my Associate's Degree in electronics at a community college. A company immediately hired me and I have worked there ever since. I worked there as a technician for a few years before transferring to management. At the same time, e-mail changed the way companies are run. I now spend half my workday writing.*
>
> Bruce Fugere
> Repair Shop Supervisor
> Air Transport Systems
> Honeywell International

> *I've been employed by a major university for the past three years. My job is to manage the computing resources of a nonacademic department. This includes system and database administration and programming, as well as overseeing the support of desktop computing systems. I can't stress enough the value of being able to communicate effectively with the written word. Just about every task I perform, whether it's developing proposals, commenting on code, documenting procedures or collaborating via e-mail, involves writing.*
>
> Glenn Sudduth
> Technical Support Specialist

Disuse: Statistical Realities and the Practical Facts

Practice makes perfect. If writing is only an occasional activity for you as you go about your tasks at work, then writing will be a challenge because you are not using the craft enough to cultivate writing as a primary skill. If your reports seem a little ho-hum or if they seem to be created by hit-or-miss tactics, it is because you are using a secondary skill to report the findings of your primary skills in engineering or engineering technology.

In industry, most employees do not write much or often. For example, a study commissioned in the state of Washington found that employees in trade and industrial occupations spend only 20% of their communication activity using any written format. This is one reason students study writing in schools. Students seldom study "speaking" in schools apart from a public speaking course. People are masters at conversation long before the first day of their first-grade year. The skill of speaking is pervasive and more useful than writing in everyday life.

Most people can safely guess that they communicate by speaking about 99% of the time. For working writers this is a problem, particularly since the remaining 1% is likely to be important work-related writing projects. If you view your writing time as a percentage of your eight-hour workday, you will see that the percentage increases just enough to be a responsibility, and that responsibility can be a challenge because you may not write often enough to feel confident with the task.

For some writers, developing a written document is slow going because they are concerned about getting over the first hurdle—the rough draft. A common misperception that many employees have about writing is that the draft is the beginning of a writing project, which it is not. On the contrary, the rough draft is the final product of a long series of events that lead up to and generate the rough draft. It can even be argued that a draft is the *end* of most of the really hard work of writing, which is *thinking*.

Many people who write with skill have a tendency to place a great deal of emphasis on the chore of turning the first rough sketch of a project into a jewel of a finished product. Write the document. Correct it. Write it again. Correct it again—and again. So much effort and attention are devoted to this level of a writing project that it may seem logical to see the first crude but fairly complete sketch—the rough draft—as the place where it all began. In fact, the completion of the rough draft is a victorious moment regardless of what shape it is in. The first draft is the outcome of what an author has been thinking. It is an end product. Besides, a writing project is an *event* as well as an outcome. If you see writing as an activity or an *event* you will have a very different approach to the task. Writing is not a text with errors to correct. Writing is the *effort,* the time it takes to produce a document and not the document itself. The plan of action and the activities of the job of writing are important features that many writers overlook.

You actually achieve much more than you imagine when you produce a rough draft. The chances are that you have done little or no preparation to write the draft, and so it is a compliment to your skills that you create any kind of draft at all, since your way of going about it may be *the* most difficult way to get the job done. To explain these two observations I will take a close look at the tasks that come long before a writer sits down to write. Then I will identify a strategy for *quickly* producing a useful rough draft.

The Tasks of the Text

Workplace writing evolves out of your job activities; it is writing based on a certain situation or set of circumstances. You may be filing a report on a T-bar stress analysis, or you may be examining the costs of adjusting a shop floor plan to meet federal safety regulations. These are what I call "situation-based writing projects," the usual daily fare of working writers. You may need to write lab reports at work or submit estimates as a contractor or develop proposals for a firm or produce projects in your college program. A *fast* approach to getting the job done will make the *best* use of your time and will be the most appropriate way to deal with writing. Is there a system for writing that will be fast and still produce quality results? There is if you first emphasize the tasks that precede the rough draft, tasks that most working people may overlook.

There is much to be done before you start to write the draft of a project, even if the project is fairly short. Whether the project is typed out as a two-page brief or generated as a ten-page report, there is a series of stages you can go through to improve the quality of your document and to speed up your writing process.

First, for the working writer who may have some difficulty with writing, a focus on these preliminary tasks has one particular merit: you can master these activities without ever writing a sentence (almost)! Everything that comes *before* the rough draft—what is called "prewriting" or "prep"—can be handled without any particular background in writing. This means that you are about to embark on a strategy of writing that places few demands on your writing skills.

To repeat, one sure way to improve your writing is to apply yourself skillfully to the number of prewriting or prep tasks for which writing skills are not critical. This preparation, however, will result in improved writing. Parts I and II of *Basic Composition Skills* focus on these preparation and prewriting strategies.

In addition to preparation tasks, you must also seek out a method for roughing out the first draft but one that will generate the *maximum* amount of material in the *least* amount of time. Proper prep will assure that the material produced is appropriate to the project. Once the pencil is in your hand, or once the word processor is awaiting your commands, you need additional methods for drafting your project as rapidly as possible. Speed allows you to keep up with your thinking; it also allows you to meet your deadlines, and it makes writing more manageable as a job-related activity that you may otherwise find time-consuming. Part II of *Basic Composition Skills* focuses on speed and efficiency.

As you will see, however, methods for high-speed production encourage a rough product that may have errors in spelling, grammar, punctuation, and other related considerations. Although I can't go so far as to say that these errors won't be important, rapid production is far more important than correct mechanics. Traditional issues such as grammar rules and spelling simply won't matter at this early stage of the game.

To repeat, a sure way to improve your writing skills is to learn the tricks that will help you to quickly generate the material of the rough drafts. If you can learn to generate a rough draft quickly, you will later have the time to devote to polishing the product so that it conforms to the expectations of your business setting.

The efforts you make to save time will provide you with the time to edit the project properly, which, for the working writer, means making sure the end result is logical and clear and free of errors. It all matters. If the document looks good, you look good.

Part III of this text concerns the structural necessities of writing projects. There are logical architectural principles that can be built into any project that you develop, whether the product is a brief memo or a long investigative narrative based on laboratory findings. Several chapters will survey a number of these prototypes and their simple architectures. Further, inside these structures are the logic infrastructures, the paragraphs. The text will explore these fundamentals also, focusing on engineering and technical uses in particular.

Part IV of *Basic Composition Skills* is a survey of basic designs or structures for a variety of workplace writing tasks. These chapters explore the mediums you most use in the workplace: memoranda, business letters, laboratory reports, bids, and proposals. Each application is examined with reference to engineering technologies.

To help you learn how to prepare for a writing project, you could simply jump ahead to Chapter 5 and apply proven methods of developing the groundwork for a writing task, but first, it is important for you to understand *why* you should proceed in a certain manner. Later chapters will tell you *what* to do to quickly produce a rough draft, but you may not be inclined to perform a task in a certain way unless you know *why.*

The reasons for establishing a set procedure for writing will give you motives for going about the task correctly. As a result, the two brief chapters of Part I discuss three activities related to a writer's tasks: thinking, speaking, and writing. If you come to understand something of each of these activities, you will clearly see practical methods for improving your writing. If you are writing at work, remember that your company's library (if there is one) will be modest, so you must *think* your own way through your projects. Then you must *speak* your way through the projects in discussions with coworkers. Then you must *write* them.

The result of the discussions in the next chapters should help you gain confidence about writing, about yourself as a working writer, and about how to handle writing. In later chapters I will walk through a project so that you can learn how to proceed with your writing task in a way that will build on efficiency in thinking and speaking. You will see that all the techniques for developing a completed writing project are based on our discussions about thinking, speaking, and writing.

NOTE: You will occasionally see a disk icon and a text box containing basic tips. The directions concern Microsoft applications that usually work on both IBM and MAC platforms.

Activities Chapter 1

Select one of the following suggestions and develop a brief document to explain your experience.

1.
- *Develop a brief account of your employment history. Use full paragraphs to describe each job of significance.*
- *Explain your primary responsibilities.*
- *Identify any writing responsibilities that were part of a job.*

2.
- *Develop a brief account of your educational experiences concerning English.*
- *Explain what kinds of projects you learned to write.*
- *Discuss your strengths and weaknesses in writing.*
- *Do you anticipate having to do much writing in your technical specialty once you are in the workplace?*

These projects should be 250 to 500 words in length, typed in a memo format (see Chapter 11), and addressed to your instructor. Your instructor may request a memo in response to each of the suggestions.

Words at Work

Work in Progress

Task Definition—Getting Started

Facilitating the safety compliance process, the president, Karen Stevenson, arranged a meeting between herself, the two working supervisors, and me to discuss the extent to which this required program must be written and by what means. At this meeting we developed a loose timeline for when the program should be implemented. It was also mentioned that certain programs, such as hearing conservation and respiratory training might or might not be needed. The following was determined: I was to become the OSHA expert for the company, and using my new-found knowledge (from reading and research), I was to evaluate the company as a whole and write a proposal to be submitted to the committee that would then be read by the management team for consideration for possible adoption.

The president sent me to an OSHA certification class to determine what rules Pacific Aero Tech should follow to remain OSHA compliant. The class was one week long and extensive and often tremendously tedious. It gave a basic overview of all basic elements of OSHA's safety and health programs by utilizing all forms of visual aids. Two reference books were provided to supplement the training. The goal of this federal agency is to provide safety officers with the knowledge to know where to start and the basic material for planning a compliance program. There was a lot to learn. Considerable training must be undertaken prior to embarking on such an extensive journey. I knew that it was imperative to know the general facts first and then, if necessary, to study related state laws to determine how to apply the regulations to the needs and structure of a small company such as PAT.

While OSHA is a national safety program required of all employers, each individual state has its own, additional laws governing the health and welfare of the state's employees. Washington State has the Washington Industry Safety and Health Administration (WISHA), which works in conjunction with the Department of Labor and Industries to assure compliance with state and OSHA regulations. Usually, the state laws are created to meet or exceed OSHA standards. Being more stringent, WISHA standards are used by most companies as the regulatory guideline for safety and health among employees.

Of course, WISHA guidelines include the OSHA regulations. Sound confusing? I thought so too. OSHA has a written form of its guidelines, and the form has been updated annually in the past, but with the emergence of the Internet, OSHA and WISHA information has become easily accessible through the average business computer. So, armed with my certification books and the information super highway, I started my basic search. It was established at the safety meeting that hearing conservation, respiratory protection, and fire safety were the most important programs concerning Pacific Aero Tech, so there is where my research began. I needed to know the standards prior to evaluating PAT's work environment. I also needed to know our workplace status in order to determine the extent of my written program.

L.C.L.

The Way We Think

Many people make the intensely frustrated observation "I can think it, but I can't write it." I would like to suggest that this might just be an illusion. Thinking might seem to be clear and simple and obvious, if writers could only get it down on paper, but this is usually not the truth of the matter. Their efforts to write—in words—will be based on what *seemed* to be clear thinking, but if the thinking does not show up on paper, one reason might be that the thoughts were not created or experienced in neatly organized words or sentences—even though it all "looked" like terrific material before they begin to write.

Obviously, a great deal of thinking is conducted without much effort to think out processes with words. Any high-speed sport suggests that people don't think in words the same way they speak or write in words. The downhill skier and the Ping-Pong player must think with amazing speed, and words are only going

to hinder success. Specialists argue that slow readers are slow because they verbalize each word they read by moving muscles in the throat. The muscle movements slow the readers down. In contrast, high-speed readers demonstrate that the activity of "wording" the thought process of reading is not necessary.

If you do not make much obvious effort to think in words, you certainly do not spend much time thinking in sentences. The sentence is your primary building block for creating logical, orderly patterns of words to construct clear thoughts for other people to understand. The structure of a sentence forces your thinking into tight organization, and if you realize that you do not think in complete sentences much of the time, you can see that you might have difficulty when you try to speak or write ideas in words that are, in turn, grouped into sentences.

Half sentences or fragments of sentences are certainly much more common in thinking than complete sentences. Thought is not as coherent as everyone thinks it is. Thinking *seems* to be based on words and sentences and *seems* organized and logical—but it is not. Thought, as a matter of fact, can be surprisingly unrelated to writing, in particular.

As a result, the sentence, which is a speaking and writing tool, is a very handy test of the coherence of thinking, in the same sense that an equation is a test of the coherence of mathematical thinking. When was the last time your bank was wrong and your checkbook was right? I would imagine that you lose the argument every time. I do. What you "think" about the account conflicts with reality. Reality is the final test, and our accounts don't always balance.

It is important to realize that thoughts are prone to error. Partly, the problem is the speed at which thoughts are produced. Thought is capable of responding to immediate needs, but without generating the models needed to evolve perfected results. Resounding failures can result.

The speed of the thought process has probably helped humans survive as a species, but that speed generates a high probability of error also. Even labored thoughts are often simply wrong. The result is evident in the U.S. Patent Office, where thousands of goofy devices have never been put to use; in pet projects that become a great source of amusement to family and friends; and in any effort to speak out or write up ideas that does not succeed. Projects crash, and they crash often. Thinking is fast. Perfecting is slow—very slow.

Thinking is a marvel, but it can let you down. Thought imagines solutions to problems. When engineers act on their thinking and build their ideas in the world around them, some are successes, but, alas, some are failures. At the turn of the century, for example, many an early aviator met with calamity in machines that proved to be very dangerous. Exploring thoughts can be risky. Words invent, and efforts to build with them may be brilliant at one moment and hot air the next.

In summary, thinking builds your world by inventing and constructing whatever you need (to assure well-being and security and so on), but ideas do not always work. Admitting the cleverness of thought, you must be aware of the flaw.

Aeronautical engineering is an engineering field in which design problems are commonplace and failures are well documented for posterity. Because aviation is a product of recent times, photographers captured many delightful moments in the early years when inventors, and experimenters—and a few explorers—sought fame and fortune in the hopes of being the first to fly. There are many amusing engineering examples of design and testing—the expression text and the reality check.

Octave Chanute's "Katydid" glider is perched on the sand dunes of Lake Michigan. Chanute was a civil engineer by trade, but he published aeronautical engineering articles in the 1890s and experimented with gliders. (Photograph courtesy of Heritage Center, University of Wyoming)

Reality Checks

Because thinking is rapid and fragmented, things can go wrong when you try to convert thoughts into logical and orderly systems such as language and math. Two problems are usually obvious: (1) You may not be able to express a certain sentence or equation—you "can't get it down on paper." (2) Even if you can think clearly enough to create a logical and orderly sentence or equation, it *still* may be inadequate because it may be *false*. In other words, even after all the work, you may simply be wrong anyway.

Everyone tests their ideas constantly. Being able to properly state a thought is one test. Once it is stated, on paper, for example, it is out there in the world and is usually either going to be right or wrong. Reality, then, becomes the second test of ideas. In sum, these two challenges, the *expression test* and the *reality check,* can show you whether you are thinking with organization and accuracy. In fact, if you cannot produce the expression of an idea, then you cannot check the accuracy of it.

Word constructions may not be the clearest example for engineers and technicians to consider. Consider the ceiling over your head. The room you are sitting in has endured a specific test of reality in converting thoughts into mass. The angle of the wall in relationship to the ceiling is a perfect 90°. And the relationship between the wall and the floor is also exactly 90°. The right-angle post and beam construction passed the reality test thousands of years ago. Thus, any 90° corner of your room represents a very old concept; it is an idea that has assumed reality. The energy of thought has been applied to mass, and the reality test has checked out okay. It works. Of course, many other ideas are destined for failure, as you will note among the wild flying machines that are illustrated here.

Putting thoughts into words, as in speaking, is always a test. A great deal of thought has been given mass right here on this page, although the tool of language is more difficult to examine. In as many ways as I choose to combine twenty-six symbols, you must scan each cluster of them with your eyes, recombine them as a set of symbols that refer to some larger symbols English speakers have been taught to share—S-H-A-R-E = SHARE—and then you must recall the meanings or things to which the words refer. These units, in turn, are combined into sentence units, and your mind must race through endless functions to pull them all together. Then, you test the material. If you understand me, I have passed the expression test. If perhaps you agree with me, I have passed the reality check according to what I set out to accomplish. My wall will stand strong.

The primitive obsidian stone that has been chipped to razor sharpness for use as a primitive tool, the corner of the room, and the words on this page are all thoughts that you can see. These are thoughts that are solidified, and fairly permanent. First, you must express thought in the world outside your mind if you are to see whether an idea is clear. Then, if it is clear and complete, you must allow it to stand, or fall, as a result of giving it reality.

Obviously, then, you cannot sit down to a piece of paper and trust that what appears to be the crystal clarity of mind is easily rendered *or* very real. The fellow who mentally pictures a garage he would like to build may think that the project is complete and organized in his mind, but he will run into many difficulties when he begins the project without drawings. You have seen the result.

The test of words can succeed or fail by degrees, but as you have seen, two basic judgments can be made about any effort you make to speak or write what you are thinking. First, if you cannot get the ideas out in words or down on paper, then your thinking probably fooled you. It was not clear or complete. Second, if you do get your ideas transferred into words but you prove to be wrong anyway, your thinking was either not complete or was, in reality, incorrect. *Expression* is necessary for a reality test in the world of ideas. Your ideas either line up and defend themselves or else they demonstrate errors. They can, of course, be corrected! There is nothing about being wrong that you should see negatively. You are just as likely to be correct in your thinking. The world of ideas is a world of internal realities, and when that world confronts the external world there are bound to be disjunctions.

For most working writers who are professionals in their fields, being wrong is probably a less likely problem than expression. The problem is sitting down to write and not being able to write what seemed clear to you. You are trying to build the garage without plans. It is risky, and certainly it will be slow.

All writing pits your wits against reality. Writing is simply a way to get ideas out into the world to see if they will hold up. In this respect, writing can be a challenge for everyone. Working with words is an exciting way to explore ideas. From my perspective, writing shares a fundamental trait with engineering: both activities build models. An author creates models of ideas with words just as engineers create models of ideas with drawings or constructions. The models—be they words or flying machines—may or may not work. The accompanying illustrations represent a pursuit that writing and engineering share: the perfection of ideas. Remember that the error factor is very high, and these flops are certainly not specific to writing.

The 1908 multiplane designed by the French Marquis d'Equevilly. It never left the ground. (Photograph courtesy of Hulton Getty/Liaison Agency)

Getting a Handle on It All

Whether it is a report on proton magnetometry, a repair manual for a '55 Chevy, a letter from your cousin, or the shopping list posted on your refrigerator, the mission of each document is the same. Writing is above all the effort to record and communicate thought. It is thinking on paper.

If writing is thinking on paper, then why isn't writing easy to do? There appear to be two basic difficulties. One is the complexity of the way people think, and the other is the inability to focus that process for any length of time, in order to write a paper, for example.

You need to understand one basic notion from psychology, which is that thought is your total history, and that total history is what you work with when you think and speak and write. Your experience, which is ever growing, is your working material for thought. Understanding this fact, you need to examine the basic mechanical functions and habits of the thought process that may influence writing skills.

When you sit down to write, it is hard to focus yourself temporarily and maintain a "one-track mind." Sitting down with you is your entire world of active thoughts in process—words, half sentences, a maze of somewhat organized logical patterns and feelings, your responses to your senses, and your memory of everything. All these elements are intermixed, criss-crossing one another, coming and going. (Take heart; everyone is in the same situation.)

If thought is erratic, unstructured, fragmented, nonverbal, and vague, you can see that thinking is not easy to transfer into writing, which should have none of these characteristics—certainly not in engineering and technological fields. If you have an insight of some sort, what seems like a set of marvelous ideas will need expression and some reality checking. You must build the ideas outside yourself with lumber, with chips, with reagents, with calculus, with words. Thoughts involve complicated systems that hardly speak in words, much less in concrete or steel or integrated circuits. The effort to externalize ideas is a demanding task, since the mind's processing systems are enormously complex.

A working writer should not simply sit in front of the keyboard and expect miracles. It is popular to think of the human mind as a magnificent machine akin to an old timepiece with polished brass wheels and remarkable, logical inner workings that systematically click to the rhythms of well-organized ideas in an orderly fashion.

When you sit down to write and command your mind to produce, you are largely perceiving your mind in these simple mechanical terms. This is a classic mistake for the working writer. Yes, there may be similarities between a clock or a computer and the mind, but the mind is infinitely more complex and obscure. The mind created the clock and the computer, but they are not models of the mind. You must clearly understand that the mind you ask to write your project is amazingly creative; it is not a mechanical assembly of gears and springs or chips and bytes.

Quality ideas will not automatically take shape when you sit down and tell yourself to write a project—perhaps an engineering report—preferably in the shortest length of time. If you believe that your thoughts have failed you because you cannot express them, you have misunderstood the relationship between *thinking* out your technical interests and *writing* them out.

The mind can be stubborn and self-serving. When you want to focus intensely, it may be difficult because of such mental activities as your emotions. Emotions can override logic processes. When someone goes through intense emotional experiences, it is very difficult to lock in on job performance and do well. Emotions force people astray and interfere all day long like static on a radio. Any type of emotion can easily distract your thoughts, and writing is an easy victim of any form of distraction.

You must make an effort to *focus* on your work. During your workday, the actual length of time you can maintain your attention at a high performance level is largely a matter of how much pleasure your work gives you. Obviously, if you want to enjoy your work, you have to seek out a field you find interesting, but that may not be all there is to your ability to generate a top performance. Elements of an otherwise exciting job may slow you down, and one of the problem areas can be writing. The working writer may be a top-performing employee who cannot focus on writing demands. If the employee has writing problems, he or she will not focus attention on writing chores with much vigor. The maze of the mind becomes too great a challenge.

Yet another pesky little problem is the tendency for thinking to drift. People have an amazing skill for using their mental resources for purposes other than the task at hand. In particular, daydreaming, a great creative tool that is at once the measure of your hopes and dreams, is also a clever little escape artist. Don't overlook your uncanny ability to stray.

Good sense should tell every writer that there is little point in writing beyond the moment of distraction, but most working writers will *force* themselves to sit and produce. Of course, they don't always produce a lot of quality writing, but they are trying. I suggest that you try to write up to the end of your attention span, and then work on increasing your attention span.

As you progress try to increase your writing attention span from five or twenty minutes to thirty or sixty minutes or possibly hours. Time management is a key element in writing. Since you are your own boss while working on a document, only you can judge how you manage time. Avoid wasted writing sessions. False starts waste quality time. At the college where I am based, I see the problem frequently. Students will sit in the library, dutifully open their books, and that's all. It looks like serious business, but, alas, little gets done. They have produced only an "image" of work. This is a very serious problem because it is a self-deception. If the students do not call their minds to attention, there will be no useful results.

One way to increase your attention time is to increase your skill in writing. This book is intended to get you to the point of maximum efficiency by helping you to focus attention on the engineering or technical *subject* of the writing project (which is the actual focus of your interest and pleasure). You also want to increase your speed at producing a written project. Interest and speed will lead you to another goal: increased quality in the finished product!

The French Vuitton-Huber helicopter dates from 1909. It never left the shop room floor. (Photograph courtesy of Roger Violett)

The Way We Speak

The act of speaking is a projection of thought. It is, therefore, an expression of ourselves. Language of one sort or another is necessary if people are going to share their thoughts. Speaking allows thought to take on a kind of reality. A speaker takes the unseen world of his or her mind and condenses it into structure in language. Speaking takes on dimensionality. It becomes the energy of sound, and you can hear it. Sound gives thinking a firm and definite reality.*

Because language tends to force the random fragments of thinking into more rigidly defined patterns—into vocabulary and into sentences—the result is that you tend to improve your thinking by talking. You can communicate with someone under only one condition: you both must agree on the symbols you use and the order in which you use them. Put another way, the great difference between thinking and speaking is that you now have an audience, and this means you have a responsibility to be orderly and clear. The erratic, unstructured, and fragmented nature of thought needs to take on order and clarity.

Spoken communication brings your thinking into an external world. Speaking out, then, becomes a test of communication. To the extent to which people share language standards, they will understand each other. You must now *precisely* verbalize the thoughts into *precise* structures—structures that make sense to other people.

When you speak, you shape or control your thoughts, although you don't always speak what you think. You carefully shape messages for several reasons. One reason you control your comments is because they are usually contributions to a conversation and you want to shape a logical discussion out of the *interaction*. For example, I don't interrupt a conversation to say I have an itch. I *do* think about the itch, perhaps even to the point of distraction, at which time it becomes a greater focus of my attention than the conversation. But I maintain the *image* of being attentive and I try to share the logical pattern of conversation that develops between myself and those around me. I abide by certain rules of order in a conversation. Conversations have a certain neatness and orderliness that probably doesn't much reflect the real, or at least the total, thoughts of the individual speakers in a conversation.

When I speak, I also begin to build an image of myself for others. Speakers are quite like chameleons in this activity. I will project myself in one way to my family, another way to my coworkers, another way to my boss, another way to my old friends. People vary their speech to selectively project an image of themselves, yet the adaptation is also motivated by the desire to communicate. Like everyone else, I want to meet people on terms I think they will understand. If they swear, I might also. Speakers suit words to circumstances. They *adapt* their communication so that they can communicate more successfully. They

** Voice is sound and sound is energy, but unlike light, which can traverse a vacuum, sound depends on mass to be conveyed. In this sense your voice has a mass that you can share with others.*

usually will use a vocabulary that will allow people to meet them on common ground. The topics of conversation will also be those they can share.

To the extent to which speaking organizes thought, the medium is a great assist to your everyday efforts to build on your ideas. By the same token, speaking may also help you translate your ideas into writing by helping you create a preliminary organization out of the jumble of your genius. If you can speak an idea, you are halfway to being able to write an idea. In addition, since you have a gift for adapting yourself to communication needs, you should be able to extend that communication skill to writing also. You should, for example, be able to develop a sensitivity for what a reader needs from you. You should be able to judge what data, what terminology, or what organization a reader needs. You carefully gauge these matters in conversation, so you should be able to apply the same skill to writing when you try to understand your reader's needs.

Speaking Out

There are a number of communication systems at work in a conversation, and the language of words is only one of them. A conversation is constructed with words, but there is another conversation going on in the *way* people speak and act out comments.

Tones speak. Obviously, you can tone your way along with great ease. Words such as "*Hey,*" "*Yeah,*" and "*Ha ha ha*" can be meant in many different ways—that is, until they are spoken with a particular tone; then one specific meaning is clear: "Hey, that was a great catch!" "Hey, get off my foot!" "Hey, calm down. I'm sure he didn't mean it that way."

For you, as a working writer, this ability to endow words with endless meanings should hint at a major problem authors have in writing. Writing, even among the skilled, is fairly dreary stuff compared with the drama and direct impact of the spoken word. Writing has little of the pizzazz of the language everyone hurls into the air with infinite tone and pitch variations. In fact, there are only three major ways to tone written words.

You can capitalize, you can highlight (underline, *italicize,* and **boldface**), you can add an exclamation point. Contrast these modest variations with your spoken language, and you begin to see an enormous writer's problem. Yes, your vocabulary is quite limited when you write, but for a particular reason: writing is rather like speaking without a voice.

Pace adds meaning to your speech patterns, as well. Consider the simple device of pausing. At work your supervisor asks you to write an unusual report, and you are willing to do the job but you don't know whether you can handle it. You might quickly say, "Okay." Or you might decide to communicate your insecurity and place a long pause in front of a slowly stated, ". . . okay." Obviously, in the second case you are voicing (and partly nonvoicing) some caution about taking on the job. The supervisor will understand the meaning of the pause and assure you that you have the skills to perform the task. Ob-

The art of writing is quite similar to engineering designwork, particularly if the writing itself has an engineering intent. Writing is constructed, and the end result is a construct—a document. Both a design and a written document will test your ability to express thought. You can go an additional step and try to build the design or the document in a model or a full-scale prototype. Then the reality of the thought goes a step beyond the conceptual shaping of drawing and writing. The model or prototype is the final moment of its originator's thinking, and it is ready for the test. Will it work in the real world? In the case of the would-be aviators depicted in this chapter, ideas often did not translate into engineering achievements.

An unknown experimenter designed this motorized passenger kite. The vehicle was not successful. (Photograph courtesy of Dover Publications, Inc.)

serve that, apart from pace, our example suggests that silence speaks. Tone, pace—and even silence—add an entirely new dimension to the words you speak. But when you write, all is silent.

As mentioned earlier in the chapter, people often say, "I can think it but I can't write it." Equally common is the observation "I can say it, but I can't write it." There is no doubt some truth in this comment. Writing functions without voicing, and voicing is a key element of the ability to speak with precision; it also greatly enlarges the spoken "vocabulary." When you write you are robbed of this entire voicing structure. Voicing becomes a lost language.

There is yet another major way of speaking that adds to the use of words and voicing: body language. You are probably familiar with the term, but you may not realize the extent to which body language shapes and colors your way of speaking.

For our purposes here, we are concerned only with the idea that body language is a "vocabulary." People are extraordinarily sensitive to what they perceive in the body behavior of others. They spend their lives "reading" other people and creating behavior for them to read in turn. Movements act as a language.

When you speak, a number of language systems come together to create "sentences" where there are often no sentences. There are words or word clusters that do not form *traditional* sentences, but they become "completed thoughts" because of the addition of voicing and body movement. For example, I might say, "Well . . ." and nod my head in agreement with a comment. The thought is complete: I agree.

Speakers also use another device I call the "partial response." Because two or more speakers share a conversation, they can be somewhat lazy about their sentences, since they rely on the context, or situation, of the conversation to give meaning to short comments. Any one-word response to a question is an example. Will you graduate in June? Your response will be "yes" or "no". The response completes a statement because you join your partial comment to the former comment. The thought is complete, which happens to be one of the traditional definitions of a sentence. In fact, you could simply use body language to say yes or no with a nod of the head.

In writing, there is no voicing, no body movement, and no occasion for partial responses. Three of the "language" systems are missing when writers write. Many people don't realize why they are at a loss if they can't write with skill when they are perfectly at ease while communicating with voice and gesture. In many respects, writing is another language altogether.

When working writers don't understand why they can be excellent conversationalists but weak writers, they are basing their observations on one similarity between speaking and writing: both systems use words. "If I can speak with skill, why can't I write with skill?" As you can see, the answer has less to do with words than that speaking is essentially a dramatic medium. You speak well because you are a fine actor or actress, not because of your ability to manage words. Speaking is acting. Writing is not.

Speaking is, however, the ideal helpmate for a writer. Given the skills that you have for speaking your ideas, you should use conversation as a convenience to help express ideas. Conversation is the easiest place to begin. You can write the ideas after you see if you can speak them.

Conversation as a Contact Sport

Think out a project by trying to see drawbacks and shortcomings and other defects in your thinking. You can develop compromises on your own, but you are limited by your own capacities. However, in the process of explaining a confusing project to fellow employees, you might see a solution. You need to *react* to conversation. You can greatly enlarge the analytic process by talking with people, and conversation allows you to use all the "languages" you cannot use when you write. Shaping your thoughts in speech is usually easier than shaping your thoughts in writing. Speaking is a handy first test for your ability to express your ideas.

However, conversation does more than produce articulation. Conversation *builds* ideas where there may have been none. Employees and students alike tend to overlook conversation as a prime element of research; *at times dialog is the only preparation available*. This is often true at work. Dialog may also be the *best* resource.

In addition, conversation also serves as a springboard. Many working writers say they do not know where to begin. Getting started is difficult in almost any job—the problem is not exclusive to writing. Many people seem to need a kick start. Talk out your task and you will quickly lose the fear that you don't know where to begin. Through the easy pleasure of conversation, you will find yourself in the thick of your writing project before you know it. It will have begun when you weren't looking. There is no easier way to start.

I enjoy talking. I enjoy listening. I enjoy learning. And I enjoy a good argument. These are, or *should* be, the tools of a writer. You will find that the shortest distance between you and a completed writing project will be the time it takes you to talk it out to get it going. Nothing helps a writer to write faster than conversation.

Library research, I might add, can serve a similar function that is often overlooked. Books are simply voices to listen to and react to and respond to. Dialog is dialog. But at work, the problems, as you will see, often fail to be resolved with "book learning." Employees learn from the circumstances and from each other. People learn from people in a workplace environment.

This airplane was called the Geary Circular Triplane. The device didn't work. (Photograph courtesy of Dover Publications Inc.)

The education system may inadvertently cause some misconceptions by directing students to "do their own work." This experience leaves its mark, and people do not learn to collaborate and help each other at work. As a result, they are likely to assume that a lab report or a proposal or any other project has to be thought out and written by the author alone. In the workplace, team spirit is often an important tool for project development.

There is also another function of conversation that you should note. Some authorities argue that a unique phenomenon occurs if a group of people form a team and set to work on a task. In theory, the group will produce better results as a team than they would as a sum of individuals. In other words, a five-member *team* will produce more or better results than will the five individual employees. This phenomenon is called *synergy.*

Specifically, you have experienced exciting conversations in which ideas take shape and build empires in the air. What happens? Your own ideas become clear. Some of your ideas are disproven in conversation, whereas others are proven to be true. Yet others are supplemented and expanded by the ideas of the other people involved in the conversation. In other words, conversation demands that you "structure" your reasoning; that, in itself, is a valuable outcome.

You could just say that two heads are better than one, but the concept of synergy assumes that *three* heads result from the two. How can two people in conversation exceed the sum of their combined respective capabilities? Communications theorists have a number of explanations for the phenomenon.

According to synergistic theory, group members experience more than a synthesis of ideas. Rather than simply an analytic force emerging in an analytic solution, an additional personality starts to emerge from the people involved. In this case a *third* perspective evolves from a discussion: yours, that of the person you speak to, and the results of your discussion together. The new ideas are not your ideas. They are not the other person's ideas. They are *additional* ideas. I would argue that in every situation of this kind you improve and build on your ideas because of *interaction.*

The function of an encouraging, provocative, or logical force—for example, someone who simply asks questions— is an indisputable complement to your thinking. Your ideas develop if you talk them out.

Use conversation as a tool. You should seek out all the help you can get. You must realize that teamwork is the hallmark of the corporate world, and the strategy of teamwork applies to a great deal of work, including writing. Teamwork brings together what each member of the team knows. Your writing task may be your own, but you are part of a team, so use it as a resource.

Conversation is your preferred medium. It is your gift. Dialog should become an integral part of your writing process.

Working in Words

Now that we have examined the activities of thinking and speaking, the best way to relate the two systems to writing is to see how the three activities connect—or don't connect. I noted that speaking involves a language of coloration called "voicing." The elements of pitch and tone and pace are lost in writing, and so, too, you lose your body language and every other aspect of your potential for animation, except your ability to push a pencil or tap on a word processor. The result is that your communication is greatly restricted by a medium that is functioning in words alone. Further, writing is, indeed, black and white, that is, without color. It is also linear, a two-dimensional medium as flat as this piece of paper.

Of course, writing operates in its own ways to compensate for the missing dimensions of speech. Writing is an exacting and exciting tool that is based on a different set of principles. It can be as precise as a math system or as magical as the fictional world of a novel. How useful is the manual you just used to rebuild your engine? How stimulating was the trip to the stars in your last science fiction novel? You will find that writing *can* be a match for the color and flexibility of your speaking habits, but in its own fashion.

In general, the final distinction between thinking, speaking, and writing is to be found on the tip of your pen. There is a funnel effect of diminishing freedoms that you experience as you move down the spectrum of expression from thought to writing. The available methods of expression become fewer and fewer. The demands placed on what you can do become increasingly rigid, but there is something to be gained from the tight controls of writing: structure.

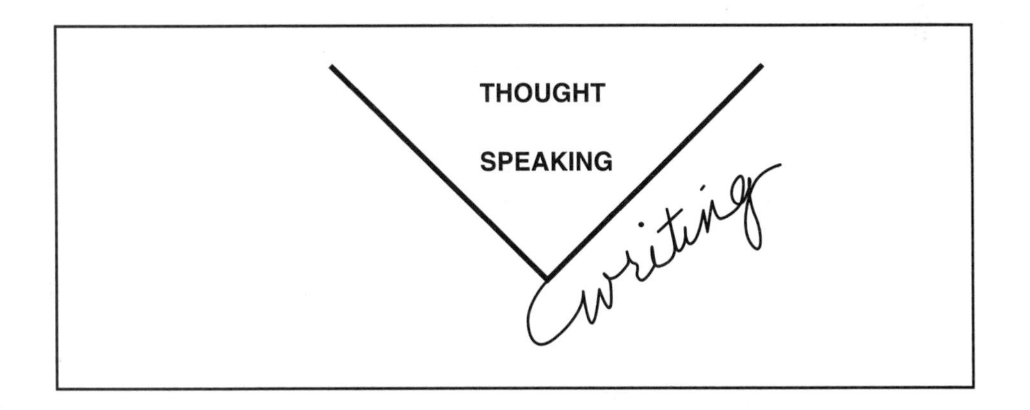

This twelve-winged plane was built in 1912 by Howard Huntington of Queens, New York. The additional wings didn't help. (Photograph courtesy of Dover Publications Inc.)

The world of thought is a fairly unstructured, fragmented, multidimensional world of freedom expressed in all dimensions of your experience. To communicate thoughts, you speak. In speaking you don't have as much freedom as you do in thinking, because in speaking, you have to express your ideas in logical order using the vehicles of speech, voicing, animation, and partial responses. In turn, when you write, you narrow down the available vehicles to words, and you are forced to develop a highly controlled method of expression. The language of writing is also tightly organized. Talk rambles all over the place, and writing should not. Writing should stay put.

Let's try to represent these differences graphically by simply looking at the control of organization. Assume that the following dotted line represents an ideal logical progression of thought that will put your dinner on the table tonight.

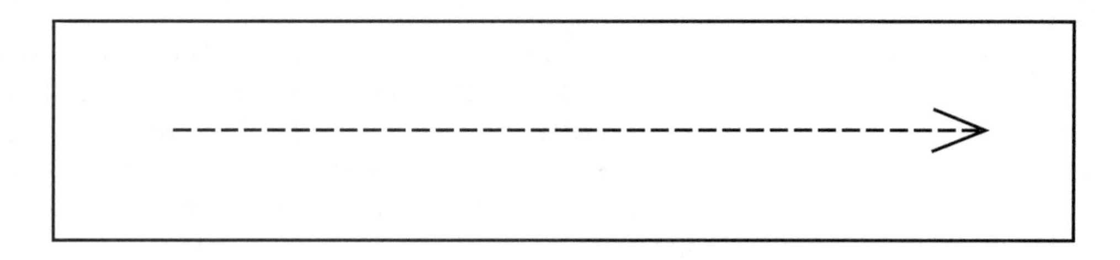

The actual thought process that accompanies the activity, however, might look something like this solid line:

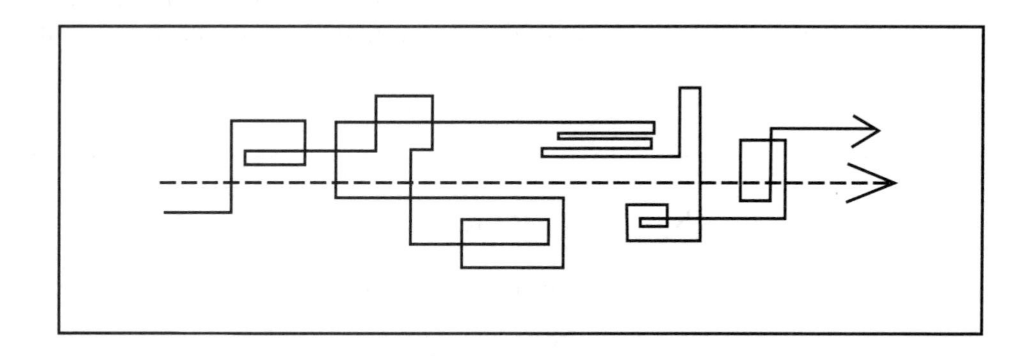

This drawing could just as well be the pattern of your activity back and forth in the kitchen, couldn't it? In your thinking, you might be going back and forth in your mind in a similar way. If you were now to explain to me how you went about the creation of the dinner, the demands of communication would force you to be orderly in your words, your sentence construction, and the logic of your overall explanation. Even then, I would have to ask you questions at, perhaps, four points. You would obviously use toning and gestures to assist what you say. You would also alter the organization to simplify or clarify or order the activities. The result would be a pretty odd jumble of logic, tone, gesture, phrasing, and so on. However, I would probably understand what you mean, and you would have constructed the first step in the task of explaining the activity in an orderly way.

You should realize that, regardless of the haphazard results, conversation is the ideal medium for preparing yourself to write. It is, after all, an articulation of ideas. It is a starting point. Again, to use the initial drawings, if your thinking looks something like this,

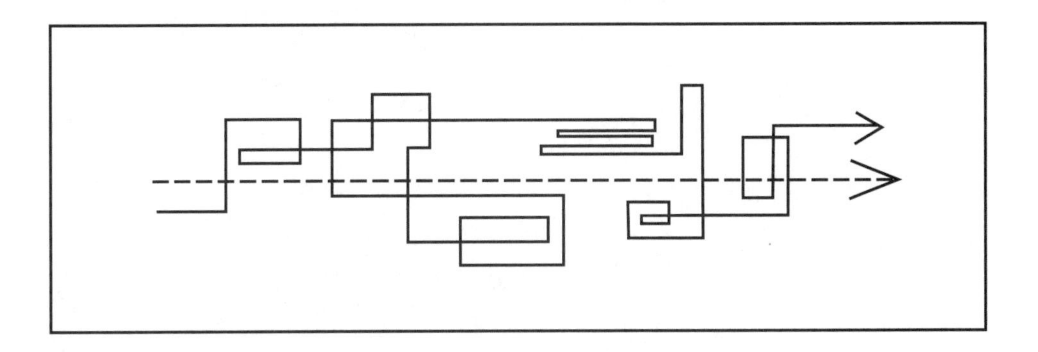

and your supervisor wants a report that generates the straight-line logic of this,

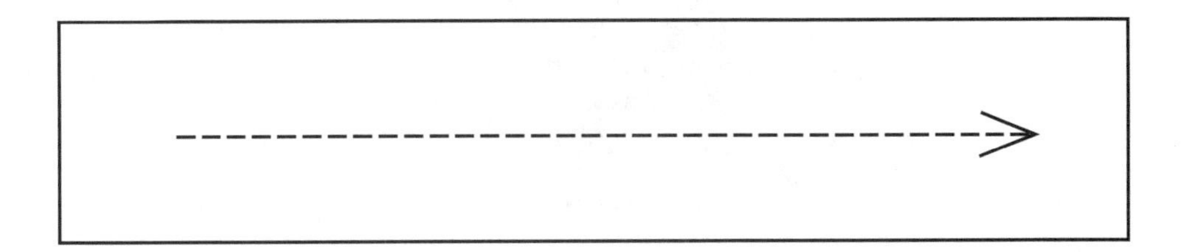

the obvious key is to discuss the issue first. You have to start talking to add organization to your thinking. Otherwise, the task of straightening out the maze of thought will be slow and hard. If you transfer your ideas directly from thought to paper without trying to structure them in conversation, the written result may not meet the challenge; it will not be straight-line logic, and good writing is clear from the first word to the last.

If you are to write well, you must exercise enormous control over the activity. You have to funnel thought into the narrow channel of orderly logic, and you have to funnel all expression into one mode: words. Clearly, you are looking at the hurdles that create the challenge of writing. *The challenges are word-based language and the linear thinking of straight-line logic.* In sum, writing has little to do with the language we speak, but the language we speak is often the most convenient test of our thinking. In this respect, these two distinct processes can work hand in hand to achieve our goals as writers, particularly in the workplace.

Summary

- Thinking is a complex and rapid process.
- Words are one path to invention, and they demonstrate thought. Language slows down the thought process and can increase the accuracy of our ideas.
- The *expression test* is the ability to articulate or construct an idea.
- The *reality test* is the measure of the truth or accuracy of the idea in the world around us.
- Mental processes are complex and tend to be erratic, unstructured, fragmented, nonverbal, and vague.
- It can be a challenge to channel thought into a highly focused activity such as writing.
- Conversation is extremely important as a method of channeling and shaping ideas.
- Speaking involves a number of language systems that are not part of the world of writing: tones, body language, partial responses, and other elements.
- Conversation is an ideal medium for fine-tuning ideas because dialog is inspired by challenge, improved by synergy, and perfected by interaction.
- Writing is unique because of the straight-line logic that is used for document organization.
- Conversation is an important interim phase that can help an author prepare to write.

Among the colorful machines you see here, perhaps the least likely candidate for flight is this design, a glider invented by the German Otto Lilienthal. It was, however, the only machine of the group that could take off on a briefly sustained flight. Hang glider enthusiasts will recognize the secret of Lilienthal's success: gliders are light enough to be carried aloft if the air currents are correct. Unfortunately, as Lilienthal discovered, the challenge of flight was only the first test. The design problems of flight control and landing were the next challenges. Lilienthal was killed on a flight in 1896. The design problems began to take a deadly toll in the 1910s once flight was sufficiently understood and machines could climb hundreds of feet in the air.

German aviation pioneer Otto Lilienthal built and flew a variety of gliders between 1891 and 1896. (Photography courtesy of Dover Publications Inc.)

Activities Chapter 2

Select one of the following suggestions and develop a brief document to explain your experience.

The expression test *and the* reality check *are equivalent to your ability to design and build an idea. Briefly describe a success or failure you have experienced in the process of "constructing" an idea. Did the design phase work? If so, did the building phase work?*

The expression test and the reality check can also be used to assess a company's success or failure with design ideas and their construction. Briefly describe a work-related situation in which you experienced or currently see a problem in carrying an idea through to reality.

The use of speaking as preparation for a writing task is most obvious in work-related environments, particularly when a committee is involved. Have you ever been part of a task group that drafted a document? Briefly explain the history of the project and explain the role of discussion in relation to the written product that the group created.

These projects should be 250 to 500 words in length, typed in a memo format (see Chapter 11), and addressed to your instructor.

The Wright brothers, who were Dayton, Ohio, bicycle manufacturers by trade, developed and flew the first successful airplane in 1903—after only four years of diligent research and experimentation. They built gliders first in order to learn from what Lilienthal and others had discovered. They communicated with other experimenters such as Chanute. They tested glider after glider in the winds of Kitty Hawk in South Carolina, conducting hundreds of flights each autumn. The rest of the year they would argue and discuss and build and test. They constructed the world's first wind tunnel in Dayton so that they could test hundreds of small models without wasting time on the full-scale prototypes. Talk, design, build, test—they were relentless in their efforts.

Wilbur Wright flying a glider in 1902. (Photograph courtesy of Dover Publications Inc.)

Work in Progress

The Proposal

I knew that my ultimate goal was to write a safety implementation program for the workers at Pacific Aero Tech, but as I began gathering information I realized that there was one set of regulatory information for employers and another body of information concerning employees. Since the company did not have any written plan, I needed to expand my research to include the employee's responsibilities as well as the employer's responsibilities for safety and health in the workplace.

Gathering the information was simple—I printed it off the Internet, but the problem was sorting and finally compiling the material to tailor it to the audience that was going to read it. At this time I understood that two different types of programs had to be written. First, I had to compose a comprehensive workplace safety and health proposal that introduced OSHA and WISHA and their compliance guidelines. Also, I had to stress that the programs would be tailored to PAT, yet within OSHA standards. Second, I had to scale the written programs—which encompass a myriad of safety topics—to one or two practical, easy-to-use documents for each topic to avoid employee confusion. I also had to be careful not to let the suggested safety training process impede work production.

In essence, the challenge was to develop the details of a safety program that would clearly communicate PAT's intended programs to supervisors, employees and regulators. Management commitment and employee participation is vital to the success of PAT's safety program. To gain the President's acceptance of the proposal, I needed to recommend suggestions as well as write the required programs. In order to complete my first task, I had to observe my fellow employees at work. This took some time and was not without some negative responses. As with any new program, fear of the unknown as well as breaking long established habits were problems. Also, management cannot stop production, which was an obstacle and a hassle at first.

The proposal I developed was written in general terms by highlighting the needs and priorities of the company, including all required OSHA, WISHA standards and the employer's general responsibilities. It included a summary of a 1996 industrial hygiene walk through, my recommendations, and future sources and benefits. I inserted additional reference material at the end in a supplement that would not distract the President's attention from the proposal. I struggled to keep the proposal as short as I could without cutting out important information that she would need.

Since I intended to "sell" my safety ideas to Karen, I packaged the proposal in a company three-ring binder. I inserted a complete copy of the 1996 hygiene walk around, placed my individually labeled reference materials in the back, and put a cover letter, briefly explaining the proposal, in the very front to set the stage. I wanted the President to see a few highlights of the proposal in the letter. But first I wanted several other employees to review the package. I had put a lot of time and effort into researching, compiling, and writing this proposal, so I wanted everything to look like my best effort. It was seventeen pages long without the supplements.

L.C.L

The Third Medium

Writing is surprisingly disconnected from thinking habits and speaking habits. I observed that talking out a matter seems to be a happy medium between chaos and order. This is an important point. The mind is quite a maze, but writing demands a great deal by asking for nearly an ideal logical presentation.

The mere fact that writing does not involve voicing and animation does not explain why it has to be so perfect. There is another reason. Suppose you are working a swing shift at a lab or assembly plant. Your company uses "shop sheets" on clipboards to record what has and has not been done.* When you arrive each night, you discuss ongoing projects with the technicians on the earlier shift. If you are late and the other crew is gone, you read the shop sheets. But if the sheets are not clear, what do you do? Specifically, whom do you ask for clarifications? No one. This is the dilemma of writing. Writing represents people in their absence.

People usually write because they cannot be at a certain place at a certain time. In engineering and industrial settings, people also write to control the sheer complexity of their work by keeping the technological data ordered and recorded. If they are not there to answer for their writing, it must be letter-perfect before they dispatch it. *Writing talks, but only once.*

Yes, you can reread a document and learn more from it, but the document will never say more than it said the first time. If there are any errors or omissions or unclear passages, you cannot address the writer. Speaking is much different because the speaker is always where he or she is speaking. You can talk out the evening lab activities and ask all the questions necessary for clarification. Writers tend to disappear when you need them the most! When you read a document, the speaker is speaking, but the author is not there to answer for his or her ideas— especially when something you read is not clear. Thus, as a writer, you must do your best in writing so that there will be no questions.

Having to write in such precise terms is complicated by another fact. In the same sense that a reader has no author to address, a writer actually has no one to talk to when he or she is writing. There is only an anticipated audience.

Writing anticipates a conversation of some kind, but the writer and the reader never really come together, except through the thoughts that are left on the written document. *My* half of the conversation on this page was written a while ago; you read the page and activate this conversation for your use at a later date—this very moment. You then try to understand my ideas and judge them, which is *your* half of the written conversation. My task as a writer is a complex one. I must second-guess your needs so that I will write in such a way as to make you understand me without question—literally without questions.

** The Boeing Corporation is based here in Seattle where I work. Shop sheets are their traditional choice for shop floor messaging because this world leader in aviation uses around-the-clock shifts to gain maximum utilization of space and time. Without accurate shop sheets, workers on one shift would not know how activities were handled by the shift before them. It is important to realize that the modest notes of mechanics and technicians are important signals in the production of some of the world's most complex machines.*

The task is not always easy, nor is it natural. Speaking across vast spaces does not seem that mysterious to anyone in the age of telecommunications, but speaking across time remains the stuff of science fiction writers. People simply take writing for granted and fail to realize that it is the ultimate time machine, one that has been around for thousands of years. What, after all, is the mission of writing? The stock answer is one word: "communication." Not so. People *never* needed writing to communicate, at least not until they found a need to communicate through time and space. One employee writes to send a mail order to some distant city, bridging space. Another employee writes to record the readings on a complex instrument so that he or she can refer to the data at a later date, bridging time.

Talking To Walls

Your role in sharing this unique phenomenon called writing brings us back to the practical issues. Because writing is intended for travel, into the future or across the landscape through time or space, having no one to talk to when you write becomes a major obstacle. People who do a great deal of writing become accustomed to talking to no one, or talking to themselves, when they write. For you, this imaginary effort may be awkward. Writing puts you in a situation that you might call a monolog; you must speak to yourself. Your preferred medium is obviously dialog, or conversation, an environment in which you can speak to someone else. As a result of this preference or convenience, writing seems awkward. You have to talk, but there is no one to talk to!

The key distinction between monolog and dialog is the presence of someone, the interaction of conversation. Dialogs push your thinking because more than one mind is involved in the topic of conversation. Your mind will be creative in conversation, but it will also be highly reactive, which makes you even more creative. Your ideas grow and emerge as a reaction to the other person in a conversation. The act of writing is not very reactive, and as a result, the output can be a challenge to produce.

Combining the two major conditions of writing puts a writer in a bind. Since the reader cannot talk to you about your writing, you must create a perfect document that consists of your best effort to create what I have called straight-line logic. If there is one time when you need help communicating it is precisely when you write. However, since you, as a writer, have nobody to talk to as you write, you are in exactly the situation that is least likely to be creative or linear—or even productive. The result is one that some writers know well: after a few hours of staring at blank paper, they experience a meltdown.

You now know what is causing the problem, and you can try to take control. Put yourself back in the situation of dialog; you must keep the writing projects in the mode of conversation. Writing seems unnatural to many people because it is like talking to a wall. They will speak of writing as "difficult," but that is not an accurate reflection of the challenge.

Once they understood the aerodynamics sufficiently, Orville and Wilbur designed a 160-pound engine that could generate 12 horsepower. The first "Flyer" weighed about 750 pounds with a pilot and flew on December 17, 1903, a distance of 852 feet in 59 seconds in the best of four flights that day. (The first flight, on December 13, was not very successful.) There were many hundreds of flights under power in the following years as the brothers sought to perfect the Flyer. They were extremely methodical and inventive and built and tested improvement after improvement.

The first motorized flight. It was 10:35 A.M., December 13, 1903. The plane hovered over 120 feet of ground. (Photograph courtesy of Dover Publications Inc.)

We simply live in a dialog world and we are, usually, unfamiliar with monolog communication. In addition, the business of making writing letter-perfect is a contradiction to human preference for the lazy rambling of conversation where people can be utterly confused and still talk themselves to clarity if the lunch hour is long enough. You need to realize that you can make much constructive use of conversation in ways that can assist a writing project.

One old trick is to write all your work-related projects *to* someone. For example, when you write a personal letter you can usually produce material fairly rapidly because you can somewhat imagine how to go about talking to your friend. Similarly, at work you can help yourself along by writing projects with a very specific person in mind. Your supervisor is a likely candidate. This practice can be helpful, particularly if the document does not seem to have a specific audience, or if many people will read the project. It is hard to write for no one, and it is hard to write for a group—or everyone—so select a representative person (preferably someone you know fairly well), and write to that person.

In two years they were back at Kitty Hawk with Flyer II and flew for 39 minutes and 24 miles! The reality checks were a huge success. By 1909 their flights were a sensation during demonstration tours in Europe. In 1911, pilot Harry Atwood completed a daring 461-mile flight from Boston to the White House lawn! The higher, the faster, the greater the risks. Although Orville and Wilbur were lucky, Wilbur was seriously injured in a test demonstration for the U.S. Army. The army representative on board was the first passenger to be killed in flight.

Wilbur Wright at the controls of one of the "Flyers" during his successful demonstration tour of Europe in 1909—only five years after the first flight. (Photograph courtesy of /Hulton Getty/Liaison Agency)

Reading The Fine Print

Because writing has physical reality, the medium can be particularly risky—especially if your name is on the document. This risk is the underlying concern that many people have about writing: writing can reveal them in devious ways and may cross them unexpectedly.

In contrast to writing, people can be fearless in their thinking because they alone sit in judgment of themselves. It is interesting to note that people usually do not lack confidence about speaking either, even though speech is clearly public. Partly, their confidence in speaking has to do with the sphere of trust in which they operate in conversations. They know the people to whom they are talking, and they can measure what can and cannot be said. Writing, however, ends up out of an author's control once a document is in circulation.

Of course, speakers handle conversation with such confidence that there must be more to the issue than simply trust in their friends. One consideration is that no one can *see* the spelling or the punctuation or a host of other errors that can be seen in writing. The fact that you cannot *hear* writing (no pitch or tone) is one problem we observed earlier, but the fact that you cannot *see* speaking is probably one reason you take such pleasure in it! You are confident in speaking partly because it does not have permanence.

Obviously bad grammar *can* be heard. If your supervisor asks you if you had luck in completing a difficult analysis in the lab and you say, "There's still three pages to write," the poor English is evident (since the verb and subject do not agree). Note, however, that the words vanished. Your boss has no evidence that the sentence ever occurred. Speaking is a pleasure because it is transient. It comes and goes in an ebb and flow that guarantees its impermanence. This means that everyone can bungle along fearlessly because they leave no trace—and bungle they do.

The usual patterns of spoken English are full of errors that people either do not notice or ignore—either as speaker or listener. The spoken language is outright sloppy if you contrast it with writing. In other words, the ground rules are flexible, probably because speaking is so highly improvised and spontaneous that you cannot demand your best grammar from such a rapid medium. Even if listeners hear errors, they are not likely to pay much attention to them. I would argue that not very many errors are even apparent to most listeners. A listener will be listening for content; a listener will be focused on what you say and not the way you phrase it. To notice the errors, the listener would almost have to transcribe your speech, which would reveal all the grammatical problems.

The transience of voice also results in another phenomenon: a conversation is usually not remembered by details but by the general outcomes of the dialog. The listener walks away with the gist of what was said. I rarely remember particular comments, and those I do remember are likely to disappear in a day or two. The nature of memory contributes to the transience of speech, and the tendency to recall a conversation with an overview or general recollection tends to eliminate a focus on particulars. In fact, if you ask someone to recall your exact preceding sentence in a conversation, odds are your listener cannot do it.

As for writing, the basic problem is very elementary: writing is permanent. Writers can make ashes out of embarrassing love letters by throwing them in the fireplace, and they can make confetti out of incriminating corporate records by feeding them through a shredder, but they may not always have access to the documents they write. After all, the idea is to write something to *send* it somewhere.

In the last analysis, much of what concerns people about writing has less to do with not being able to spell, or not remembering how to write after being out of school for years, or similar familiar ideas than with self-preservation. Everyone wants to preserve their reputation and image. They know that the language will be judged, and they would rather avoid the judgment. The physical fact that writing exists on its own is their greatest concern. Writers desire to have control over their communication so that they can have confidence in the results. Acknowledging this unspoken apprehension, observe that there is a bright side to the issue.

Blackboard and Chalk

Consider a mathematician. She speaks a unique language when she is at work. She cannot easily *think* long and complex theoretical combinations of mathematical symbols. Nor can she easily *speak* them. She must *write* them. Seldom will you see a photograph of a professor of mathematics without a trusty blackboard covered with calculations. She cannot easily communicate the thinking of mathematics to a class of students without writing the calculations. Speaking the calculations is too difficult—and imprecise. *Writing makes perfect.* Plenty of chalk and plenty of erasers help. *Revision makes perfect.*

Precisely because writing has a physical reality means that the medium can be *worked.* The *content* can be *reworked.* The *form* can be *reworked.* The mathematician uses the physical reality of writing to her advantage. She can build and build and build on what she constructs on the blackboard. She can, by virtue of the fixed reality of writing, make thinking as perfect as possible. So too, the poet can work words in endless pursuit of a certain form so that the words achieve a certain grace and meaning. So too, the engineer can design and redesign a concept until it is perfected.

To the extent to which your technological specialization demands precision, writing alone will stand by you in your efforts to record and communicate complex data. You can always make sure that a written report contains the correct data. You can construct a discussion that is written in precise straight-line logic. You can remove every imaginable error! You can perform these tasks precisely because of the physical existence of the medium.

I first urged you to use conversation with people to encourage you into your projects. Then I suggested that you write *to* someone, to a person you know. Writing to someone helps create a kind of dialog. Now I would suggest that you talk to yourself to add yet another conversation to the project. When you work on a manuscript or some document you have developed, you *are* conducting a conversation. You are talking to yourself when you revise material. Anything you write takes on its own existence and becomes an older version of your thinking by the next morning. It is history.

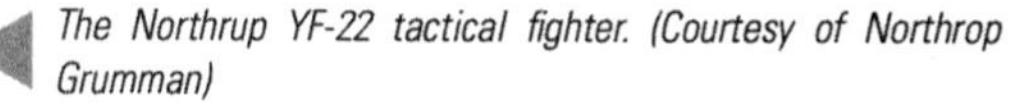

The Northrup YF-22 tactical fighter. (Courtesy of Northrop Grumman)

No matter what you write, the writing becomes historical; it becomes the historical "you." The conversation you have with yourself when you revise your work is an extremely valuable dialog. No one is writing the document you are composing except you, so the best person to talk it over with is yourself. No one else is quite as involved. Your revisions create this important dimension of internal conversation. Any book on writing will tell you to revise—and revise and revise. The best revisions, however, are not simply acts of polishing. Conversation is the secret. Conversation constructs. Conversation builds. When you revise your work you build a response to *yourself.*

Talk to coworkers.

Talk to the reader (write for someone).

Talk to yourself (edit your drafts).

Talk the project into existence.

In a little more than ten years after the first flight, aviation became a military tool in the First World War, the first war of the fighter pilots. Since that time aeronautics has become a science, aviation has won and lost wars, and air passenger transportation is a global reality for millions. In less than a century this most adventurous and daring of the engineering fields has evolved into the aircraft illustrated here and on p. 52. Politicians lost no time in observing that aviation was, indeed, beginning to look like Star Wars.

The B2A Stealth Bomber (photograph courtesy of Northrop Grumman)

The Challenge Of Perfection

As you have observed, there are several reasons why writing is a fussy medium. First, you use the medium of writing to fine-tune your ideas to perfect your thinking. Second, you transmit the medium and will not be around to answer for your work if someone else is puzzled by the document. As a result, to perfect your ideas and to transmit them perfectly, you must do your best to generate an ideal document that is itself perfect. It must be perfect in form (spelling, grammar, punctuation, format, and so on). It must be perfect in content (straight-line logic, evidence, discussion, paragraph development, clarity of ideas, and so on). Of course, any such hope is terribly optimistic; nothing *ever* seems to be perfect. The goal of perfection at least means that the *effort* to reach it should produce informative and effective material. That is all we can ask. You can be sure that the book you are holding is not letter perfect no matter how much effort went in it. However, the effort you make will produce the desired results—your *best* work.

Your mission is to produce an orderly and logical project that your coworkers or your supervisor or your clients will understand without difficulty. Thinking is quite spontaneous and often disorderly. Writing is highly organized. You have to be able to get a round world onto a flat map. Mercator did it. So can you. You can achieve this mission by carefully constructing the units of logic you find in the sentence, in the clusters of organized sentences you develop into paragraphs, and in the clusters of paragraphs you group into the overall organization of any project you have to generate.

Then you go to work on the cosmetic touches to make the vehicle of writing as readable as possible by making certain that there are no mechanical errors in your efforts. This means you finally have to get *their* and *there* straight. You must look for any blunders that can confuse or annoy your intended audience. If you attend to correct spelling, proper word choice, appropriate punctuation, and standard grammar, you will create a smooth and easy-to-read final product, a near-perfect map.

The need for mechanical accuracy in writing is most easily seen in the basic logic unit of the sentence. Formal writing of the sort you generate at work is usually composed of whole, or complete, sentences. In other words, every sentence must have a subject and a verb, an actor and an action. If either the actor or the action is missing, you have a fragment of a sentence. More precisely, a fragment of a sentence leaves your reader with a fragment of your logic, and this is a problem in writing, since a reader cannot call you for clarification.

To anticipate every question from the reader head-on, you use complete sentences to complete all the logic of your thinking. It is the best you can do to be thorough and to assure yourself that you are transmitting your meaning accurately and completely.

Then, too, the details of a sentence, such as punctuation, help create the precision of your logic. You must select these symbols as precisely as you choose a math function symbol. Every symbol in the sentence couts. I left out the letter *n* in the word *counts,* and you are instantly a little confused. Writing demands that you be precise in the use of symbols because errors in communication are difficult to correct in the world of written communication. The author and the reader come in contact only through the written document. They *must* expect precision from each other.

Communication involves a transmitter, a medium, and a receiver. Errors can show up at any stage. An author can control at least two of the problem areas, since the author is the transmitter and controls the medium. Precision will help eliminate any misperceptions a reader might have in the reception or understanding of the message (see the illustration).

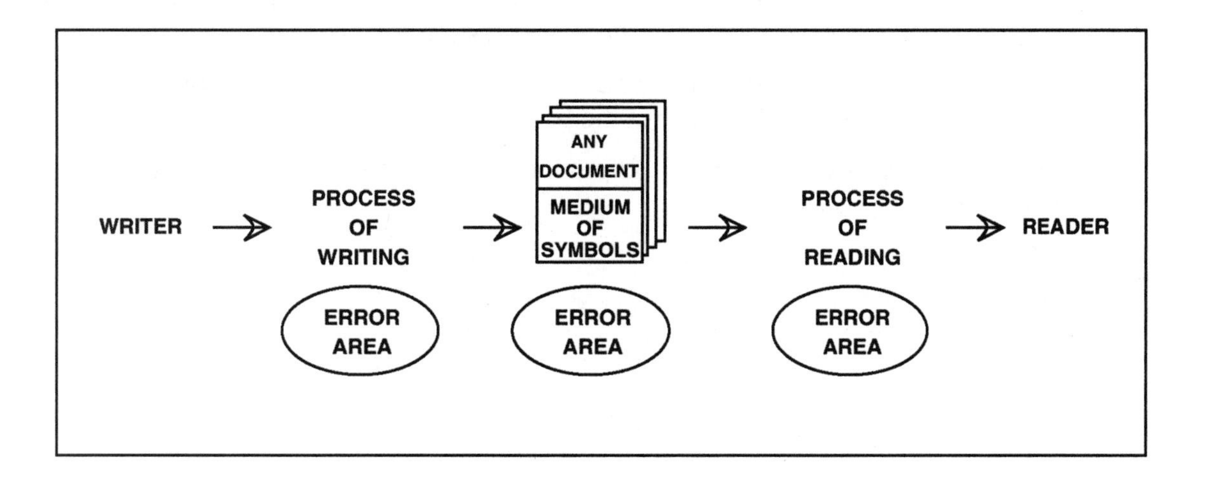

If an author can skillfully control the process of writing and the development of the document, he or she will effectively eliminate the probability of error for the reader, which is the third problem area. The document should be easy reading; then, the only question is, as the popular saying goes, Will the project fly? First, will it be understood? The expression test. Second, will it be accepted? Approved? Endorsed? The reality check.

Summary

- Writing allows an author to perfect the ideas embodied in a document.
- A reader depends on the accuracy of writing because the document stands alone as the representation of the author's ideas.
- Writers should use conversation to prepare for a writing project, partly because the dialog helps state and explore ideas, and partly because the *dialog* helps prepare the author for writing, which is similar to a *monolog*.
- Writing to a specific person is helpful, particularly if the document has a large audience or otherwise lacks a specific reader.
- Because writing is a relatively permanent medium, authors are often cautious about writing activities.
- Writing creates a historical document that can be revised again and again. Each revision is a dialog between today's author and yesterday's author.
- The goal of a *perfect* document is one a writer may seldom achieve, but the effort to perfect a document will bring out the *best* the author has to offer.
- Revisions are signs of dedication and proof of professionalism. The revisions do not indicate writing problems.

Activities Chapter 3

Select one of the following suggestions and develop a brief document to explain your experience.

Are you familiar with any activity that involves considerable preparation and precisely detailed development? Schematics, structural drawings, and legal documents are three products of this sort of activity. How are these or similar products perfected? Explain the process.

Have you had a run-in with a poorly developed set of instructions? Or have you found yourself on the phone with the 800-help number provided by your computer or software manufacturer? Explain what the problem was and why the instructions or manuals or software instructions were not clear.

If you can go directly to the instructions in the preceding exercise, why not make the appropriate revisions? Rewrite the document so that it reflects the precision it should have had in the original version.

These projects should be 250–500 words in length, typed in a memo format (see Chapter 11), and addressed to your instructor.

PART II

The Workplace Writing Process

Work in Progress

Fits and Starts

The president of Pacific Aero Tech and I established a tentative proposal deadline of one month from the last day of my OSHA training class. My final draft was the result of many "fits and starts" that caused me to have much trepidation and resulted in several short drafts, two of which I completely trashed. Each time I started over from scratch.

My method of writing included choosing a topic, usually by order of importance, reading all the material I could find, and taking copious notes. The research notes allowed me to think through how I would like to present each individual section. After mulling the topic and my notes in my head for a while, I would sit down at the computer and just type. My goal was to get the information down on paper first and worry about editing the grammar and the format later. This procedure required many revisions, and a lot of time and effort, yet it yielded a solid written presentation—which was the whole point.

When I felt comfortable with my first draft, I set it aside for a few days to rid myself of the "halo effect" a writer often gets when critiquing his or her own work. I asked Gloria, our receptionist, to proofread the draft and document her findings. I asked her to concentrate more on the clarity and layout instead of the content. I also had one of my fellow technicians, Don, analyze the proposal from a technical standpoint. Both made good suggestions.

I completed the final draft of the proposal on time, and the project ran ten written pages—this was not including the supplemental references that I included to support my findings. This draft was the condensed version from my original draft of seventeen pages. Remarkably, this proposal just glanced over the main issues and was very general in nature! It was only the beginning. With the material finished, I took some extra time to make the proposal presentable. Since the proposal represented me, I wanted it to look sharp as well as read well. After placing the revised document back in the three-ring binder, I wrote a cover letter, in lieu of an introduction, to briefly describe to Karen, the president, the contents of the revised proposal and all the supplemental material included.

Tight time management was needed to meet my deadline. There was so much information that needed to be researched, compiled, written, proofread, and rewritten, that no time could be wasted. I did get into a pattern that eventually helped me reach my deadline goal—which was to put out a good product in the desired time.

L.C.L.

Deadlining: How Not to Do It

The following chapters focus on seeing and removing as many obstacles to writing as possible. The appropriate way to begin a discussion of strategies is to deal with all the wrong methods, so let's look at how *not* to write.

There are popular misconceptions about writers and how they go about their tasks. Movies or television programs that portray writers are usually action-packed dramas about journalists writing madly to meet a dateline. Everyone has seen a few of these romantic portrayals on television, but few viewers have ever seen *real* writers at work.

Professional writers cannot quickly assemble a project without a major effort. I have only once been asked to do a rush feature—complete with photography. The occasion was very special because an author had withdrawn an article from a magazine that was soon going to press. Such a hasty process (two days to write an article) is not very realistic. Nevertheless, the notion that a writer can sit down and grind out a writing project in one intense burst of activity remains an attractive idea to many working writers. These writers persist in sitting down to the task of putting pen to paper with the idea that they should be able to get the job done in one sitting.

Procrastination

For many people, writing tasks create a predictable response: the more documents they have to write, the less they want to write. College students, for example, rarely sit down to write with a head of steam; to the contrary, they often procrastinate and stall until the tensions are overwhelming. There is a certain wisdom in this practice. Since students may not have a real desire to write, they cleverly wait for a real drive to get going: deadlines. They finally act out of panic. I call this technique the "crisis model." Panic is, after all, a prime mover. Fear of the wrath of supervisors will force everyone to produce. Like college students, many employees settle in to the task at the last minute—fully expecting results. The method is risky, even if students and employees can cleverly motivate themselves in this manner.

The problems that result from deadlining are many, but the first problem—and the most serious—is that if you are willing to deadline, you are assuming you *can* meet the challenge at the last minute. I suggest that there is a curious vanity in this idea. Professional writers work slowly and persistently on a task. Obviously, if the professionals do not feel comfortable with deadlining, the logic of deadlining for an amateur is not very convincing, is it? The results are likely to be destructive.

The practice of deadlining is widespread, and, to compound the problem, there are often times, particularly at work, when you will have no time to leisurely pace yourself through a writing project even if you would like to. The fundamental dilemma, however, is that the working writers who are most inclined to deadline are often the writers with the least skill for the task. Most of these writers will not be able to match the challenge. Their procrastination puts them on a collision course with their project datelines.

Then, the threatening hands of the clock finally motivate the deadliners. They begin to write and promptly encounter the next problem: preparation—or the lack of it. If they are really skilled procrastinators, they will have avoided adequate preparation for the writing task. "Out of sight, out of mind," the saying goes. Besides, the media's false image suggests that a writer sits down and then thinks of something to write. Unfortunately, this is not the procedure an experienced writer uses. You should sit down to write because you have something to say. Thinking precedes writing. You cannot tackle the difficult chore of thinking out the very ideas you want to write at the very moment you want to write them. Yes, you shape and refine your ideas as you write, but you must have a sense of what you intend to do before you can proceed.

Without preparation, the results are predictable. At worst, the result will be white-paper shock. The author sits for hours and draws a blank. Hours may pass before a trickle of ideas start to take shape on paper. The wasted hours could have been used to write the entire project if the author had known *beforehand* what was to be said on paper! Most people sit down to write and *then* start to think. The process is backward. You should be finished thinking about a topic before you pick up a pencil or turn on the computer.

At best, the final product will be a quickly produced text by an experienced deadliner who is also sufficiently skilled as a writer to know how to write his or her way into saying something. In the end, the product is submitted on time, but there is little coherence and polish to the project. It will ramble and sputter because of the haste that produced it. Everything goes wrong when writers deadline. They can hope only that their supervisors do not read the material very closely.

Front-to-Back Writing

Even with adequate preparation, procrastinators face another huge task. To write the project in the least amount of time, they reason that it must be written straight through from the first paragraph to the last paragraph. There is little time for corrections, much less major text revisions, so they try to produce a text that they envision as an end result.

The idea, of course, is to avoid subsequent drafts. This is the "front-to-back" method. Using this method, an author starts with the first paragraph of the introduction and writes through to the other end. I suppose this tactic seems logical enough. After all, documents are *read* from front to back, so it seems logical to assume that they are written that way also.

Sometimes a professional writer's project falls into place in one writing effort. If a professional writer knows what to write, she has the training and the years of practice to generate a front-to-back text. However, trying to write a document from front to back is a very difficult method of writing, and it may be the *most* difficult for the working writer. Besides, the front-to-back writer is likely to be motivated by lack of time. In other words, the crisis model leads to the front-to-back method of writing, and now the writer's hands are *really* full.

Imagine what you are asking yourself to do if you sit down to create a front-to-back document. You are asking your mind to organize an overview of the project, control the logical continuity of the entire sequence of points to be made, and systematically develop each point one after the other. In addition, if you are deadlining, you will try to create a text that will not need corrections, and so each sentence must be *the* sentence and it must be rendered up as a finished product that reads with clarity and contains no errors. Each paragraph must be *the* paragraph, and as a group the paragraphs must logically run from one to the next from the first page to the last page.

Recall the discussion of the nature of the thought process. It is little wonder that most working writers have difficulty writing and assume they are not cut out for such work. Inefficient writing habits accidentally create impossible self-expectations and rigorous demands that would alarm even a professional writer. Yet many working writers do procrastinate and inadvertently place themselves in the most clumsy of writing situations.

If through sheer courage you complete a task under these circumstances, you run into a third set of problems when you look at the outcome. Suppose you stay up all night and get a job completed. The project is in the supervisor's mailbox. What your manager then reads is the result of a rushed job that will contain several predictable problems. First, the overall presentation is likely to be hard to follow; the logic will ramble. Second, the important opening paragraphs will probably lead the reader to expect one set of outcomes, whereas the actual text will no doubt yield a somewhat different set of results.

A poor introduction is something of a trademark of front-to-back writing. Since the writer composes the introduction first, it promises a bill of goods that may or may not ever actually be delivered. A document under construction tends to go its own way. The introduction reflects *intention*. The rest of the document reflects *reality*—what actually got written. Usually there is no time to rewrite the introduction after the paper is composed because the author is up against a deadline. The result is that the important opening paragraphs will be misleading. The reader uses the introduction to understand an author's intentions. If the introduction is off the mark, the reader will not understand much of the report. It will read like what it is: a last-minute, hurry-up-and-get-it-done production.

The final predictable problem involves the cosmetic appearance of the document, particularly typographical errors and spelling errors. If you deadline a writing project, many little errors creep into the finished product, and the project will lack polish. We might simply blame the rush job, but there is more to the problem.

Deadline Fatigue

Rushing leads to errors, but so does fatigue. Errors increase due to exhaustion. You cannot maintain the pace, particularly in one sitting. In addition, there is another dilemma that you, as a working writer, must be aware of: memorization of patterns. No doubt you have put together projects in circumstances that allowed you to be fairly careful about errors only to be surprised at how many errors you missed anyway. Upon reviewing the document a week later, every error leaps out at you. How could you have overlooked obvious typos and similar errors? Rushing and fatigue are contributors, but in addition, you simply did not *see* all the errors.

The culprit is your memory. If you loom over a project all day, you will memorize your text. Editing for errors is something of a disaster at this time because you will not see half the errors. Since the entire text is imprinted in your mind, your mind tells you what *should* be on the page rather than allow you to read what you actually wrote. The first word or two of each line serves as a cue for your memory, which then recalls the rest of the line for you. You are blinded by your memory. You cannot edit because you are not seeing the manuscript. It is not there on the page. It is in your mind.

There is almost no recourse once you are in this situation. There is only one old trick for deadline editing. Journalists used to read a text *backward* to correct spelling and typos. In other words, go to the bottom of the page and read from right to left, bottom up. The text then becomes incoherent and your mind cannot simply read the text from memory. You will clearly see each word, and you will find almost every typo and many spelling errors. But other than using this one device, you are on your own.

There is no cure for this kind of memorization. You can read backward for typos and spelling errors because you do not need logical continuity to catch the errors. You can also run the document through the spelling checker on your computer, but you must read forward in the natural order of the text to edit for any other sort of error. All the other editing problems will depend on reading the text properly, and you will have trouble because you have two texts: the one on paper and the one in your mind. At work you can, of course, have someone else read the document for errors, and this is a common practice, but in the wee hours of the morning, there will be no volunteers. In addition, having another employee do your editing may not be a cost-effective use of time if the job was yours to do.

The final solution is to avoid deadlining. If you can clear your mind of your text by waiting several days to edit the document, you can continue with your work properly. The editing process should then be undertaken with care. Neither the writing nor the editing can be rushed if quality is the goal.

The Front-to-Back Method

Since the front-to-back method is a temptation for everyone, I have to be realistic and discuss it as a system for writing a draft of a project. You may have luck with the technique *if* you are well prepared as a result of fact-finding activities of the sort we will describe in the next chapter. See what happens. Begin at the beginning with the introductory paragraphs and have at it. Remember, however, that this is not the quick-solution strategy that people imagine it to be. It is basically slow and frequently shows poor quality control.

White-paper shock is very likely to occur in front-to-back writing, which is one outcome that slows down the technique. You cannot get started, or else an entire series of false starts lands in the wastebasket. Notice that the time it takes to overcome white-paper shock can almost equal the time it takes to write the entire project. You can sit for two hours until you finally hit on an opening, and then the rest of the document may fall into place in little more than the next two to four hours. You have just wasted as much as half of your time in this scenario.

Examine the problem. You are trying to write the introduction. The introduction to what? A report that is not yet written. You are trying to write a report, but you have not written it because you are trying to write an introduction to a report you have to write. Do you see the problem? You are chasing yourself by the tail. You are running in circles because you are trying to introduce a paper that is not there. You are attempting to do the impossible. However if you are intent on writing out the project as a completed document, there are a few tips that will help.

First, introductions are for readers; they are not for writers. You know what you are trying to write. Write it. Forget the introductions. The introduction of a report always comes first, but that does not mean that you have to write it that way. I wrote the *first* draft of the preface to this book *after* I had written four chapters. Why? Because no matter how prepared you are to write, an introduction is difficult to create if there is no text to introduce. *If you use the front-to-back method, I suggest that you omit the introduction and charge ahead with the main points of your project and try to write the body of the report.* If this seems to leave you without a rudder, go ahead and jot down a very sketchy introduction in no more than five or ten minutes just to get a sense of direction, but otherwise do not waste time on it. Be sure it is only a rough sketch. Do not devote time to it unless golden words simply flow across the screen of the word processor. Forge ahead. When the project is written, you can write the introduction. At that point, the task takes very little effort because the composition exists and you can clearly see the main points you may want to highlight in the introduction.

If you are fairly skilled at front-to-back writing, you may find that you can assemble an introduction before you go on to write the paper. If you take this tack, you have a special problem. You must make sure that the introduction does indeed reflect what actually happens in the project as it later evolves on paper. As I noted earlier, writers often plan on one set of outcomes and generate another set of results. *Once the project is written, go back and make any adjustments that may be necessary so that the introduction reflects the actual project you have written.*

This particular editing task is very important for two reasons. First, if the reader is going to read your report, the introduction must not be misleading or the reader will be confused by your work from the very start. The second reason is a corporate consideration. If your reader is not going to read your report closely but expects to see all the results on the first page, then you have a responsibility to mold a very precise introduction. The reader will assume that the report demonstrates or proves the findings you present on the first page or two. The second type of reader is very common in business and industry. Readers who want the documentation but who intend to read only a short briefing will depend on the accuracy of your opening remarks. Very commonly, this group will include chief executives and other managers. Their time is valuable. They expect the executive summary on the opening pages of a document to be a clear text and an accurate reflection of the entire project.

A second tip on front-to-back writing is to avoid deadlining if possible. We have examined the pitfalls of rushing the job. If you deadline, you cannot afford the luxury of writing at your very best. Time is cut short and so is your writing. However, if you want to leave enough time for several drafts, you will have to charge through a front-to-back format as fast as possible without having time to handcraft each sentence. *If you are willing to be careless about your writing, you can greatly speed up the otherwise slow process of the front-to-back method.*

I used a variation of the front-to-back method to write this chapter. I started writing as fast as I could in the front-to-back method because much of what I wanted to say was easy to state (I have said it all a hundred times) and because the logical order of the chapter was easily fashioned around what is called "temporal sequencing." All that means is that I discuss what to do first, what to do second, what to do third, and so on. In other words, I knew what I wanted to say and it was easy to organize, so I could rush through the creation of the text from the first paragraph to the last paragraph. However, unlike the typical front-to-back writer, I was *not* trying to write finished copy when I wrote the first draft. There is a big difference between the two goals, as we will see in the next chapter.

- If you use the front-to-back method to create a *rough draft,* the method is a practical tool, but it can be slow.
- If you use the front-to-back method to create a *finished product,* the method is ineffective.

I will later explain to you that you must judge the difficulty of your writing projects before you begin them. I judged this chapter to be somewhere between easy and moderately difficult to write. The work went fairly fast *because* I was reckless with my phrasing and word choice and careless with the cosmetics of spelling and punctuation and the like. The trick is to reread nothing if you can avoid it. *Just keep moving.* However, the genuine front-to-back method that many authors use to meet deadlines is quite slow because they work on continuity and cosmetics sentence by sentence as the project develops.

The final suggestion that may help you write a front-to-back project is quite simple: reading what you write wastes time. As you are writing your first draft, read back through the text only if you need to reestablish the logic of what you are trying to say. The secret is to never look back. Don't judge! The goal is never found on the line above the one you are writing. The goal is to create the line below. Do not stop the thought. By never going backward, you avoid judgments that will bring you to a halt.

When you are writing a front-to-back first draft of a project, you will confuse matters if you shift out of the role of writer and move into the role of editor or critic. Do not confuse the stages of the writing process. The creative first-draft stage is your only concern at this point. It is difficult enough to create a paper without also stopping to dot an *i* or dig through a dictionary to check on the meaning of a word.

Time Management for Writing Efficiency

A final convincing argument against deadlining has nothing to do with writing. Consider physical stamina for a moment. If you are a jogger, you know that you should prepare for your daily run with a warm-up composed of stretches to limber up the muscles, and a few vigorous exercises to get the heart prepared for the run. Then off you go.

For a few minutes the run is exhilarating, and then there is a phase during which the run becomes work and you sense the need for effort. After the body responds to the stress and you are fully up to your pace, you are on your way. Several miles later you sense that your stamina is declining and you stop at some point in response to fatigue. You finish the run by walking home to "warm down" the body. You can plot this activity on a curve.

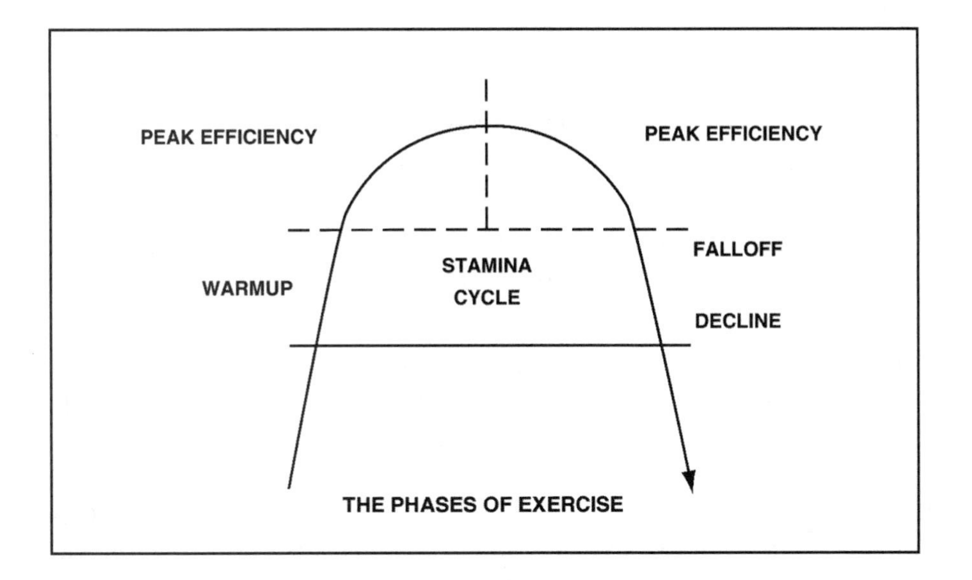

This phenomenon is an "efficiency curve." Athletes know it well, and so do coaches. A good coach knows exactly when to take a player off the field or off the court because the coach understands every player in terms of the efficiency curve. At the first sign that players are not doing their best, off to the bench they go.

Thinking is also a matter of stamina. Writing, as an effort to organize thinking, calls for a great deal of stamina. Mental stamina follows exactly the same curve. When you sit down to write, it may take an hour to "get into it." You are warming up. Then, for a time you produce with a certain amount of pep. You are moving through the peak efficiency part of your stamina cycle. Finally, the run is over; beyond a certain point your writing efforts decline. If you are deadlining, you can see a disaster in the making. Let's plot an efficiency curve on a clock with the assumption that you take an entire workday at the office to finish a project because it is due tomorrow. You start at eight in the morning and "finish" at four in the afternoon.

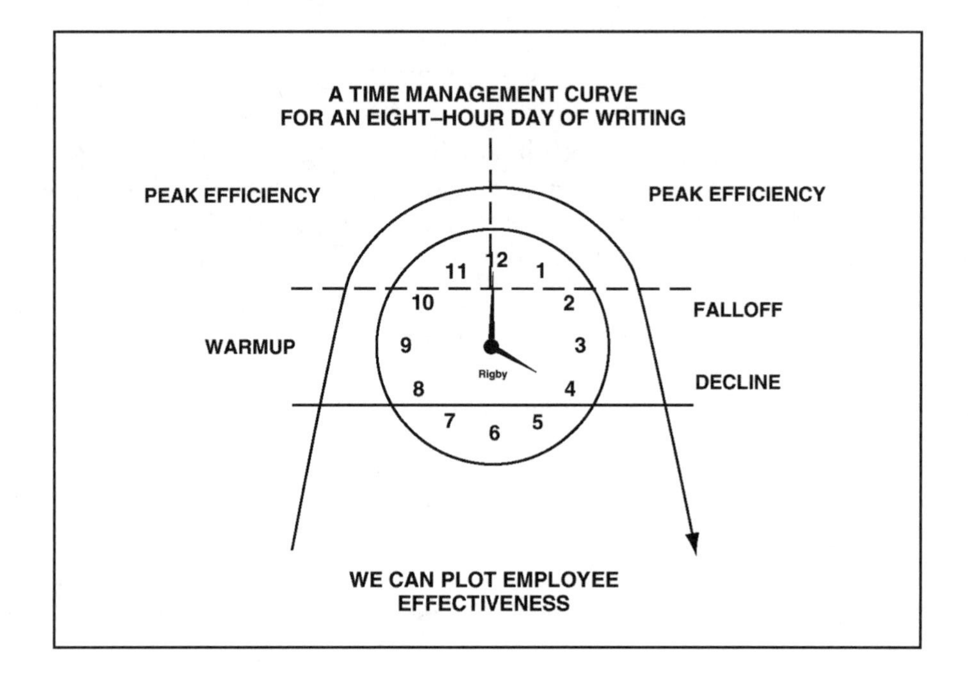

We clearly see a problem—in fact, it is a well-known national problem. The project is completed (let's say at four) when you are already tired, and it is edited (let's say from four to five) when you are at your worst!

College students are prone to an even more unmanageable pattern in the way they commonly go about writing papers by burning the midnight oil. Usually you will have time to develop writing projects over several days or longer at work, but if you are in college, you have a unique situation on your hands.

For a host of reasons, thousands of students across the country are busy writing papers every night of the year. The result is the familiar writing pattern I called the "crisis model." This particular group of writers starts to get down to business at about nine in the evening because a document of some kind, perhaps a short term paper, is due the next morning. The following drawing is the same as the preceding one except that the drawing has been amplified because students frequently *compress* the time they make available for writing projects. The efficiency curve has also been tilted to accommodate the odd hours.

The basic problem is self-explanatory, but do not overlook the following dilemmas. Remember that the efficiency curve was originally applied to *physical* stamina. At least the employee who writes the project at work has the advantage of daylight hours and an alert body.

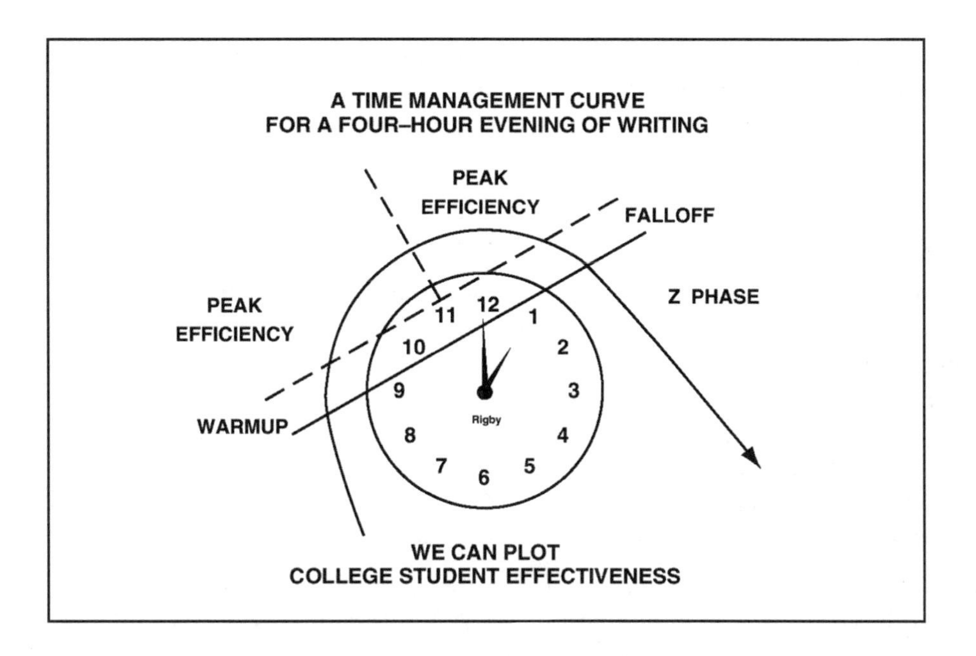

The traditional practice of writing projects "the night before" involves writing that puts the mind and the body at cross purposes. The body is in a falloff phase by midevening, just as the mind decides it is time to write. Frankly, if you have already put in a day at work or at college, the falloff began by four in the afternoon. What to do?

The answer is fairly obvious. Rather than waste four or eight or twelve hours trying to deadline a project all at once, use short sessions to increase the number of "peaks." The logic is apparent in the next illustration.

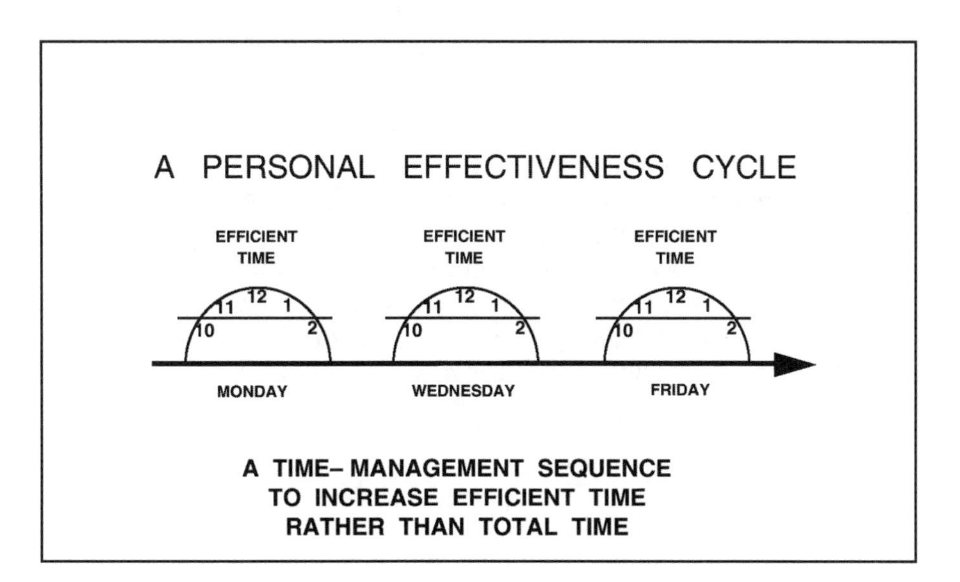

You also need a good coach. Unfortunately, you must coach yourself. Recognize "falloff" and respect it. Perhaps you run for health. It makes good sense to write with a healthy style as well. Do not turn the finish line into a deadline.

Since I am on the subject of stamina, let me say a few words about "writer's fitness" at the risk of seeming as if I am standing here in a gray sweatshirt with a whistle and a stopwatch. I shall limit myself to two key points that may help your writing.

First, I recommend that you write during the hours of the day (or the night) when you are at your best. You have to face the realities of work, but if you can schedule writing into your *best* hours, the practice may be of some help. Be at your best and you will do your best. This advice is obviously common sense, but it is also takes into account the scientific reality of circadian cycles. People can be fairly self-destructive if they are not sensitive to such obvious ups and downs. You *can* complement your skills if you perform the right tasks at the right time. Given the growing evidence from the new science of chronobiology, do not overlook the obvious.

The other little tip I suggest is that you utilize the pleasure of silence. Our homes seem to be the least likely havens of peace and quiet. The ever-present din of radio and television is hard to escape. At the office, the problem is usually the bullpen concept of open carrels that do not efficiently muffle sound.

If you write at home, be sure the television is off. It talks too much, and, besides, television is always engaging your glance because the medium is visual and so are you. Radio is not much better, since most popular music stations have talkative disc jockeys, plenty of ads, and music dominated by lyrics.

Words of any kind can distract your thought processes. They do not go in one ear and out the other no matter how accustomed you are to the medium. Words are *symbols* that go in your ear and disrupt your efforts to focus on writing. If you must listen to music, I suggest that you try your stereo and play only instrumental music. In a shop setting, oddly enough, noise may prove to be your best friend. If the noise is pleasant, it can drown out the distraction of talk.

In the office or lab, the problem is a delicate one: your coworkers. Recent findings have suggested that the bullpen concept of office design that became fashionable in the seventies is not the ideal office concept it was supposed to be. Removing office walls was supposed to produce a host of positive results in employee behavior. There is now growing evidence that open space is actually a barrier to all kinds of communication. In particular, because there is no place to go for silence, communication with yourself can be difficult if you are constantly disturbed or if you hear conversations all day long.

A marketing analyst executive for U.S. West recently told me that writing is the one chore he cannot seem to accomplish at the office. He spends his "writing days" at his home office, where his thinking is undisturbed. The complaint is common. If you find yourself dealing with the struggle to find sanctuary, see whether you can arrange for silence somewhere. If you solve this one problem, you will be one step ahead on your next writing task. Seek out the environment you need for your optimal efficiency. Know when to write; know where to write.

Summary

- Writing skills can be greatly improved by addressing undesirable habits that either make writing difficult or result in an inferior product.
- Procrastination, a well-known problem for many writers, uses deadline pressure as a task motivation, which produces rushed results and front-to-back writing.
- Avoid writing the introduction until the body of the document has been drafted.
- Front-to-back writing is popular, but the method is effective only if there is time for several drafts.
- Compose a front-to-back rough draft quickly and focus on content and not spelling and mechanical details. Address the spelling and grammatical accuracy in subsequent drafts.
- Avoid deadline fatigue, and write a project over a period of several days whenever possible so that you utilize peak-energy periods and use time efficiently rather than quantitatively.
- Write during your optimal hours and look for a setting that is quiet.

Activities Chapter 4

Select one of the following suggestions and develop a brief analysis of your writing habits.

Discuss a recent close call you have had with a project. Why did you procrastinate? Did you use the front-to-back method? Were you and your instructor or supervisor satisfied that you did your best? Discuss briefly.

If you effectively used your time on a recent project, explain why you used the planning methods you used. Explain how the project was handled from start to finish. Were you satisfied you had done your best? Did you do well in the eyes of your instructor or supervisor? Did you get a raise (or at least a high grade)?

These projects should be 250 to 500 words in length, typed in a memo format** (see **Chapter 11), and addressed to your instructor.

Work in Progress

The Project Research

Once the proposal to the President was submitted, I was relieved to have it finished. Since I was proposing an expensive, time-consuming reevaluation of my company's safety program, I believed that Karen's total acceptance of my entire proposal would be somewhat unrealistic, so I needed to get to work on what I thought she would definitely adopt. Based on our past discussions, the Respiratory Protection Program and the Hearing Conservation Program were definitely two plans that were to be implemented.

Pacific Aero Tech refurbishes, repairs, overhauls and sells almost all kinds of aircraft windows and avionics equipment. The company has a strong customer base for aircraft parts—especially windows. We are able to compete with other larger commercial companies because we do our own work. By keeping the repairs and overhauls "in house," our overhead is minimal and we operate at a lower cost. Also, because we are involved in each stage of the repair or overhaul, quality workmanship is guaranteed. PAT takes pride in its professional work. Specially trained workers have been hired to be a part of this team that can produce quality products. To accommodate these specialists, unusual tools and special work spaces are required for their unique skills. For example, we have technicians who remove scratches from passenger windows and others who polish and buff these windows to a sheen. Others take the windows apart, remove the old paint, install all new hardware and new glass and repaint the window frames.

In the process of getting these windows finished properly, and on time, certain safety precautions were identified to protect the employee from all recognized safety and health hazards. There are several potential health hazards—and associated regulations—that have been identified at PAT. These include several problem areas:

- *Airborne particulates*
- *Airborne organic vapors*
- *Noise*

As a result of my research, I could identify and explain the regulations concerning these matters.

PAT creates "nuisance" dust as a result of industrial processes. Nuisance dusts are nonfibrogenic particulates (they do not cause scarring of the lungs) and so no PELs (Permissible Exposure Limits) have been assigned. However, the total particulate exposure should not exceed Washington Industrial Safety and Health Act (WISHA) limits for total dust and for respirable dust. Also, these dusts are not permitted to accumulate on the floor or on other surfaces.

PELs for organic vapors are specific for the vapor of concern. Acetone, toluene, isopropyl alcohol, and other potentially dangerous chemicals are used at PAT.

Personal Protective Equipment (PPE) issues also needed to be addressed. Respiratory, hearing, eye and body protection practices are required by law if the hazards exist. Chapter 296-24 WAC contains regulations that pertain to the general health and safety standards.

The hazard assessments had to be documented. I needed to assess and document what specifically each employee in his or her working environment required. What I found was astonishing, but more surprisingly, the research created a very negative response from the employees.

L.C.L

Before You Write: Preparation Strategies

Since you now know that you face a unique *engineering* task in getting thought to paper, you must develop strategies that will complement and utilize your thinking, speaking, and writing habits so that you proceed with a writing project in a way that will produce the best possible results. If deadlining is hard going and if front-to-back projects are going to be less than the best, you need alternatives.

Your first task is to have some subject to write about. I noted that sitting down to think of *something* to say and also to think of *how* to say it is too demanding. This approach is starting too late to produce quality work. Preparation is a must. Writing should not merely be a task you have to do because you have been told to do it; you should write because you have something to say. An assigned writing task is not just make-work. Engineers write reports because they need to record circumstances of one sort or another. If the results are going to be worthwhile and useful, you must accept the task of first thinking out the project with care.

The actual number of stages that an experienced writer will go through in a writing project may come as a surprise. Many people think that writing should be a swift process and that they are poor writers because they do three drafts and make endless changes. The truth is that writing many drafts and making lots of changes are signs of professional behavior and professional responsibility. In reality, a project that is developed properly goes through at least ten stages of development. Because of unique situations that characterize corporate documentation, technical and engineering projects will go through a dozen or more stages of development. Notice the documentation schedule on p. 85.

The work schedule we see in the table is designed to imitate the "critical path" scheduling system that is used in many engineering firms. The third column shows the starting point of a given activity; there the length of the activity is indicated in the overall scheme. It is a rude awakening to compare this method with the typical one-night turnaround time that many college students depend on for important projects.

Although the schedule in the table is intended to reflect standard procedures for workplace projects, it also reflects two additional realities that are worth noting. At the outset of this text, I explored thought and speaking and writing as three discrete skills that should be at work in the writing process. This fundamental reality is at the heart of proper writing strategies in the workplace. The table reflects stages of the writing process that are well outside of a simple perception of "writing" as just writing. Several important stages of development, for example, are driven by conversation (Stages 1 and 5 on the chart). Several others are driven by the activity of coordinating thought (Stages 3, 4, and 6). Also observe that it might appear as though most of the later activities (Stages 6 through 12) are aspects of the writing process, but this is deceptive in a workplace situation. There may be a good deal of employee interaction, particularly in Stage 12, where there is a need for peer review. Since work-related writing is often collaborative, the table must include this final provision for peer review. Fundamentally, any review of a manuscript by coworkers is an invitation for dialog. Editing is a version of dialog, particularly when the editor is not the author. If a document involves teamwork input, then conversation is implicit in the later draft stages (Stages 9, 10), and indeed, a substantial amount of corporate work is handled with some level of team involvement.

In addition, engineering documentation calls for a special variety of peer review in which you invite someone to read through your work for *technical* accuracy. You need this specialized review as a double check or a fail-safe edit. It could, of course, occur earlier. This stage is an important feature of technical writing. Cost-accounting features are the other major area of concern in which you want to be certain that the text is technically accurate.

I will now walk through a typical project and talk about planning and executing a document. In this way you will experience the situation vividly. The table is not self-explanatory, since it represents at least a week or two of typical work-related processes. To see the table at work you need to hear the clock ticking and experience the hurdles of the job. Let's start on a Monday morning.

DOCUMENTATION WORK SCHEDULE

	Phase	Project Input	Path Duration (accrued)	Project Output
1	Project is Initiated	task memo		perception of project
2	Critical Path Analysis	agenda for the activities		calendar of the events
3	Project Sketches	author's notes		key positions and supportive logic
4	Research	scientific data, developmental data, cost data		statistical findings for support
5	Coworker Responses	critical dialog		recommendations, emendations, negations
6	Form and Content Analysis	logic architecture for the document		document outline
7	Development One	creation of initial document parts		rough copy of components
8	Development Two	creation of organized whole		rough draft copy of full document
9	Revision One	copy modifications		second draft
10	Revision Two (option)	second copy modifications		third draft
11	Final generation edit one	"clean" copy detailing		model document
12	Peer Review	engineering/ technical edits		ready document

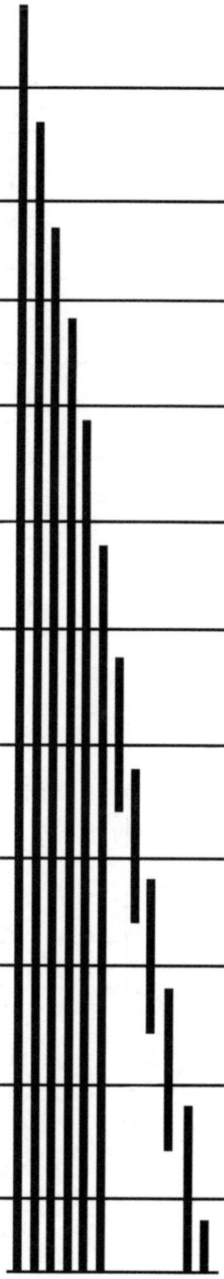

Stage One

Sketch of the Project

Try to get the tasks defined in writing even if you are the person in charge of the guidelines. Make the task real; make it visible. Sketches, notes to yourself, a rough outline, or a formal outline will do the job. Then discuss the plan with your supervisor before going ahead with the project.

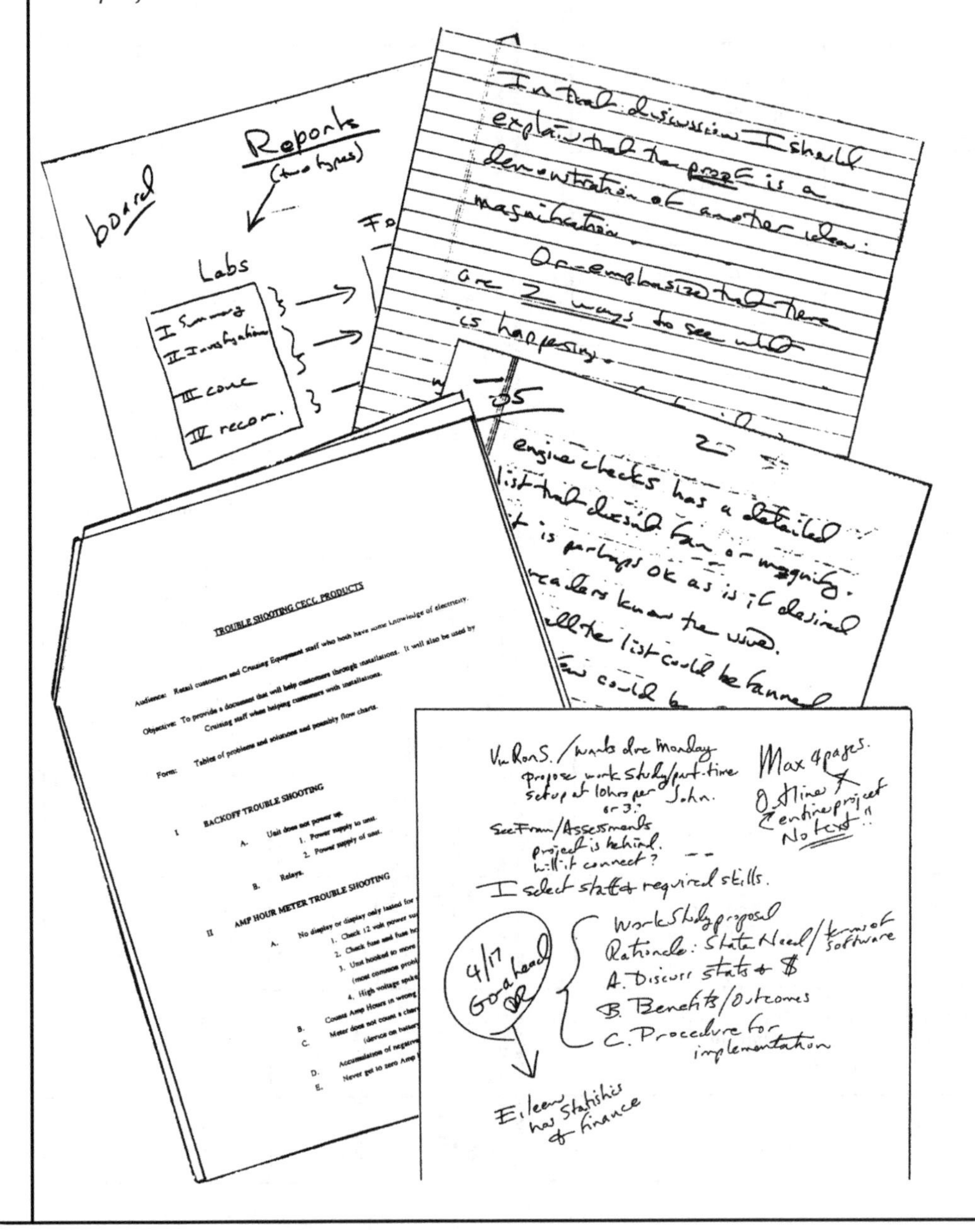

Project Evaluation

To gauge your involvement, your first job is to determine the *task*. If the writing is requested during the course of conversation in a lab or office setting where there is a lot of activity underway, do not hesitate to look for the supervisor at a later hour when you can calmly follow up the initial assignment with questions so that you know how to proceed. Be sure you understand what you are being asked to do. Writing takes time, and unless you know exactly what you have been asked to produce, writing can waste time. This is the first step in your game plan.

DOCUMENTATION WORK SCHEDULE

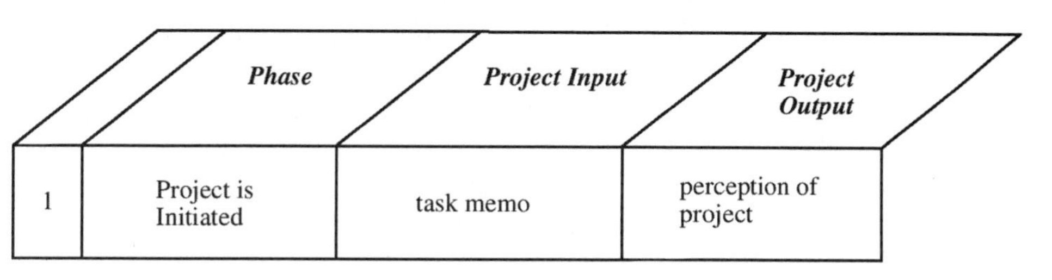

	Phase	*Project Input*	*Project Output*
1	Project is Initiated	task memo	perception of project

It is helpful to have a memo explaining any writing project you must produce. If you have a sketch of the tasks, you can calendar the events necessary to complete the presentation. You may be lucky enough to receive such a memo. If not, write it yourself. In other words, note on paper exactly what you have been asked to develop. Discuss this sketch of the project with the supervisor before you go any further. In effect, you will produce a memo to yourself, and it is a logical place to start. I refer to this document as a "task memo." I see it as quite distinct from the content outline that we will discuss later.

The idea of a task memo is to tell yourself what to do. Even routine writing tasks can be analyzed in this way. If you have a variety of writing chores that are everyday occurrences, there will usually be no memo to explain them. You can scrutinize any of these tasks and write a short note to yourself about how each effort is to be handled.

Once you sketch the tasks you are expected to develop for a writing project, look for some way to measure the degree of difficulty. Do not look at length as the key; besides, you will not usually know how long a document will be until you complete it. Instead, look, for example, at the extent to which your thinking out the issue must precede writing out the issue.

Calculate your preparation time first. You can measure a project as least difficult or most difficult simply by determining the extent to which it is or is not investigative. You then can judge your calendar accordingly. An easy project, for example, will take a minimum amount of lab time or office activities.

Another way to judge the difficulty of a project is to look at the measure of difficult thinking involved. It is important to realize that your judgments are involved in every type of analysis you might conduct. In the technological environment this fact is not always clear. I can divide engineering and technological documentation into several types to examine your role in creating them. Suppose you are troubleshooting a defective mainframe computer for a banking institution. Your task is simply to get the device back on line. You keep a checklist on your bench and a product evaluation report form for detailing your findings. Because this particular mainframe model has proved to be systematically defective in many banks, your firm is compiling a report for the manufacturer. Suddenly, you find you must write out full reports on each incident of a malfunction for this particular computer. Basically, the task is *descriptive:* you describe what went wrong.

But suppose that the report requires *more* of you. If the supervisor wants the report to go beyond an explanation of the defect, and he requests your analysis of the *cause* of the problem, you would observe that your perceptions are assuming greater importance in the project. Your *judgment* is now definitely involved in the work requested. In the process of determining the trouble with the computer you can apply more than your know-how. You can go beyond the repair and offer an analysis of the problem.

We can go a step further. You may be requested to find the fault and examine the cause, but you may also be asked to recommend *solutions* to the problem. At this point, of course, there is little doubt in your mind that all your critical and professional skills must go to work to develop the report. Most technical documentation is likely to fall into one of these three categories of involvement: *(1) observations, (2) observations with analysis, (3) observations, analysis, and recommendations.* The more demanding tasks involve your evaluations. Your thoughts on the matter become more important as the category of involvement shifts to analysis and then to recommendations.

The approach you take for developing a report must first be measured in terms of your involvement. Yes, you have to write it anyway, but the more a project depends on your *thinking* skills—from observation to analysis to evaluation—the more demanding the project will be to write. If the project is going to involve background data, an analysis of problems, and an analysis of solutions or a presentation of recommendations, then get ready for the challenge. Your mental efforts will be at a maximum level of involvement. Of course, the document will be longer as well.

Stage Two

Calendar of Tasks

I like to schedule my writing projects ahead of time. A month-by-month calendar is handy for time management. Divide your time into units of preparation, units of writing, and units of wrapping up the job. The number of units is based on the situation. A business letter takes few; a "presentation" may take a dozen.

JANUARY 2000

SUN	MON	TUE	WED	THU	FRI	SAT
Osmiote Balanced Test CallSRS/via?		International documents There are 2 papers evaluations > judgmental description				1 1/365 New Year's Day
2 2/364	3 3/363 New Year's Bank Holiday (Scotland) $750	4 4/362 Vernschys reservations Rosebk	5 5/361 Banklive	6 6/360 Call Anderson Rob	7 7/359 Groc. Lumber for Salt.	8 8/358
9 9/357	10 10/356 Dept Meeting 9:00	11 11/355 call DyNam	12 12/354 Apr 28 Mariners game 7:00/call for	13 13/353 Tom 1:00	14 14/352 Div. Meeting 1:00	15 15/351
16 16/350 Keep open — Article week!	17 17/349 Martin Luther King, Jr. Day (US) Scott Report due!	18 18/348 Do egg's Fiorini Sketch	19 19/347 DATA? See Sue Steve Frank	20 20/346 1st DRAFT	21 21/345 FINISH DOC. IF POSS	22 22/344 WRAPUP by noon?
23 23/343	24 24/342 PROJECT DUE	25 25/341	26 26/340 Australia Day (Australia) call John for Friday	27 27/339	28 28/338 John 2:00 reviewed approval	29 29/337
30 30/336 Plane 5:45	31 31/335					

Your Calendar

Next, you should look at your calendar to schedule the stages of the project. Once you know *what* to do, you can turn to the calendar and plot your tactics. For now, let me simply remind you that calendaring a writing project signals that you understand the actual nature of the chores. Writing is not a product; it is an event. I made this comment much earlier, but now you can see clearly the meaning of the observation. I will plot out the activities of your task and discuss them. Whether you call it time management or critical path scheduling, writing is a dollars-and-cents issue, just like every other work activity.

DOCUMENTATION WORK SCHEDULE

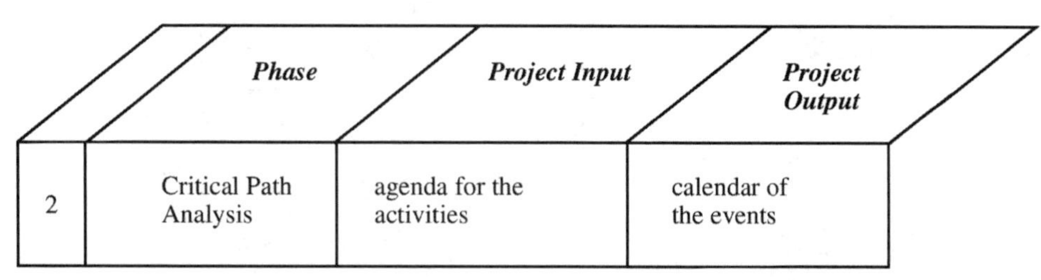

	Phase	*Project Input*	*Project Output*
2	Critical Path Analysis	agenda for the activities	calendar of the events

To develop a model writing event, let's continue with the hypothetical project. Assume that you were assigned a writing project during a weekly-update meeting held every Monday morning. You were a little hazy at that hour, and so you did not clearly understand the task. Your supervisor does not usually write memos, so you go to her office after lunch to discuss the project. You jot down the intended tasks of the project and *repeat* to her the outline of chores so that you and your boss agree on what you are about to do. Be sure to repeat the tasks as you understand them.

Back at your workstation, you assess the degree of difficulty: since the project is fictional, assume that the proposed project is very complex, so the degree of difficulty will be maximum. You stare at your calendar and write the word "due" on the Monday following. Next, develop a critical path schedule. Writing is an industry like any other. Learn to know where you plan to be tomorrow because planning will be your profit or your loss. You should buy what is called a "planner"—a roomy calendar with space for filling in your days. I like the monthly planner the best. It gives me an overview of four weeks of whatever I am working on, which is plenty.

With the task memo in hand, or at least some kind of written sketch of the tasks, begin to tentatively assign days to the activities. The actual time frames will vary depending on the nature of your work. Research, for example, may be only a printout away, or it may take days or weeks. With practice, you can make the calendar synchronize your work. The challenge that remains is to follow the calendar, if possible.

Stage Three

Mind Time

Carry a notebook for jotting down thoughts. I carry a small notebook that fits in my shirt pocket, and I try to capture ideas as they come and go. The following samples do not show the grocery lists, phone numbers, and endless reminders I maintain along with my project ideas. The small items are yellow sticky tabs from my dashboard. If you do not write it down, you can write it off.

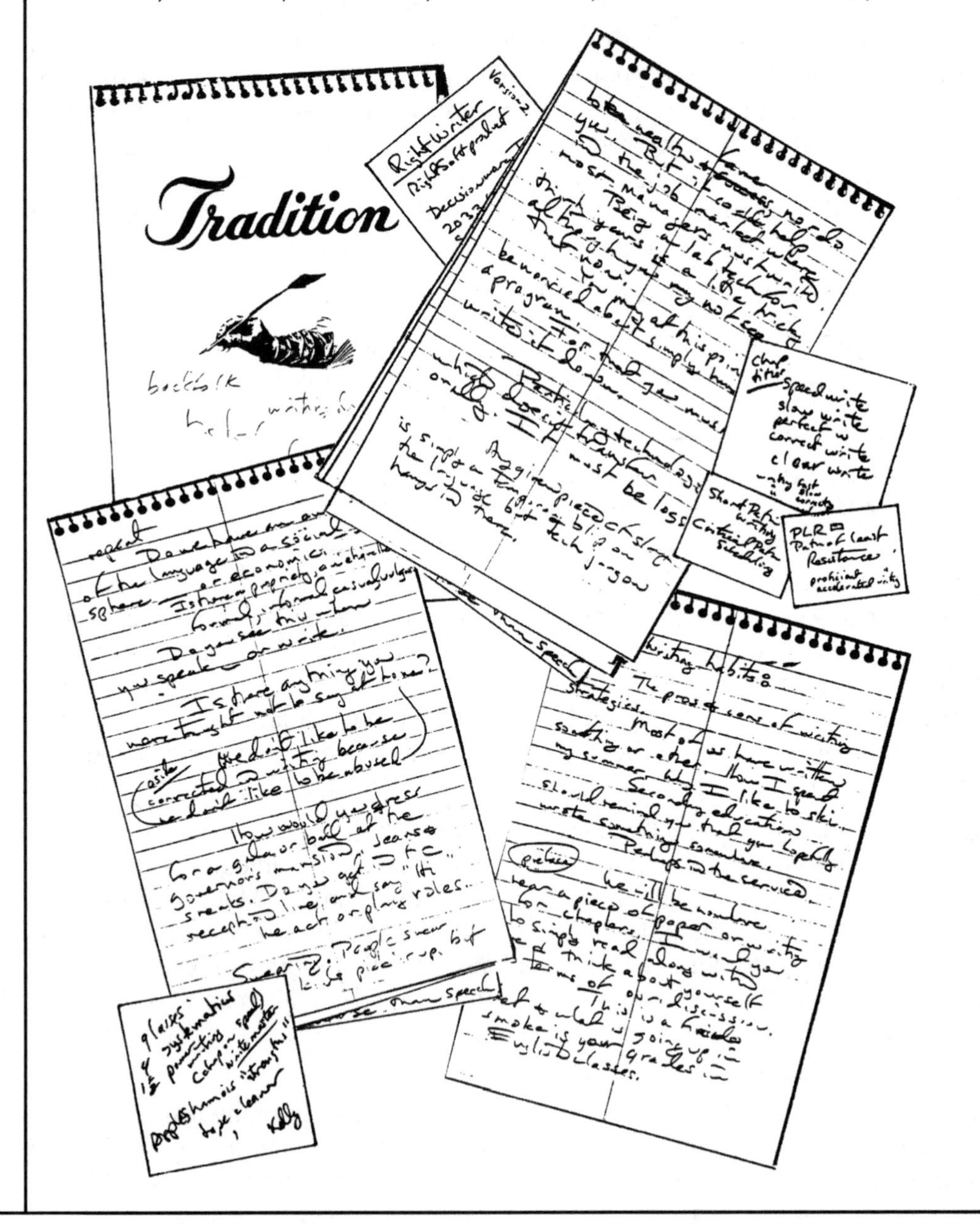

Your Log

Your mental commitment has begun, and you want to keep track of your ideas in a logbook. Do not think of this item as a journal or a diary. My experience at keeping a "journal" has been unsuccessful. I have three empty books stored away, each of which contains about five pages of very complex writing and several hundred blank pages. I am not a person of leisure, and I do not have time set aside to think and write page after page of material. Besides, I found a journal too challenging because I felt I should say something fairly profound every day, and nothing very profound happens to me or occurs to me on most days. I abandoned every one of my journals. Logs, however, I fill one after another! Let me explain.

DOCUMENTATION WORK SCHEDULE

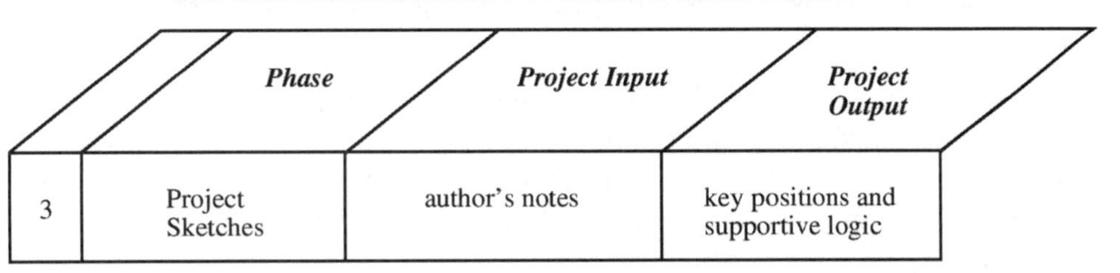

	Phase	*Project Input*	*Project Output*
3	Project Sketches	author's notes	key positions and supportive logic

I am a busy person and I am certain that you also are busy. You have a lot on your mind, and so do I. It is very hard to keep up with life's details. If I go to the grocery store for three items, I will certainly forget one. It is a joke in my family. I do not remember phone numbers, and I am hopeless when it comes to birth dates and anniversary dates. How in the world am I going to remember serious matters concerning my work if I can not even remember to buy coffee? I could argue that I should remember work-related issues because serious matters are more important than grocery lists, but there are, in quantity, a great many such serious matters, and they can be very complex. People forget, or get confused, because they do not develop the habit of tracking their thinking.

Most people do not write things down. Forgetfulness is the best reason in the world for keeping a logbook. And, of course, to say "I forgot" means that your supervisor may be a little forgetful when the next promotion comes around. The log is not a hefty journal composed of great mental moments, because most of our thoughts are less than monumental. The log book is nothing but a small notebook that can be kept on the lab bench or an office desk. It must be portable so that it is with you at all times. Use it for phone numbers, grocery lists, recipes, appointments, all manner of daily reminders. *And use it for ideas.* Develop the habit of logging your thinking. If you do not, your best ideas will never see the light of day.

Start off by kicking around ideas. Set aside time for the activity. If you use the hypothetical model of a report that is due in about eight days, perhaps you can allow three days to shape the needs and demands of your document while you decide about the lab work or fieldwork that will have to go into the project as your support data. However, keep your logbook with you. If you allow a half-hour each day for thinking about the project, your mind will start to take up the challenge, and your thoughts may start to percolate at odd moments.

You must be keenly aware of the use of setting aside a little time to face the project and briefly writing down a sketch of your ideas. You must also, then, be prepared to act fast if your mind takes up the task.

The only way to deal with the mind is to take it on its own terms. Try to keep up with it. Write down what you may not remember. Follow your ideas. Perhaps the most frustrating event is the occasional perfect sentence that will occur. The expression is what seems perfect. The thought rings clear because of the *way* it is phrased. But it will be gone in *one* minute if it is not written down. I take the problem so seriously that I will pull off the road to write down ideas.

Sometimes I am only trying to write down a few words, but words I had never before thought of and might lose in a minute. I also keep yellow sticky tabs on my dashboard so I can write at traffic signals, which is where I thought of my trademark, *Wordworks*. Many ideas for *Wordworks* were spontaneous ones that I placed in my log.

You might think that this is going too far, but remember, you will have some of your best ideas at odd moments—in the cafeteria, in the parking lot, at the wheel of the car, at home. Do not let them go. Do not forget the discussion of the way people think. It is a little busy up there. *It is easier to write down what comes to you than it is to sit down and try to think out what you want to write.* Do not discriminate, by the way. The log is for your convenience; it is *not* for your writing project. It must be a *habit;* it must be your method of being efficient, your method of making the most of your thinking skills and daily tasks. I use mine almost every day and at any time. You cannot assume that your work environment is the only setting for your thinking time. You are as likely to have your best ideas as you sit in your car stuck in the middle of the five o'clock traffic on the freeway. I have often stopped one of my lectures to write something down.

You know from the analysis of the thought process that you need to somehow deal with the wild free spirit of the mind. There are only two options: either you force the mind into productive channels, or else you tag along and watch yourself think without restrictions. People tend to underutilize this second strategy, or they do not record the results.

If you are in the middle of a very important meeting, and you remember that you need antifreeze for the car, log it. If you hear an engineer explain a useful idea, log it. If you do not agree with the opinion of your boss, think out your objection and log it. Use the log to keep up with your mind. To the extent to which you develop the practice, you will keep up with your writing projects as well. Memory does not serve you well, but writing can hold on to every thought.

Under the usual circumstances of a workday, your thought processes do amazingly well even without time to "sit and think"—but you have to be fast. You have to catch yourself problem solving in line at the cafeteria and log any quality ideas. Inventive though people may be, imagine the incalculable loss they experience out of self-doubt and a little laziness because they let ideas vanish.

As you try to apply your thinking skills to the project, be on the lookout for a misleading preparation habit. Worrying about *having to do* the project does not move you one step closer to completion. You can be sure that anyone who is assigned a worrisome task is going to be thinking about it *constantly,* but thinking about the matter *constructively* is a different issue. Thinking about having to write a paper is not the same as thinking out the project. Worrying is not very helpful.

Try to distinguish between productive thinking and plain old anxiety. You can easily waste time worrying. Everybody does. The point is to avoid both wasted time and needless worry. If you would rather be done, then manage your time. Begin by managing your thinking.

Stage Four

Research

You may need some sort of research to support your project. The situation may call for no data, or nothing but data. Usually the sources will provide economic, statistical, or scientific findings to substantiate the project.

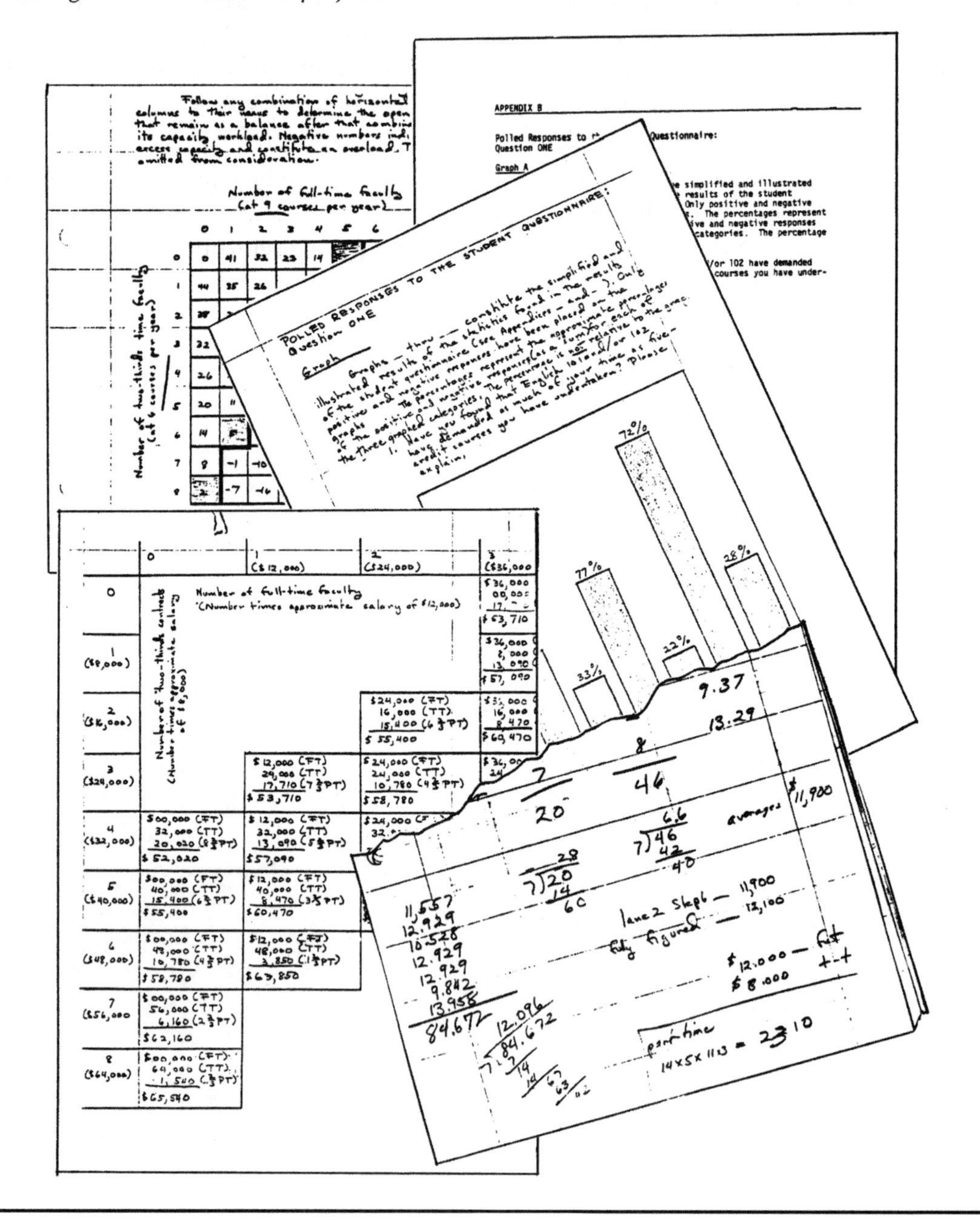

The Fact-Finding Phase

One of the determining factors that will shape a company document will be the research. Although I am not looking at this theoretical writing activity primarily as a "research project," most technical and engineering documentation will depend on some measure of input that supports the overall presentation. The context in which any particular project operates is so enormously variable that I need only make a few obvious comments at this point.

DOCUMENTATION WORK SCHEDULE

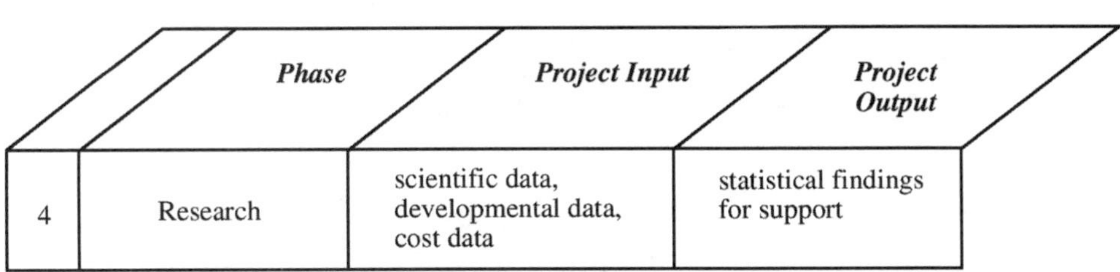

	Phase	*Project Input*	*Project Output*
4	Research	scientific data, developmental data, cost data	statistical findings for support

Since you are a student, I will define resources in terms that I call "primary" and "secondary." A primary source is any research that involves direct, original findings that support your project. Laboratory research is one obvious support that will assist you as a primary resource.

A secondary source is research that is reported by other companies or other researchers. University library facilities and the Internet are the likely points of contact for secondary resources. College students rely heavily on secondary research, which is the historical reason for a strong academic emphasis on knowing how to document sources with reference notations and bibliographies.

You may have access to one or both types of resources. The corporate employee will draw on company resources. College students will do research in the library stacks or use search engines on the Web. The former uses primary sources; the latter use secondary sources. In truth, however, both situations use primary and secondary sources. Upper-division students are involved in their own primary source laboratory activities, and corporate employees usually know what the competition is up to if the information is available in published reports.

Fact-finding lays the groundwork for a project, but it is not, as we will see, the exclusive resource. There are additional supports that you can draw on to establish the foundation of your projects.

Stage Five

Notes From Conversations

These samples were tabulated, but casual conversation with your coworkers and clients is a more common source for the information you might need. Write the comments down, or they will be lost.

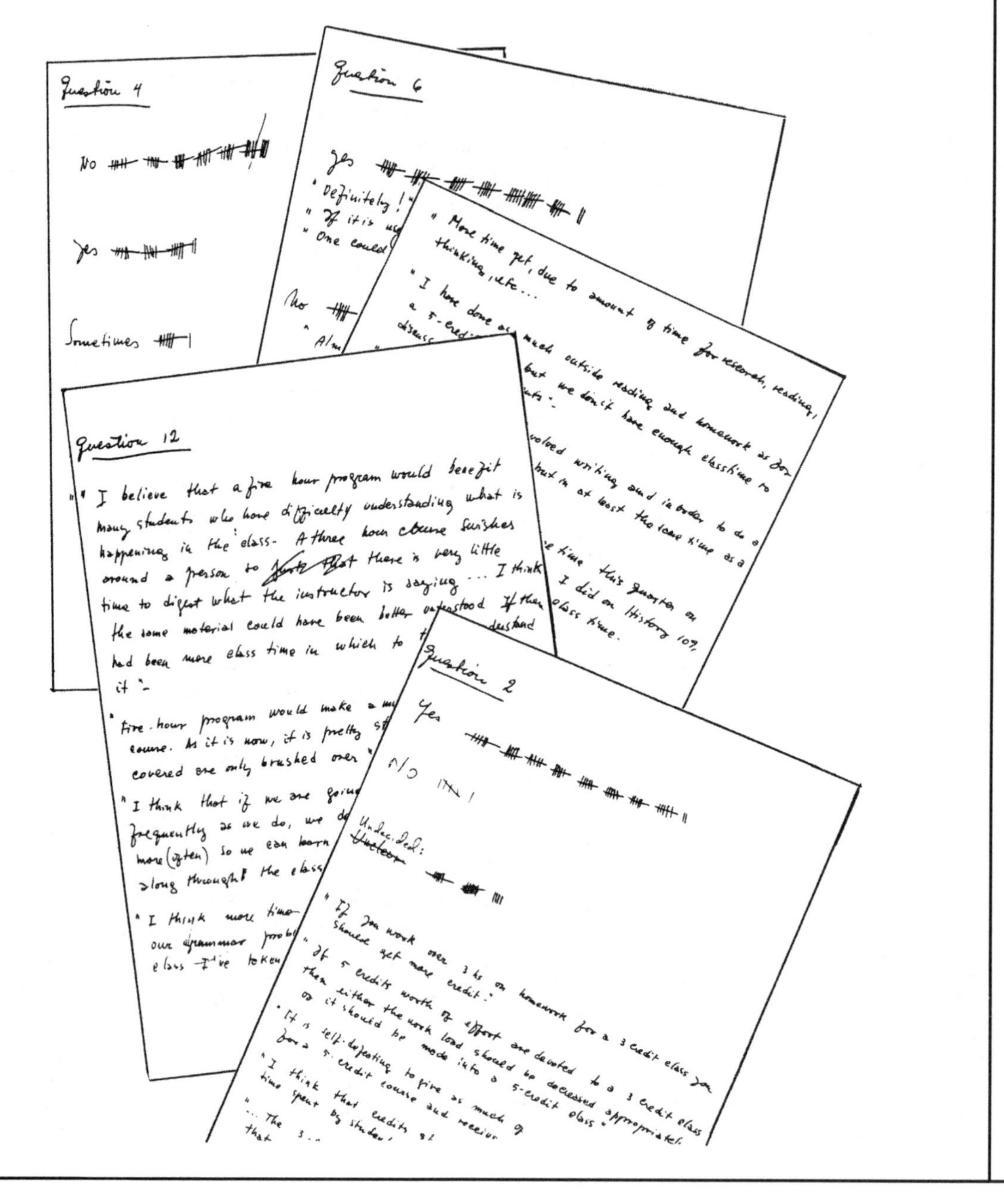

Stopping to Listen

Both the task outline and the logbook involve you in conversation. The dialog began when you discussed your project with the supervisor. In the case of your logbook, you are talking to yourself. Since conversation is a favorite pastime, turn it to your advantage.

DOCUMENTATION WORK SCHEDULE

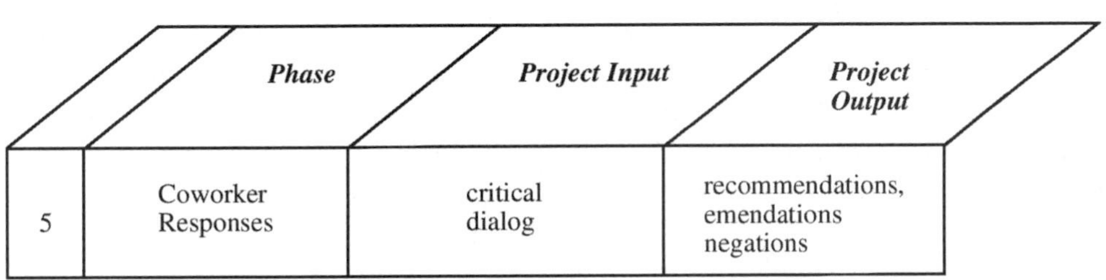

	Phase	*Project Input*	*Project Output*
5	Coworker Responses	critical dialog	recommendations, emendations negations

Inputs

As we discussed earlier, conversation is your preferred medium. Do not hide away to write your project. There are private moments to be sure; when you want to think out matters or when you want to write, you may not want to be distracted. Do not view these private activities as the only way of getting a project completed.

Conversation is a great help. The notion that writers must go about the craft of writing by sitting at a lonely desk in a lonely room and looking at a lonely wall is not accurate, and the behavior is not necessary. There are stages of a writing task that are quite public—or at least should be. This is particularly true in a company setting where teamwork is the norm.

Look for help—before you write, rather than after. Many working writers probably have the habit, or at least the hope, of asking secretaries to correct all the errors in their projects after the papers are written. Of course, *depending* on someone else's skills may not be an ideal form of teamwork, but this example demonstrates the reality of work. Of far more value to you than the secretarial help would be the use of *conversations* to help you gather ideas and to help you collect your own thoughts *before* you begin the writing. In a "situation-based" project, such as you might find in your workplace, background investigation is very likely to consist primarily of talking out the issues.

College writing projects always involve research, or so it seems. Your company is not likely to have a library for research work. Conversation is a useful resource in such a case. The use of conversation can run from a formal interview to the most casual conversations on a coffee break. Chase after the ideas. The utility of dialog is apparent. First, conversations provide information. Second, as you learned in the earlier discussions, conversations force people to express their ideas. You need to talk out your ideas before you write them down. If they are not going to pass the expression test and the reality check, you want to know that well before you begin to write a report.

Collaboration

In Chapter 3 I explained a unique feature of dialog that takes shape in a give-and-take interaction. Conversation has a push-and-pull effect on the logic under discussion. Ideas constantly gain and lose ground in a dialog, but if all goes well, the result will be new and improved ideas. To illustrate the concept, let's look at a hypothetical situation that demonstrates the principle involved so that you can examine the logical analysis that unfolds as people discuss each other's ideas.

Assume you are engineering a heating and air conditioning system for a building. Your task is to present a proposal to the architect for the HVAC system of an assembly plant. During lunch at your office, you tell a member of your design team that you think you should go with electric heat in the proposal. She suggests that you should choose gas. The discussion then evolves. Look closely at what is happening. Assume you represent a positive force. Your coworker means well, but in terms of your proposal, her observation represents a negative force. In the collision of the two ideas there is possible third or alternative option: compromise.

You might convince her that electricity is the way to go. Or else she might convince you that gas is the way to go. Or the two of you might emerge with an alternative plan that has resulted from sharing ideas. (You might write down the arguments that each of you develops.) If, in the first case, you convince your coworker of the appropriateness of your ideas, she has helped you explain those ideas. In the second case, she may convince you of an alternative that you had not fully considered. In the third case, you work out a proposal that evolves out of the best ideas from the two of you. *Your coworker helps in every case,* and in the event of a strongly shared effort to work out a problem, the two of you can exceed the sum of your contributions (recall the discussion of synergy) by developing a superior plan that neither of you anticipated. This is the value of collaboration.

Reactions

Since you want to try to transcribe or log the important aspects of key conversations, observe one problem about taking notes in a log. I have noticed that many people seem to have trouble evaluating what is or is not important in a conversation. Many men and women have trouble determining the key points in situations such as lectures, which *should* be clear cut. The person behind the podium is the authority, and listeners write down the primary and secondary points. They judge the importance by the emphasis that

ideas are given by the speaker. It sounds easy, but many people seem not to have mastered the skill of taking notes.

Committees are at the other end of the spectrum. *Everyone* is talking, ideas ramble, noisy voices may be unimportant, important voices may go unrecognized, and there may be little sign of an agenda. Trying to take notes on committee meetings can be very difficult. Open conversations in a group can be a difficult situation to analyze.

Conversations between two or three people probably fall somewhere between the extremes. It may take practice to learn how to judge the importance of comments that emerge in a conversation. Try to write down what *you* think is important. And do not forget to take notes on *yourself* in a conversation. This point is easy to overlook when, in fact, your own ideas are the most important element of a dialog since *you* have the responsibility for writing the report. *As you react to the ideas of a conversation, be sure to log your own reactions that you consider to be important.*

You can put great store in people as a resource. Of course, in situation-based writing projects of the sort employees develop at work, the "research" available to them may consist of little more than putting their heads together anyway. For the HVAC proposal, you have only the architectural drawings, your engineering tables, and your engineering know-how. The rest is up to you—and your team members.

The team is important. People are not always very imaginative when it comes to self-correction. Encourage the ideas of coworkers. Count on the extra input. I sometimes think that genius is mostly the ability to independently build a process for self-correction so that a mind can, *on its own,* exceed its limits. Conversation is a valuable assist, and for working writers, a little help from friends allows them to be at their best on paper. In sum,

Identify the task.

Think about the task.

Talk about the task.

Log the results.

Summary

- If you are asked to develop a document, request a task memo that precisely defines what is expected of you.
- As an alternative, discuss the project with your supervisor and write down the expectations.
- Measure the task. Will it involve observations? Analysis? Recommendations?
- Develop a calendar of the activities associated with the project.
- Think of the project in terms of your mission and begin to keep a log of your ideas.
- Conduct appropriate research as needed.
- If the fact-finding phases involve frequent conversations with appropriate parties, use coworkers as a resource. Take notes.
- Field your own ideas in your discussions because coworkers can be helpful in other ways. They help develop important ideas and perceptions that will assist you.
- Keep a record of your reactions to constructive conversations.

Activities Chapter 5

Select one of the following suggestions and develop a brief analysis of the stages of your writing procedure for a project.

Study the documentation work schedule on p. 85. You have examined the first five stages of these project phases. Use a recent lab project to see if you are utilitizing the steps you have analyzed. Briefly explain the history of the laboratory report in terms of how the project was initiated, how you handled timelines, how you handled preparatory notes, how you conducted research, and how you used coworker input.

Consider the same issues as in the previous activity, but discuss a paper you have written recently. Explain any difficulties you experienced during the preparation of the paper.

After you have written your first paper for this course, write a brief analysis of the first five project stages.

These projects should be 250 to 500 words in length, typed in a memo format and addressed to your instructor. The third activity can be attached, as a cover memo, to the first essay project you develop for this course.

Work in Progress

Employee Reactions

After word of the OSHA proposal got around the company, I expected some initial resistance from the employees because most of the staff had established a routine without concern for personal protective equipment, but I was amazed that the supervisors also initially balked at my suggestions. At this juncture there wasn't any concrete "go ahead as planned" from the President, so I used the lag time to gather more information.

Unfortunately, this phase required me to talk, observe, and document the employees in their natural working environment. This meant getting in their way, asking questions, and being a general pest. After a while, I had the impression that I was the enemy invading their sacred space, but I persevered. Armed with a notebook, dosimeter (to measure noise), and a healthy resolve to make a quick surgical strike, gather the information I was seeking, and get out, I entered the warehouse determined to not let the employees' lack of enthusiasm mar my confidence.

The window shop is a particularly tight group of males who excel in their particular craft in the avionics window repair industry. Most of the company's income is generated from the window shop's work, so time, production, and costs are foremost on the minds of the supervisor, Matthew, as well as the window shop employees. They felt that I was going overboard and Matthew didn't even have time to see me. In fact he flat out stated that a respiratory program was not needed; it would cost too much money and waste too much of his staff's valuable time. With that resounding defeat, I decided to abandon the respiratory program temporarily and explore the company's view on hearing conservation.

The majority of the employees are subjected to intermittent noise exposure. The noise is considered intermittent if the operator is exposed for two or more segments at different levels during the day. Denis, Oliver, Diego, and Russell work predominantly in the polish room. Peter, James, and Albert work in the final assembly area, and Roger and Mark rotate from station to station as floaters to facilitate production where needed. They work four days of ten-hour shifts overhauling passenger and cockpit windows for airlines such as Reno Air, Alaska Airlines, and Southwest Airlines. A considerable amount of noise is generated in several of the rooms. My dosimeter, an analog sound measuring device, registered serious noise levels.

In the polish room, when all four polishers are operating and the radio is cranked up so it can be heard over the din, the decibel level in the room is continuously over 104 dB except for breaks. In the final assembly room, high pressure air hoses are used and they create considerable impact noise, up to 98 dB, when used. Various areas in the warehouse were over 90 dB, at different locations and times during the day.

I knew from my research that when hearing is lost because of noise exposure, it cannot be restored. By law, companies whose workers are exposed to high noise levels must have an active program for protecting their employee's hearing. The hearing conservation standard is 85 dB for a set exposure limit. Finally I had hard evidence with which to initiate PAT's safety program. All I needed then was approval from Karen to go ahead with the changes—although that meant more writing because, in the next phase, the proposal had to be written as procedures. L.C.L.

The Production Quota

You were assigned the writing project on Monday. You "cleared" the task sketch of your responsibilities with your boss on Monday afternoon. For several days you performed the lab work on the computer. In the meantime you set aside a little time to sketch out your basic game plan to establish a precise use of your time—the critical path schedule. The schedule has been on your calendar. The project has been on your mind. You have kept a log of your ideas as they occurred to you. The concept of a document is beginning to take shape in your mind.

You have talked with employees about the project and you have listened to their views. You have logged useful ideas from them. You have also noted to yourself all your objections to their ideas. It is Wednesday afternoon. You are slowly beginning to worry about your deadline, and you want to combine your lab results with your recommendations and get the report written. You are ready to write the first or rough draft.

You need a writing strategy that is appropriate to a workplace setting. *First, learn every successful writer's best kept secret: your goal at this stage is production.* Get the job done. Remember the goal. The white paper—blankness—is the challenge. Fill it. The challenge is *quantity* not quality. Produce. Produce some more. Make the task a *production* goal. Quality control is not important yet. There must be no nitpicking. The moment you stop to evaluate, the pen or processor stops, and you cannot win the race by sitting in the pits. I cannot place enough emphasis on this point. Just remember that the perfectionist is hardly the best of employees. The boss wants the job *done.* She does not particularly want it to be *perfect.*

A perfect job that is on time is ideal. A good job that is on time is the typical goal. A late job will not do. If a project is late, an employee has failed, perfectionist or not. Perfectionists cannot see that the greatest flaw among their skills is perfectionism itself. Perfection is nice, but not if the project fails to meet the midnight deadline of a bid on a federal contract worth millions, and not if every other employee completes a good job while the perfectionist is slowly coming along on a perfect job. Yes, writing demands a certain amount of perfection, but you will worry about that later. The goal at this moment is quantity. The goal is speed.

The point is that if you are very judgmental, you will write slowly, if at all. Accept what you think are your second-rate ideas. You can build them into superior ideas later, whereas if you never get the ideas down on paper, you have no material to work with. You need raw material and plenty of it. Your first go-around on paper will be rough, indeed, but it is the only way to get the material you need.

You cannot build the structure until you have the building blocks. Be particularly oblivious to spelling, word choice, and all matters that have little to do with logical continuity and content. Everyone has been overly trained to worry about these matters. These are finishing touches. Spelling is not writing. These matters are of no concern to you whatever. You are the author; your only concern here should be the building blocks. The cosmetic effect is for the reader, and such considerations are *far* down the road for now. Deciding on the trim colors for your house has no purpose if there is no house. Do not be distracted. Focus on ideas and only the ideas. Develop them.

As a tactic, think that you are in a race with yourself. Do not waste time. If you have ever taken a written examination, you may have marveled at your output. If you had to produce the same material at your convenience in the leisure of your office, it might have taken forever. The difference between the two circumstances is a matter of pressure. When the heat is on, people produce. When the heat is off, many people just kick back and watch.

THE FIRST-DRAFT TECHNIQUE

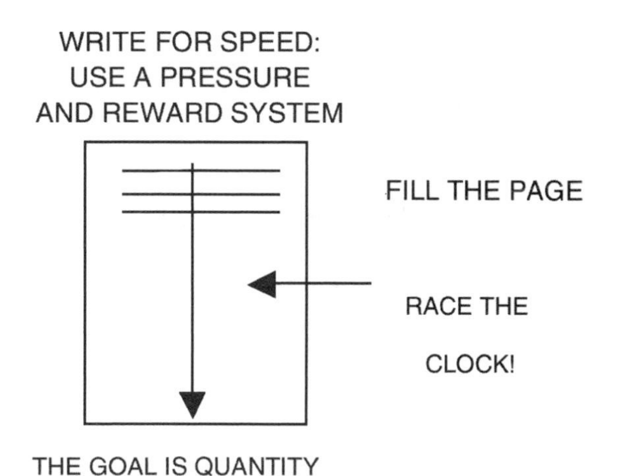
WRITE FOR SPEED:
USE A PRESSURE
AND REWARD SYSTEM
FILL THE PAGE
RACE THE
CLOCK!
THE GOAL IS QUANTITY

The problem with writing is that *you* are responsible for the activity, and if no one says "go," you will not come out of the starting gate. To give yourself the discipline to produce, use the one device coaches always use to challenge a self-paced activity: a clock. Much earlier I suggested to you that you have to be your own coach. *You* have to blow the whistle. You have the stopwatch. No one else can pace a writer but the author. Give yourself a time limit and go for broke. However, base the time factor on reasonable expectations.

Your attention span will suggest the outside limits. There is no point in writing once your attention is exhausted or distracted. Perhaps thirty minutes will do. Possibly an hour is a good target. Perhaps two or three hours are practical time frames. You decide. Do not stop for any reason during a designated session, but take frequent breaks *between* sessions because if you sit still for too long, you will become lethargic.

You can encourage the timeline game by adding another element to it. Since there is no boss to force you to challenge the clock, be your *own* boss. If you were your own boss, you would quickly provide yourself with the perks you have always deserved. Start now. Reward yourself when you write. I always do. If your bag lunch is full of treats today, write from eleven to lunch. If you work out at the gym with a group of friends after work, write from three to five. Reward yourself. Spoil the cook. However, if at quitting time you are not done, I will let you decide what you should do. *Pace yourself; race yourself.* Production and speed are the goals. You are the coach. You are the boss. The challenge is easy to prescribe:

THE FIRST-DRAFT TASKS

- **WRITE**
- **WRITE FAST**
- **DO NOT STOP**
- **DO NOT JUDGE**
- **DO NOT DOODLE**
- **WRITE ON ONE SIDE ONLY**

If you use a typewriter or write in longhand, use only one side of a sheet of paper so that all the pages can be viewed at once. I call this a "spread," and it is a vivid way to organize material. (Even the largest computer screens are not big enough to properly display multipage spreads that are readable). In addition, single-sided documents can be cut into pieces and reorganized in "pasteups" (see p. 117). A computer version of a document can easily be restructured to match the hardcopy pasteup.

The pasteup is a valuable device, particularly for developing the rough drafts of longer writing projects. The pasteup is an *editing* procedure that can promptly shape the organization of a document, no matter what method was used to get the document written. If, for example, a front-to-back composition did not yield the desired organization, the document can be reconstructed with a pair of scissors, and valuable time can be saved because rewrites are avoided. The pasteup can be used at any time and in conjunction with any method of writing. It is particularly important for reorganizing highly improvised rough drafts.

You can, of course, make all the text-shift modifications on the computer screen if you have a good visual memory for text. You will be able to see only one or two pages at a time even though you may be moving sentences and paragraphs back and forth between many pages. I find this method difficult if a project involves a substantial text.

For the *writing* procedure, you have three basic options, one of which I have explained at length. First, you can try to use the "front-to-back method," examined earlier. This technique is clearly the most popular approach, but it is the most difficult strategy for many writers. This strategy involves the heroic effort to write the project straight through to completion. The method usually asks the mind to do too much at once in the process of trying to juggle all the parts of a report, but the front-to-back method is effective for rough drafts

if the author is not expecting polished copy. You may be comfortable with this popular method, and you must recognize your ability to use whatever methods fit your needs and habits—as long as you feel that your method is the *fastest* and most effective way to go about writing the first draft of your project.

You must first determine the degree of difficulty of a writing task. You might use a front-to-back strategy for memos and business letters that are clearly easy for you to produce. The longer the document, however, and the more complex the contents, the more difficult front-to-back composition will be.

We need to look at the pluses and minuses of two additional successful strategies for handling more ambitious projects:

The outline method

The sprint method

When projects are ambitious they may involve ten, twenty, or a hundred pages. If only because of the length of such a document, the front-to-back method has its limits. Beyond a certain point the method is inadequate and will not meet the needs of the project. These other strategies conveniently meet the demands of long or complex document preparation.

Stage Six

The Formal Outline

The outline is a valuable tool for both the writer and the supervisor. These formal outlines were developed for me so that I could be certain the projects were going to be written according to plan. The formal outline is also very useful for committee or supervisor review.

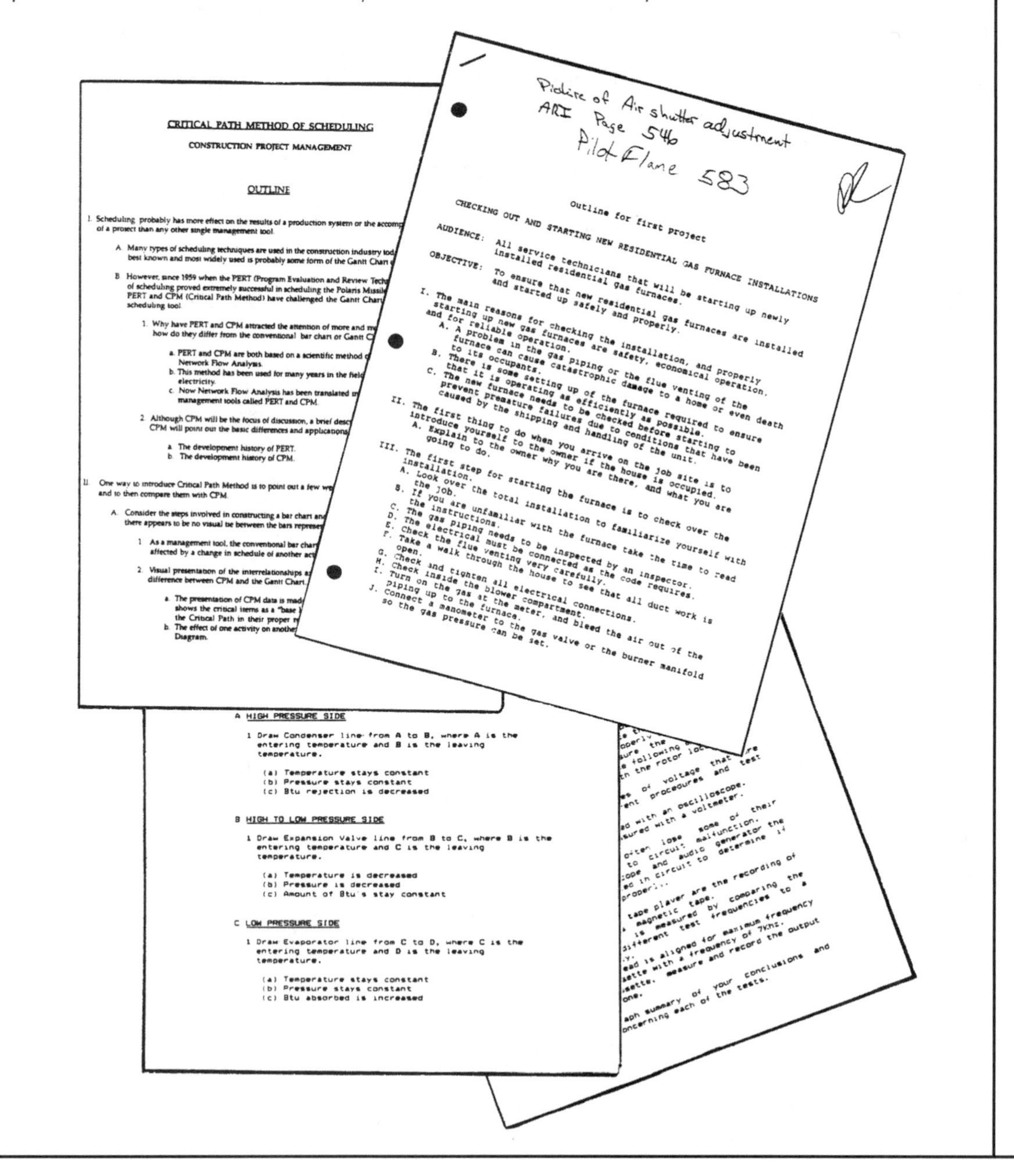

Picture of Air shutter adjustment
ARI Page 546
Pilot Flame 583

Outline for first project

CHECKING OUT AND STARTING NEW RESIDENTIAL GAS FURNACE INSTALLATIONS

AUDIENCE: All service technicians that will be starting up newly installed residential gas furnaces.

OBJECTIVE: To ensure that new residential gas furnaces are installed and started up safely and properly.

I. The main reasons for checking the installation, and properly starting up new gas furnaces are safety, economical operation, and for reliable operation.
  A. A problem in the gas piping or the flue venting of the furnace can cause catastrophic damage to a home or even death to its occupants.
  B. There is some setting up of the furnace required to ensure that it is operating as efficiently as possible.
  C. The new furnace needs to be checked before starting to prevent premature failures due to conditions that have been caused by the shipping and handling of the unit.

II. The first thing to do when you arrive on the job site is to introduce yourself to the owner if the house is occupied.
  A. Explain to the owner why you are there, and what you are going to do.

III. The first step for starting the furnace is to check over the installation.
  A. Look over the total installation to familiarize yourself with the job.
  B. If you are unfamiliar with the furnace take the time to read the instructions.
  C. The gas piping needs to be inspected by an inspector.
  D. The electrical must be connected as the code requires.
  E. Check the flue venting very carefully.
  F. Take a walk through the house to see that all duct work is open.
  G. Check and tighten all electrical connections.
  H. Check inside the blower compartment.
  I. Turn on the gas at the meter, and bleed the air out of the piping up to the furnace.
  J. Connect a manometer to the gas valve or the burner manifold so the gas pressure can be set.

CRITICAL PATH METHOD OF SCHEDULING

CONSTRUCTION PROJECT MANAGEMENT

OUTLINE

I. Scheduling probably has more effect on the results of a production system or the accomp[...] of a project than any other single management tool.
  A. Many types of scheduling techniques are used in the construction industry tod[...] best known and most widely used is probably some form of the Gantt Chart [...]
  B. However, since 1959 when the PERT (Program Evaluation and Review Tech[...] of scheduling proved extremely successful in scheduling the Polaris Missil[...] PERT and CPM (Critical Path Method) have challenged the Gantt Chart[...] scheduling tool.
    1. Why have PERT and CPM attracted the attention of more and m[...] how do they differ from the conventional bar chart or Gantt C[...]
      a. PERT and CPM are both based on a scientific method [...] Network Flow Analysis.
      b. This method has been used for many years in the fiel[...] electricity.
      c. Now Network Flow Analysis has been translated in[...] management tools called PERT and CPM.
    2. Although CPM will be the focus of discussion, a brief desc[...] CPM will point out the basic differences and applications[...]
      a. The development history of PERT.
      b. The development history of CPM.

II. One way to introduce Critical Path Method is to point out a few we[...] and to then compare them with CPM.
  A. Consider the steps involved in constructing a bar chart an[...] there appears to be no visual tie between the bars represe[...]
    1. As a management tool, the conventional bar char[...] affected by a change in schedule of another act[...]
    2. Visual presentation of the interrelationships a[...] difference between CPM and the Gantt Chart.
      a. The presentation of CPM data is mad[...] shows the critical items as a "base [...] the Critical Path in their proper [...]
      b. The effect of one activity on anothe[...] Diagram.

A HIGH PRESSURE SIDE

1 Draw Condenser line from A to B, where A is the entering temperature and B is the leaving temperature.
  (a) Temperature stays constant
  (b) Pressure stays constant
  (c) Btu rejection is decreased

B HIGH TO LOW PRESSURE SIDE

1 Draw Expansion Valve line from B to C, where B is the entering temperature and C is the leaving temperature.
  (a) Temperature is decreased
  (b) Pressure is decreased
  (c) Amount of Btu's stay constant

C LOW PRESSURE SIDE

1 Draw Evaporator line from C to D, where C is the entering temperature and D is the leaving temperature.
  (a) Temperature stays constant
  (b) Pressure stays constant
  (c) Btu absorbed is increased

WEDNESDAY

At the Starting Gate: The Outline Method

One method commonly used to write the rough draft of a major project is to use an outline as a guideline for the project. This method is particularly useful for long projects ("long" by your own standard). A great many men and women write by using the outline method. Usually they did not happen upon the system; they were taught the method in writing classes. The outline is tried and true; I use it often, and I will explain its use as a strategy for writing technical papers.

DOCUMENTATION WORK SCHEDULE

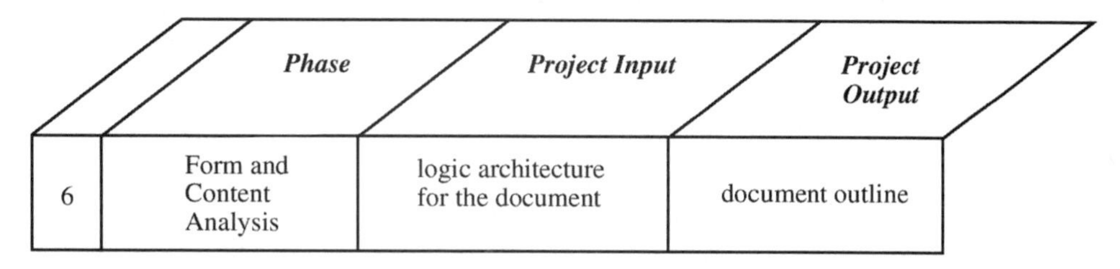

	Phase	*Project Input*	*Project Output*
6	Form and Content Analysis	logic architecture for the document	document outline

The larger the composition the more complex the organization becomes— or to state the problem in practical terms, the larger the composition the more confusing the project becomes to both the writer and the reader. For very practical reasons, writers need a large number of traditional devices to help organize the maze and to help guide a reader through the document. The table of contents, the introduction to a text, chapter banners, an index, color codes, and other devices help organize the product. However, the most practical instrument—the one that will build the table of contents and the chapter headings—is the one organization tool that will never appear in the finished product: the outline.

Outlines have at least three practical functions. First, for technical documentation the outline is the best strategy to use to get a project underway. Outlines are easy to develop, they organize any body of material, and they serve as springboards to get a project started.

Second, the outline is an ideal path for communicating with coworkers or supervisors who may be involved in a project. Rather commonly, corporate writing responds to committees and reflects group activities and group judgments. In other words, a project you might develop in college will usually reflect your sole authorship. A project you might develop for a corporation, in contrast, may be an effort to organize the material of five or ten employees. The project will structure and document their thoughts in a report, and they must approve the final product. This is a tedious process that can be greatly accelerated by the use of an outline. Outlines can be drafted, discussed, and revised much more rapidly than a text. The report should be created only after the outline has unanimous approval.

Third, with only a little reworking, outlines are easily adapted to be used as a table of contents, if you need one, in the finished projects they define.

An outline is essentially the table of contents of a book without the page numbers. The outline indexes what you plan to write about. There are many approaches you can take to designing an outline, but your goal will always be the same: you want the outline to be a skeletal structure for the *logic* of the presentation that you will develop when you write your project. I suggest that you state two important points of information above the outline as basic guidelines for the outline itself (see the illustration on the following page).

First, state your *objective*. Simply tell yourself, in one sentence preferably, *what you have to do*. For example, in the HVAC project we discussed earlier, if you have to submit an estimate for an appropriate heating and ventilation system that fulfills the guidelines provided by an architectural firm, state that fact as your objective. In the case of the HVAC estimate, the objective might be to provide a cost-effective heating and cooling installation for 4000 square feet of enclosed space.

Second, state your specific *proposal,* if one is required of you. The proposal states the *recommendation* your text is intended to support and endorse. The proposal might be that a forced-air gas system costing $45,000 should be installed. The objective and the proposal then give you preliminary guidelines for the outline, which sets about to demonstrate or support the proposal. A proposal is similar to or the same as the "thesis." Both terms refer to the author's opinion or viewpoint. In industry you will never hear the word *thesis*. In college, and in composition classes, you will rarely hear the word *proposal*. The terms are shop talk and vary from business to business, but they mean the same thing.

Now you are ready to outline. The outline is divided into whatever major divisions or primary considerations are appropriate for the topic. If you have a proposal to make, the major headings explain your reasons for making the proposal. You may have three major points to make, or you may have fifteen. You may then want to develop subheadings under each heading so that you know what elements of discussion or evidence are needed to support each major issue. The outline looks something like this:

Objective: ____________________

____

Proposal: (Optional. State if desired.)

I. First Argument
   A. ____________________
   B. ____________________
II. Second Argument
   A. ____________________
   B. ____________________
III. Third Argument
   A. ____________________
   B. ____________________

If you have difficulty seeing your objective, start with a question. Make the outline respond to the question. A key question will give you clear control of the mission of the outline, and the proposition you plan to put forward will quickly take shape. However, do *not* use questions in the actual outline. The outline is a series of answers, not a series of questions.(For additional discussion see p. 219.)

Now, fit the outline into the imaginary project you were assigned on Monday. By Wednesday your lab work and your log work are moving along. You can start to rough out an outline. Do it in one sitting, or do it over a period of days if time allows. In fact, sketching out ideas in a plotted format such as an outline is a tactic many writers like to use from the very start on day one. Other writers wait until the thinking, fact-finding, and the talking phases are done. Then they develop an outline during their first real effort to start the first draft of the writing.

With the outline in hand, you can begin to write the first draft. It can be handled in a front-to-back fashion, but since the outline organization is highly structured, you can safely begin anywhere in the document. I usually write "out of order" because it is more fun to write the parts that fit my mood. In general, if you write the "easy" parts first, you will encourage yourself into the project and waste little time. Since the outline allows you to write out of sequence, it is often a faster, more efficient procedure than the front-to-back-method.

Stage Seven

Speed Sheets

The pasteup is a very handy technique for shaping ideas. Write fast. Spread all your material on a table, cut it up, and organize the document. The following longhand pasteup was used to organize one of my projects. It was good for about twelve feet of material. Pasteups are the fastest way to go when you are writing the first draft. A printout is easily reconstructed in the same fashion.

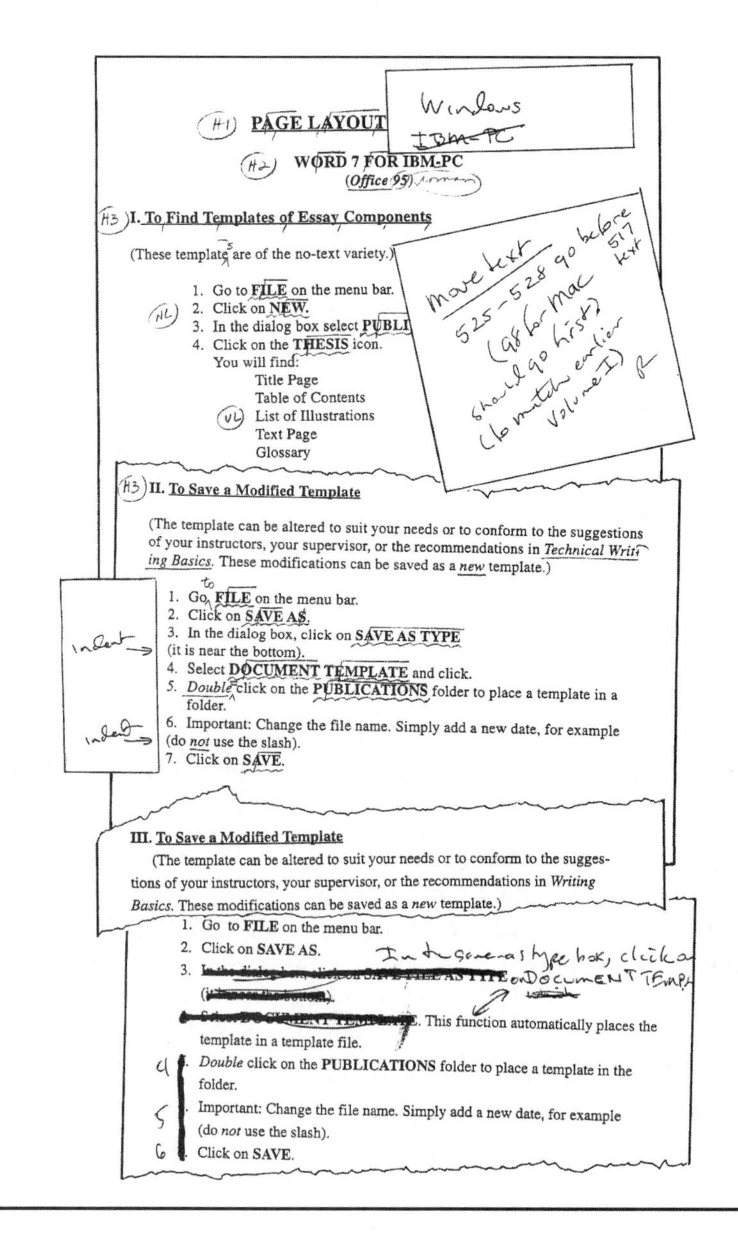

Windows

(H1) PAGE LAYOUT

(H2) WORD 7 FOR IBM-PC
(*Office 95*)

(H3) **I. To Find Templates of Essay Components**

(These templates are of the no-text variety.)

1. Go to **FILE** on the menu bar.
2. Click on **NEW.**
3. In the dialog box select **PUBLI**
4. Click on the **THESIS** icon.
   You will find:
   - Title Page
   - Table of Contents
   - List of Illustrations
   - Text Page
   - Glossary

move text 525–528 go before 517 text (98 for mac should go first) (to match earlier Volume I)

(H3) **II. To Save a Modified Template**

(The template can be altered to suit your needs or to conform to the suggestions of your instructors, your supervisor, or the recommendations in *Technical Writing Basics.* These modifications can be saved as a *new* template.)

1. Go to **FILE** on the menu bar.
2. Click on **SAVE AS.**
3. In the dialog box, click on **SAVE AS TYPE** (it is near the bottom).
4. Select **DOCUMENT TEMPLATE** and click.
5. *Double* click on the **PUBLICATIONS** folder to place a template in a folder.
6. Important: Change the file name. Simply add a new date, for example (do *not* use the slash).
7. Click on **SAVE.**

indent

III. To Save a Modified Template

(The template can be altered to suit your needs or to conform to the suggestions of your instructors, your supervisor, or the recommendations in *Writing Basics.* These modifications can be saved as a *new* template.)

1. Go to **FILE** on the menu bar.
2. Click on **SAVE AS.**
3. In the save as type box, click on DOCUMENT TEMPLATE. This function automatically places the template in a template file.
4. *Double* click on the **PUBLICATIONS** folder to place a template in the folder.
5. Important: Change the file name. Simply add a new date, for example (do *not* use the slash).
6. Click on **SAVE.**

A High-Speed Strategy: The Sprint Method

Writing a text straight through to the end is demanding. The result of deadlining a front-to-back effort is likely to be a hodgepodge of material that should be reorganized if it is not tightly structured.

DOCUMENTATION WORK SCHEDULE

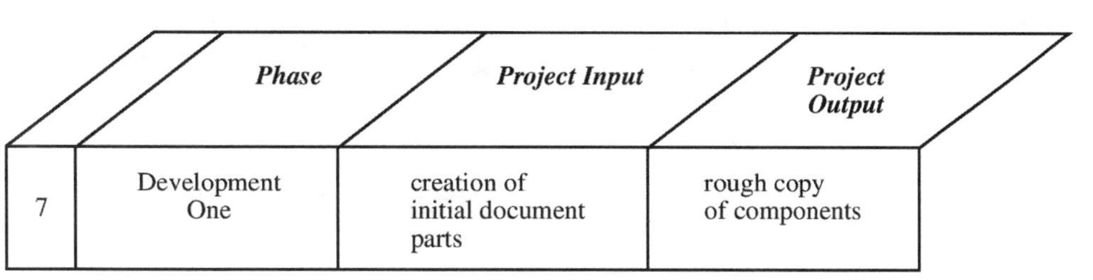

	Phase	*Project Input*	*Project Output*
7	Development One	creation of initial document parts	rough copy of components

Writers start to reorganize the many pages of their front-to-back draft with pencil lines leading here and there to reconstruct the project. There are arrows going forward, arrows going backward, arrows up, arrows down. Authors get out colored pencils to indicate what goes first, then second, and so on. It does not help if the project is composed on both sides of the sheets of paper. How are they going to indicate that paragraph two on page three now goes after paragraph one on page four? The computer screen is also a very difficult place for reorganizing, since only one or two pages are on the screen and it is hard to visualize what goes where when you cannot see the other pages. Whether the material is paper copy or screen copy, the result is a labyrinth—and just as confusing. Additionally, many an author expects someone to type up the material, as is.

There is another writing strategy. I call it the "sprint" method; it is similar to a well-known method of helping a committee to generate lots of ideas quickly by brainstorming. This writing technique, also known as "speed writing" and similar names, depends on a very lively writing session or two. There are no rules; you simply start to write whatever comes into your head. The strategy can be very effective and deserves our attention. The outline method is highly structured; the sprint method is highly spontaneous.

Recall the discussion of how the mind thinks. The *written* project a writer wants to complete is like a clock with shiny brass wheels. A pile of gears and springs is a more honest reflection of the creative *mind.* I argue that this fact suggests a tactic for writing that allows a writer to write very quickly—by letting the mind simply create. Why not forget structure altogether and produce the raw material? You can structure or organize later, especially since experience teaches most writers that they do not organize with much success during the first, or creative, stage of writing.

Sprinting

The best way to go for production is to sprint through the tasks. Put your task memo on the desk. Get out your log. Get out your lab data. Be sure to use these resources, since, in this case, there is usually no formal outline. You will improvise. Clock your effort and get your reward ready. These two time-management devices are critical to your success in speedwriting. Write on only *one* side of a piece of paper. If you know you want to use this method, it will be helpful if you maintain your log by writing on only one side of each page of the notebook. (You may also have an outline ready if you wish but do not try to follow the outline systematically. Develop the outline on one side only also.)

Start. Generate as much material as you can as fast as you can—and for as long as you can. Keep the processor chugging. Write anything. This is *not* a front-to-back model. There is no format. There is no *essay,* to put it in college terms. Just write.

There is a trick to sprinting through a paper. It is not simply high-speed writing or a game of quantity. It is more than reckless writing. The idea is to keep up with your ideas. A very common complaint is that writing a paper is frustrating because ideas can be created faster than they can be written or typed.

Research suggests that people think about four times more rapidly than they speak. For anyone other than a highly skilled typist, the mechanical differential is going to increase when you write. If you are free of the burden of your own mechanical or cosmetic corrections, you will speed up. If you abbreviate, you will speed up. If you are unconcerned about technical errors, you will speed up. Most important of all is organizational freedom. Forget any organization activities beyond the small cluster of thinking that you are structuring for a paragraph. Take each cluster of ideas and try to quickly build a paragraph. *Do not go back and reread the paragraph.*

Move on to the next paragraph. Repeat the process again and again. Do not look back. (Well, you can take a quick look if you need to figure out what to do next.) Keep the text moving up the screen and out of sight. If a paragraph does not develop, drop it. If you have sudden ideas out of nowhere, log them on an extra piece of paper or in a new paragraph in italics on the on-screen document. *Never judge, and do not delete or toss material.* I have already cut thirty pages from the book you are reading. I have saved another thirty pages and transferred the material to a file for another book. You may read the material at some point, but not in this book. I save everything. Write down anything that comes to your mind as you try to chase your ideas. No one will ever see this material. Feel the freedom. *Write. Write fast.* When you are done, conclude the sprint session.

The sprint phases are intended to develop as much creativity or originality as possible, and as much speed as possible. *Writing fast is not the secret. Writing freely—without judgment—is the key to success in this technique.* "Creative" engineering? You have to be inspired to write, and this process is "creative" regardless of your spreadsheets and stats and schematics. You are creating the document. You are building a logic structure where there was none. When you write, you want to get as much as you can out of your ideas in the least amount of time.

If you think back to the observations about the thought process, you will realize that this writing method is designed to make maximum use of the creative disorder of thought activity. You free up your thoughts and try to commit them to paper. You let your mental processes do as they will; then you build on the outcomes with structure and coherence and other features of writing that are best left to other phases of the process.

The sprint strategy is particularly handy for a difficult writing job, which usually involves complex thinking that is a maze on paper. Simply getting the material down in any way, shape, or form may be a challenge. I used the sprint method to develop the second, and third chapters you read earlier. The material was extremely difficult to pull together, and I used my log and wrote out any section that was of interest to me—pronto. Then I used the pasteup method to organize the material, slowly. I proceeded by using exactly the method I have just described. Completing the first rough draft is an excellent place to stop when you sprint.

The Pasteup

When you are out of words, put your pen down or stop typing. Print out a copy. Get out a pair of scissors and cellophane tape. Now, place every piece of paper out in front of you. Your desk may not be big enough. Use the floor if you have to. I use several large tabletops if I need them. Start to divide the material into logical groups. If you have made an outline, cut the outline into the divisions you created for the logical elements of the writing project. Under each outline division place the parts of the rough draft, lab data, and the log notes that fit the section. Structure each bit of paper under the appropriate division.

If pages must be divided because the data or the ideas go in separate divisions, *cut up the pages,* and place each part in the correct group. Then fan out each division, one by one, and organize the parts of each group. For those sections that appear to be thorough and orderly, tape the sequences together and set them aside. For those that need work, write the additional material on slips of paper, add them to the parts in the division, and adjust the logical sequence again. Tape them up.

You have just developed a pasteup. Because speed writing is highly productive but prone to disorganization, the pasteup is a critical second phase of the activity. It may come as a surprise, but a lot of people do their fastest writing with scissors—not pencils, pens, or computers—and *never* with erasers. The reconstructed mess in front of you is not an outline. You are far beyond the outline stage. You *have* the document. You have constructed the organization. You have added the data. You have developed the key ideas. The cut-and-paste process is very fast because the mind does not have to build castles in the air. You can cut and paste on your word processor without generating copy, but screen size limits your page-display capability as mentioned earlier. Use paper copy to see the *big* picture. Construct a spread. Cut and paste to see the entire castle. Then go back to the processor and click the parts around.

To Move Text

1. With the cursor, highlight the text you need to move.
2. Click in the highlighted area and drag the text to its new location.
3. As you are dragging the text, a light gray line indicates where the text will be placed.

Every scrap of paper becomes a concrete element that you need only organize; all the ideas are now on paper. But since people often think in a rather disorderly fashion, scissors will correct the problem without the endless rewrites of the front-to-back method. The result is that you correct the unavoidable disorganization of the front-to-back method. You also save yourself several drafts because the cut-and-paste technique is a shortcut to the finish line. In addition, since you have brought together a jumble of resources, you will have developed a structure based on *all* the available material.

You can use the sprint method with a formal outline (all these procedures can be mixed and matched), but there is a unique value in using speed writing in place of an outline. If all the haste and spontaneity of a sprint session results in contradictory materials, this situation is valuable. In one respect, the sprint method can be superior to the outline method because you do not force yourself to design some prearranged logical structure that you think will be appropriate. Writers tend to build outlines on the basis of what fits. They avoid conflicts by omitting material that does not conform to the logic of the outline. In speed writing a project, you try to use everything you have, and you will have to face any internal contradictions you discover. The result will be that you can uncover any number of potential problem areas as you write the report. Conflicting data or conflicting ideas will boldly identify problems you, as the author, want to resolve before you complete the project. You do not want a reader to notice conflicts once the product is complete.

Do not confuse the pasteup with the ability of a word processor to edit in and edit out materials of all kinds. I would suggest that the task memo, the log, and the scissors are irreplaceable. A printout of the rough draft you created by speeding through the material is also irreplaceable. A pasteup is like a game of solitaire; you must have all the cards on the table. You need an overview. You need hardcopy. You must see the entire project before your eyes. The day may come when a computer-screen page display will be big enough (or cheap enough) to show ten full-size pages at once, and you may be able to use some sort of supermouse to click them all together. Until that time, paper is the only way to give a physical reality to *all* of your work at once.

After the pasteup is complete you are ready for the next phase: sprint again. If you are not tired, you might proceed now. Otherwise stop for a break. There are advantages and disadvantages to stops and starts. Leaving a writing project for another day means that you must come back and pick up where you left off. That may be awkward, and it takes a little time to get back into the project, but it also gives you a fresh perspective that can be helpful.

Stage Eight

Sloppy Copy

The rough draft is the first appearance of the actual document. This sample should give you an idea of where the word rough *comes from. If you generate the rough draft on a computer, avoid deletions. Include everything in the rough draft.*

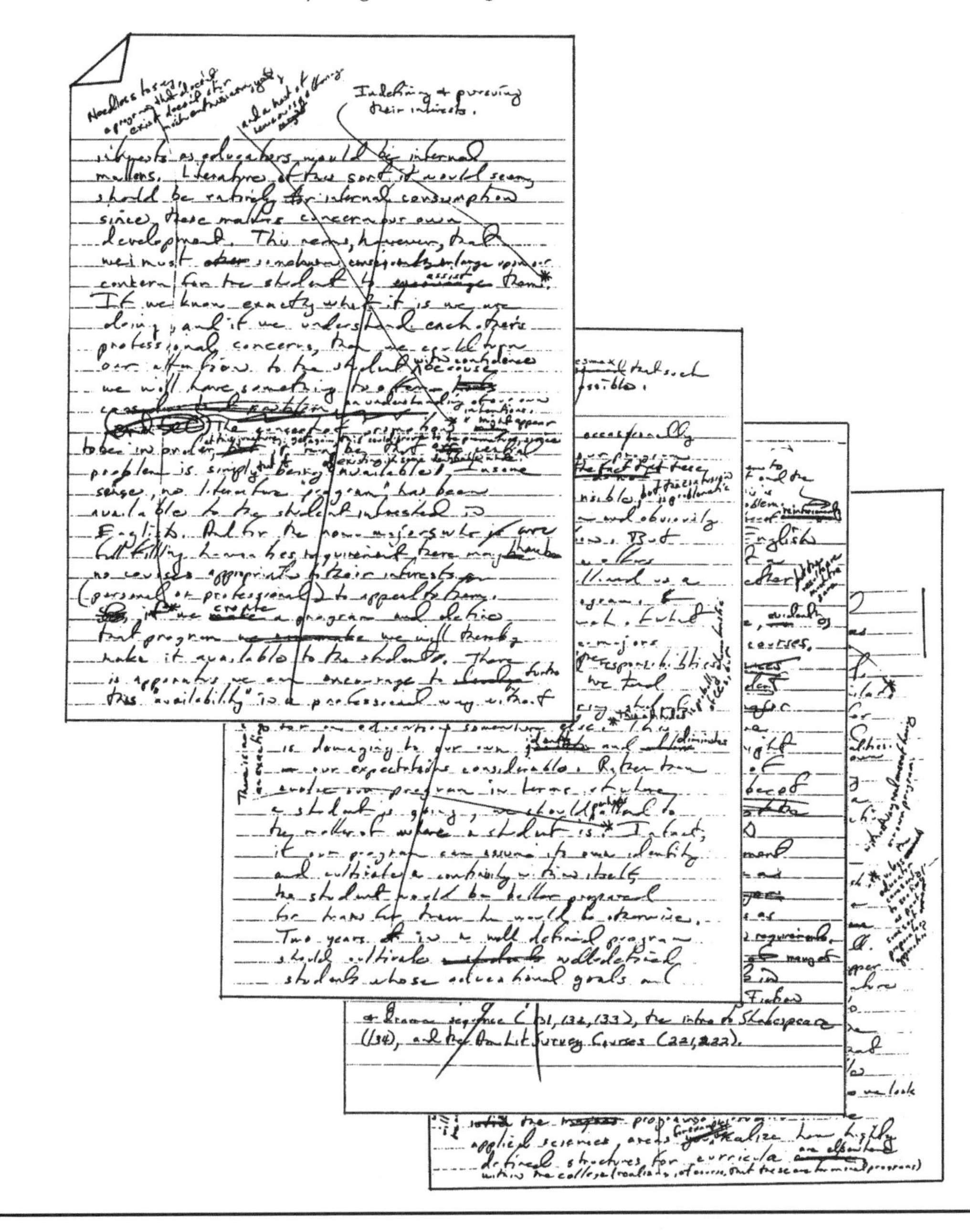

THURSDAY

The Completed Rough Draft

It is now Thursday morning and you want to try to complete the rough draft. First, you must finish any more speed writing that has to be done to finish the document. Time yourself. Place the pasteups on the desk or put the reorganized text on your screen. Try to develop each pasteup sequence into a series of paragraphs that will go into the final project. Start with any sequence you want. Do the introduction if you wish, but many writers prefer to do it last. Again, the goal is production. Forget quality. Write as much as you can as fast as you can— on one side of each sheet of paper if you are working with hard copy. Write fast. Type fast. Abbreviate all technical terms, names of devices, people, organizations, and anything else that can be shortened. If your technical data are already neatly prepared on your pasteup, or if you have the research on a printout, do not rewrite the material for the rough draft. Cut out the material and tape it in place or move it from file to file on your word processor.

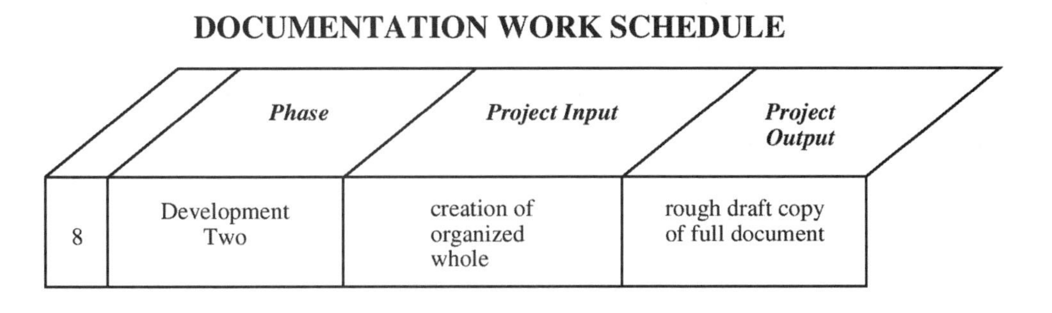

DOCUMENTATION WORK SCHEDULE

	Phase	*Project Input*	*Project Output*
8	Development Two	creation of organized whole	rough draft copy of full document

Now, go back to your pasteup. On this phase of the rough draft, focus on paragraphs and the logical continuity of the project. Develop and refine the text logic. This new draft of paragraphs now becomes another pasteup (which is why you keep writing on one side only throughout the entire event of writing). If you have the document on a word processor, print a copy. At this point, I still don't recommend trying to revise on the screen without paper copy. Once again, you need to be able to see the total of what you have, and you probably still need to cut and paste.

Do not throw out the original pasteups. In fact, *do not throw out even one scrap of paper at any point.* Save everything, but run a pencil line through every page once it is "dead copy" so that you will not be confused. Save electronic files by adding a version number each time you edit, and save it. You will know which ones are old by the datelines. File all

Sample 6A

PROBLEM 1

The computer does not boot up. When the power button is pressed to turn the PC on nothing happens: the computer appears "dead." The most common source of this kind of problem is connectivity. Below are a few troubleshooting techniques to help you solve this problem.

POSSIBLE SOLUTIONS

- ➢ Verify the power cable is secure in both the computer and wall outlet (or in most cases the surge protector).
- ➢ Check to see that all switches are turned on. Computer? Monitor? Surge Protector? Uninterruptible power supply?
- ➢ Is the computer in Doze or Sleep Mode? If so, striking any key on the keyboard or moving the mouse will "wake up" the PC.
- ➢ Make sure the wall outlet and surge protector are functioning by plugging a domestic appliance (such as a lamp, radio, etc.) into the outlet.
- ➢ Replace the power cord with a new one, or one that you know works.
- ➢ Check to see that the data cable and power cable are secure and properly connected from the hard drive to the motherboard and power supply.
- ➢ Use a multimeter to test the power supply to see if that is causing the problem.
- ➢ If the PC still will not boot, try the following procedures in order:
    - ➢ Reconnect or swap the drive data cable for a new one.
    - ➢ Reseat or exchange the drive adapter card.
    - ➢ Exchange the hard drive with one you know works.
    - ➢ Replace the motherboard.

the material in the event that you need any of the old versions. If, for example, you deleted a paragraph or a page from the text, you can recover the material from an old file. You can calculate an appropriate length of time for saving the rough copy, and when you do throw it out a month later, be sure the final document is in your files. Keep a floppy backup copy. I keep two backups, and a zip disk goes to the bank deposit vault.

When you shift your attention to pasteups, you more or less stop the creative phase and begin the editing phase. Move back and forth from one to the other. In the editing phase, you are not trying to produce; you are trying to organize and make sense of the material. The pen and the keyboard are the tools of the creator. The editor uses scissors and tape, at least for the first major go-around of editing content. Delete. Save. Move the parts around. Cut. Paste. (It is a tradition to say "paste," but everyone uses tape.)

Do not rush the job by trying to jump from a hasty pasteup to the finished project. There are no shortcuts.

Organize the entire text into a structure that is appropriate. The document may look good already. If you were lucky, and adequately prepared, the text may need to have only a few more paragraphs cut out and relocated. On the other hand, you may see areas in which you need to add material. You may see sections that are unclear or that do not seem to quite fit the logic sequence of your text. In this case, you may want to do a little more speed writing to create the missing sections or to make other major changes. Simply repeat the process, although you should focus only on the areas that need work.

The example on the left is an end product of four drafts (see Sample 6A). Technical texts have to be very precise. The author's rough draft did not identify which switches had to be checked (the second bullet), nor had he included the last five procedures. Because of the precision and thoroughness involved, the text was revised repeatedly.

Stage Nine

A Computer-Generated Second Draft

Notice that I use the pasteup technique repeatedly. You can perform these tasks electronically, but I want to see the overall product. I do a printout, and then I cut and paste on a table where I see the overall organization take shape.

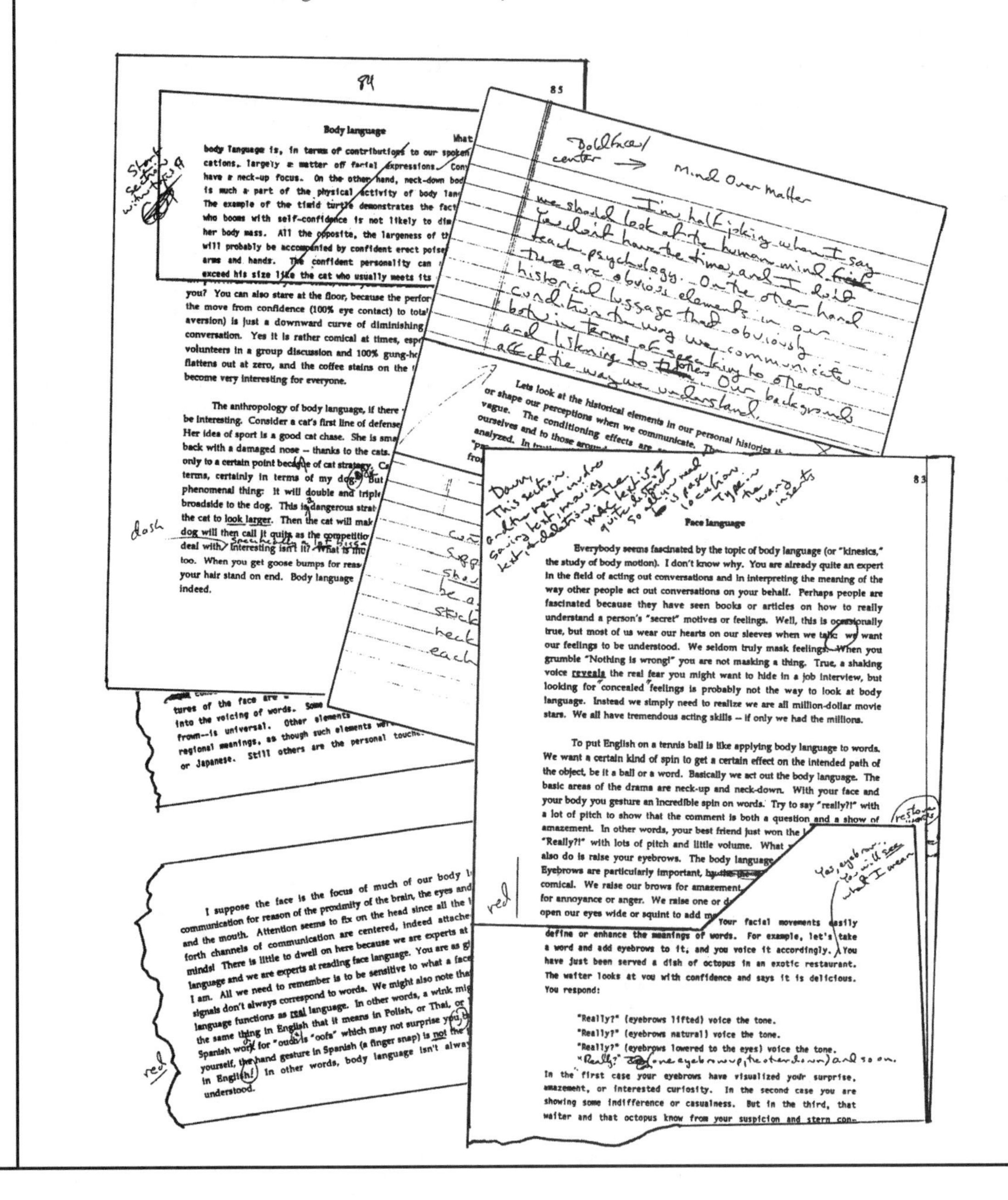

Body language

body language is, in terms of contributions to our spoken ... cations, largely a matter of facial expressions. ... have a neck-up focus. On the other hand, neck-down bod... is much a part of the physical activity of body lan... The example of the timid turtle demonstrates the fact ... who booms with self-confidence is not likely to dim... her body mass. All the opposite, the largeness of th... will probably be accompanied by confident erect poise... arms and hands. The confident personality can ... exceed his size like the cat who usually meets its ...

you? You can also stare at the floor, because the perfor... the move from confidence (100% eye contact) to total ... aversion) is just a downward curve of diminishing ... conversation. Yes it is rather comical at times, esp... volunteers in a group discussion and 100% gung-ho ... flattens out at zero, and the coffee stains on the ... become very interesting for everyone.

The anthropology of body language, if there ... be interesting. Consider a cat's first line of defense... Her idea of sport is a good cat chase. She is ama... back with a damaged nose -- thanks to the cats. ... only to a certain point because of cat strategy. ... terms, certainly in terms of my dog. But ... phenomenal thing: It will double and triple ... broadside to the dog. This is dangerous strat... the cat to look larger. Then the cat will mak... dog will then call it quits as the competitio... deal with. Interesting isn't it? What is ... too. When you get goose bumps for reas... your hair stand on end. Body language ... indeed.

Lets look at the historical elements in our personal histories ... or shape our perceptions when we communicate. ... vague. The conditioning effects are ... ourselves and to those around ... analyzed. ...

... tures of the face are ... into the voicing of words. Some ... from--is universal. Other elements ... regional meanings, as though such elements were ... or Japanese. Still others are the personal touche...

I suppose the face is the focus of much of our body l... communication for reason of the proximity of the brain, the eyes and ... and the mouth. Attention seems to fix on the head since all the ... forth channels of communication are centered, indeed attache... minds! There is little to dwell on here because we are experts at ... language and we are experts at reading face language. You are as g... I am. All we need to remember is to be sensitive to what a face ... signals don't always correspond to words. We might also note tha... language functions as real language. In other words, a wink mi... the same thing in English that it means in Polish, or Thai, or ... Spanish word for "oops" is "oofs" which may not surprise you ... yourself, the hand gesture in Spanish (a finger snap) is not the ... in English. In other words, body language isn't alway... understood.

Face language

Everybody seems fascinated by the topic of body language (or "kinesics," the study of body motion). I don't know why. You are already quite an expert in the field of acting out conversations and in interpreting the meaning of the way other people act out conversations on your behalf. Perhaps people are fascinated because they have seen books or articles on how to really understand a person's "secret" motives or feelings. Well, this is occasionally true, but most of us wear our hearts on our sleeves when we talk: we want our feelings to be understood. We seldom truly mask feelings. When you grumble "Nothing is wrong!" you are not masking a thing. True, a shaking voice reveals the real fear you might want to hide in a job interview, but looking for "concealed" feelings is probably not the way to look at body language. Instead we simply need to realize we are all million-dollar movie stars. We all have tremendous acting skills -- if only we had the millions.

To put English on a tennis ball is like applying body language to words. We want a certain kind of spin to get a certain effect on the intended path of the object, be it a ball or a word. Basically we act out the body language. The basic areas of the drama are neck-up and neck-down. With your face and your body you gesture an incredible spin on words. Try to say "really?!" with a lot of pitch to show that the comment is both a question and a show of amazement. In other words, your best friend just won the ... "Really?!" with lots of pitch and little volume. What ... also do is raise your eyebrows. The body language ... Eyebrows are particularly important, ... comical. We raise our brows for amazement, ... for annoyance or anger. We raise one or ... open our eyes wide or squint to add m... Your facial movements easily define or enhance the meanings of words. For example, let's take a word and add eyebrows to it, and you voice it accordingly. You have just been served a dish of octopus in an exotic restaurant. The waiter looks at you with confidence and says it is delicious. You respond:

"Really?" (eyebrows lifted) voice the tone.
"Really?" (eyebrows natural) voice the tone.
"Really?" (eyebrows lowered to the eyes) voice the tone.

In the first case your eyebrows have visualized your surprise, amazement, or interested curiosity. In the second case you are showing some indifference or casualness. But in the third, that waiter and that octopus know from your suspicion and stern con-

Assuming that you now have a full-scale draft either as a pasteup or as an edited printout, turn to the matter of editing the rough draft to create the next draft. It is Friday afternoon. Revise the project now and create the second major draft. You will see many areas that you will want to revise. Attention must now shift from quantity to quality.

DOCUMENTATION WORK SCHEDULE

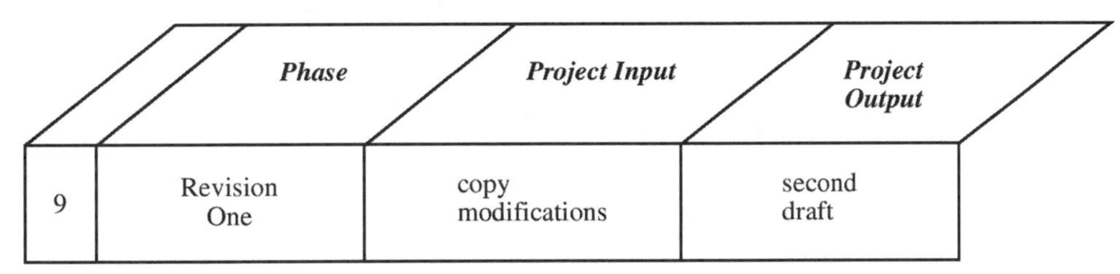

	Phase	*Project Input*	*Project Output*
9	Revision One	copy modifications	second draft

In an earlier discussion, I observed that writing survives beyond the time of its origin. Everything you write becomes, an old or historical version of your thinking the very next day. This is probably the main reason you can perfect your thinking in writing. Each time you return to a writing project, you look back over the historical selves who have gotten involved in the writing project you have in progress. Put another way, each day you add a new "you" to the operation. You should encourage as many of *you* as you can to help you write your projects.

The person you were on Monday is not quite the person you are on Friday. Frankly, since Friday certainly is not Monday, your mood, at least, should be quite different. I mentioned earlier that everyone has had the experience of seeing errors in a writing project that surely were not there when he or she last saw the document. In that discussion I blamed the problem on memorization. There is also another force at work in your ability to see errors where you did not see them earlier. As you put time between your writing and yourself an interesting phenomenon occurs.

As I also explained earlier, anything that you write becomes history the moment you write it. As you move further and further away from a document, you come to look at your work from the perspective of a different person. This well-known phenomenon is called *psychological distancing.* In the case of writing, this ability to change, and to change your mind, is a valuable tool.

A deadlined project suffers from having only one author. If you space out your project over a period of days or weeks you should be able to collect quite a group of authors to put it together. Of course, every one of the authors is you. Monday's you. Tuesday's you. Wednesday's you, and so on. For a working writer, the effect is significant. Herding as many of you into the paper as time allows will greatly improve your work. The following figure illustrates the concept.

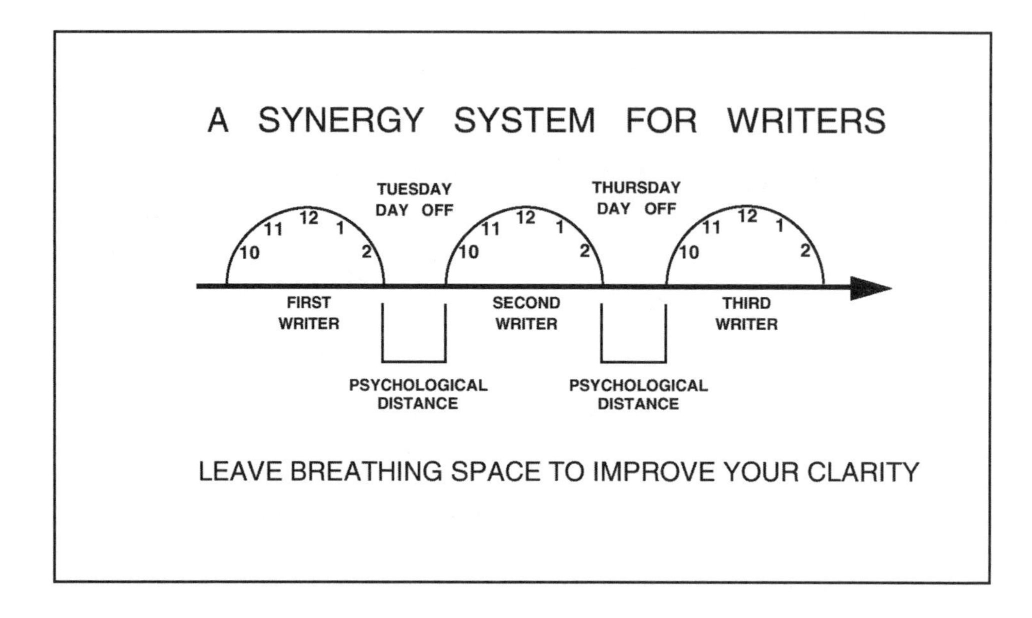

Writers are their own best critics *if* they can gain distance so that they see themselves clearly. I observed in Chapter 4 that writing gives an author an excellent opportunity to return again and again to his or her earlier perceptions to perfect and develop superior ideas. The tactic is not a difficult one. You need only budget your time so that the project is completed in phases. *Your perceptions change, and your ability to be critical of yourself is best served by time.* Time is your best corrective. Divide the project into phases that keep renewing your involvement in your project over a period of days or weeks. In this way, there may be very few errors to find three weeks later.*

You must be well aware by now that I encourage you to involve yourself in conversations throughout the project. Your log represents your effort to record your own thoughts and the thoughts of others. Essentially, you hold conversations with yourself and you hold conversations with other people. In phasing a writing project, you will see a similar activity. Each time you return to your project and continue to work on a new phase, you are beginning another kind of dialog. This time, each conversation is between your *historical*

** A very practical application of psychological distancing is never to send out any writing you do until the following day. I use this strategy as a rule of thumb with everything, including short memos. The next day you see your errors, notice any tone problems, and identify any confusion in the content.*

self, represented by the ideas on paper, and your *present* self, represented by the new ideas you bring to the project during each encounter. Each dialog constructs a process in which you can build ideas. Little wonder that writers who stare at walls and deadline a project have little luck with writing. There is no conversation.

The following diagram shows a practical use of distancing applied to a simple three-day writing project. Notice the progression of the manuscript.

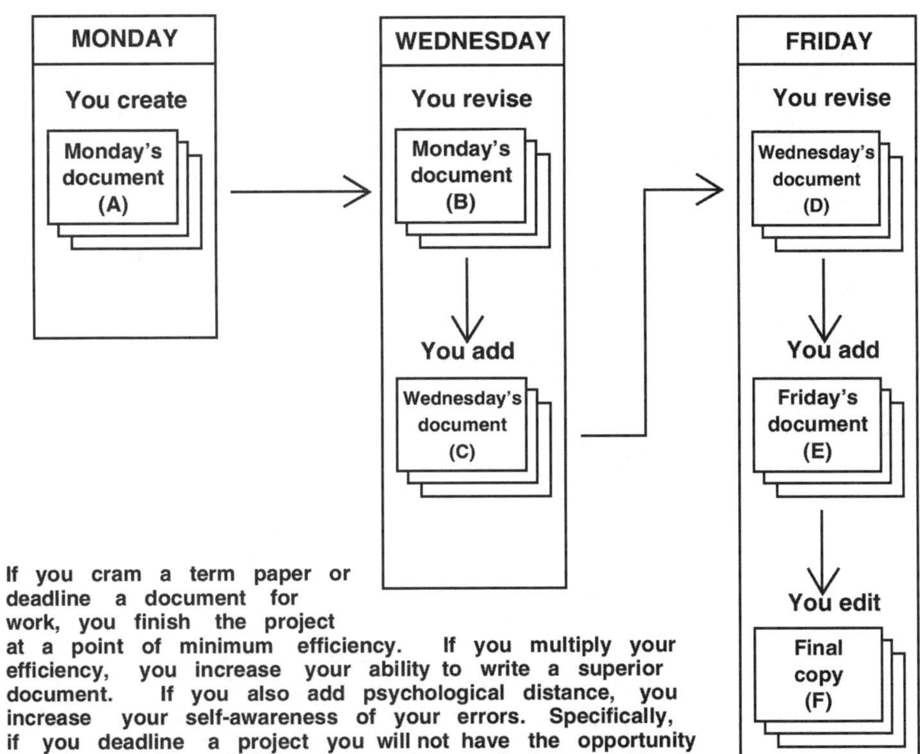
A WRITER'S PRODUCTION CYCLE
FOR A THREE- DAY SEQUENCE
MONDAY
You create
Monday's document (A)
WEDNESDAY
You revise
Monday's document (B)
You add
Wednesday's document (C)
FRIDAY
You revise
Wednesday's document (D)
You add
Friday's document (E)
You edit
Final copy (F)
If you cram a term paper or deadline a document for work, you finish the project at a point of minimum efficiency. If you multiply your efficiency, you increase your ability to write a superior document. If you also add psychological distance, you increase your self-awareness of your errors. Specifically, if you deadline a project you will not have the opportunity for the revisions (B, D) we see above. And you will not progress with quite the same new material we find in C and E.

Stage Ten

A Computer-Generated Third Draft

Some projects do not come easy. Once I got this one organized, I still wanted to make lots of changes. Fortunately, additional drafts are easy to create on a computer.

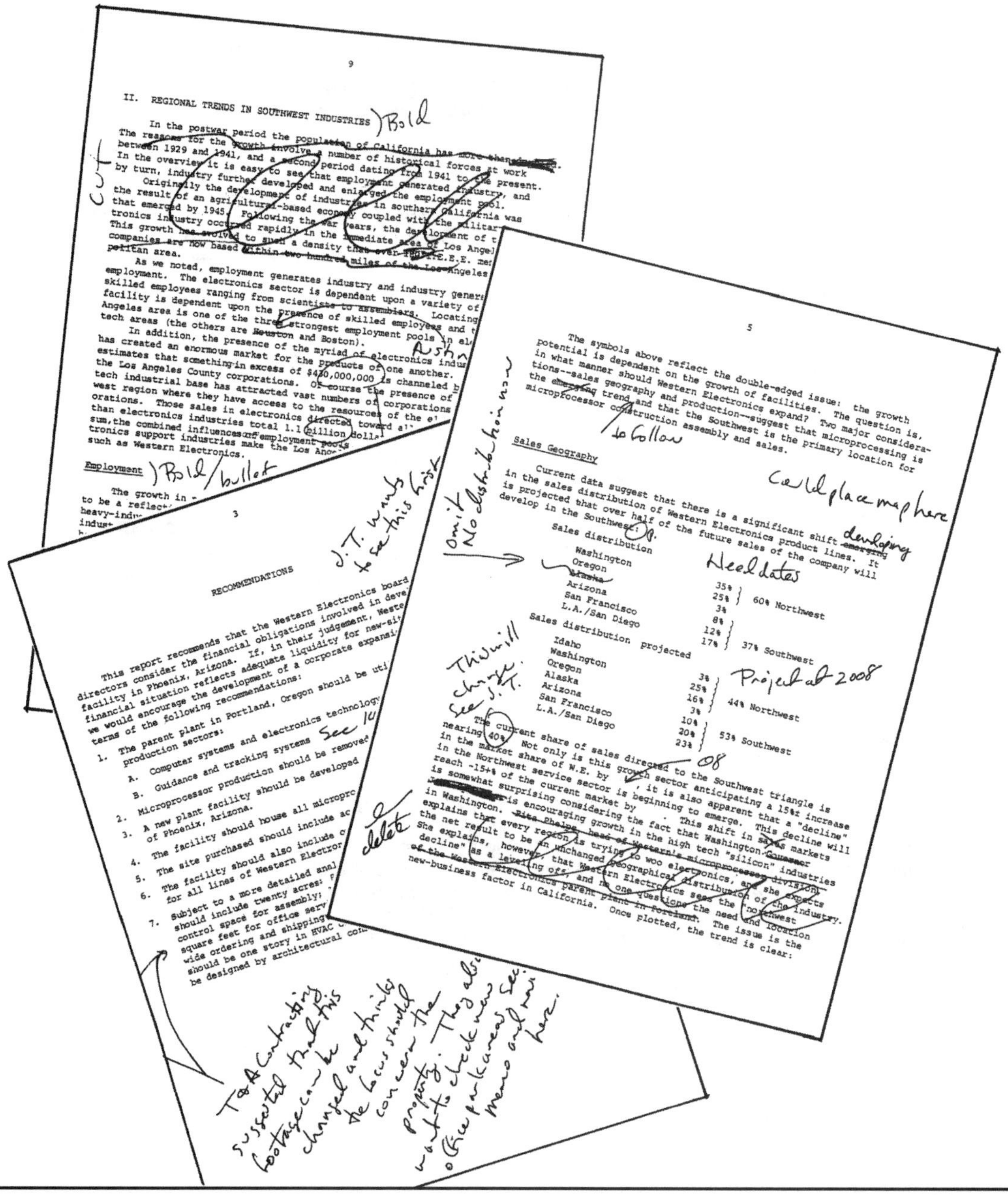

SATURDAY

Editing the Final Product

Writing a project in a series of stages—what I call "phasing"—is the secret to perfecting your own ideas, and phasing is also the secret to editing your own work. All the strategies I have explained should be carefully organized as a process to be undertaken over a period of time—days or weeks. The event of writing is best served by a series of discrete phases. You have spent five days developing the task memo, lab work, thinking, conversations, log activity, a document outline (if you wanted one), pasteups and the rough draft, and the second draft. Continue phasing when you edit for the final draft.

DOCUMENTATION WORK SCHEDULE

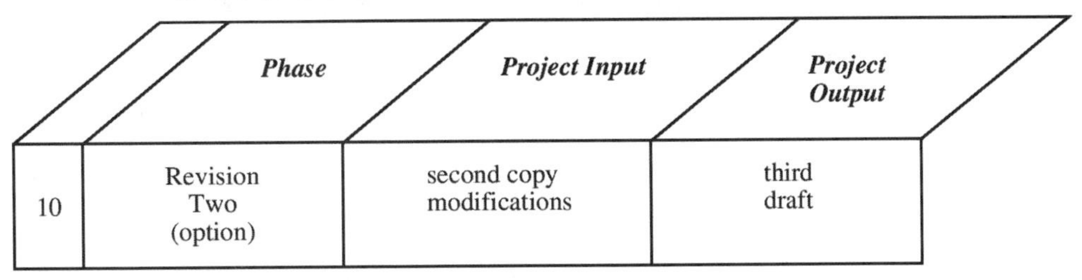

	Phase	*Project Input*	*Project Output*
10	Revision Two (option)	second copy modifications	third draft

Most writers cut the editing short because of timeline problems. You may not have the opportunity to be leisurely in editing for corrections, but try not to edit by doing a rushed job jammed between the rough draft and the final typing or printout. *Never leave the editing as a chore to be done as you type up the final copy.* You will really shoot yourself in the foot with that old trick. Type and repair. Repair and then type. Do not write and edit at the same time.

On Saturday—you arrive at a reasonable hour, let's say ten. You decided to come back to the office on Saturday for a few hours. If you had stayed late on Friday, you could have edited the text, but you would not have had the psychological distancing that can help you clean up your work. Saturday becomes another phase; let's get at it and be done as quickly as possible. So far you have a second draft, and it looks the way you want the project to look. (Some writers need another draft or two to reach this point of development.) It is time for final edits.

Divide the editing into several tasks. In order to create a sharp finished product, it is important not to edit for everything at once. Trying to edit for continuity, paragraph development, spelling, punctuation, and a host of other problems is not easy if you try to go about the tasks all at once. Let your mind focus on one set of problems at a time.

Sample 6A

THE NEED FOR SPEED:
OPTIMIZING YOUR HOME COMPUTING EXPERIENCE

We live in an era where time has become ever-increasingly valuable. In many cases it is the greatest measure of a thing's worth, performance-wise at least. We don't want to wait. So, we have fast microwave ovens, fast cars, fast planes, fast computers, and . . . slow internet connections; the weak link in the otherwise rapid computer experience.

Even with a 56Kbps (kilobits per second) modem, any serious web browsing is hampered by the amount of data to be downloaded, as people and businesses are designing their web pages using more and more multimedia techniques. Web pages can easily contain hundreds of kilobytes (8 kilobits per kilobyte) of data, and often exceed a megabyte (1024 kilobytes). Software downloads, pictures, and music files can be enormous multi-megabyte affairs that can take hours to download. A one megabyte file equals 1024 kilobytes, which equals over 8000 kilobits, and if your 56Kbps modem could actually achieve a 56Kbps rate, it would take nearly two and one half minutes to download that one megabyte file to your system.

If the file is the data for a web page encountered while browing, the computer user is likely to become frustrated and perhaps even cancel the page download. Software downloads can be huge. The application that installs Netscape version 4.7, for example, is an 18 megabyte file. With a 56Kbps modem it takes about an hour to download, depending on actual connection speed.

And then there's the problem of getting data from one computer to another. At first there was only one computer in the house and this was not a problem. But now there are three computers, and another on the way. Sneaker net, the practice of transferring data by way of a floppy disk and your Nikes, works well but gets old. And if the files are on the large side it becomes more difficult to use floppies due to their small capacity. And usually there is one "good" printer that everyone would like to use, but that would require sneaker net and so the users generally opt for their own clunky printers instead.

Edit for Content

The first of my final edits is always dedicated to the continuity of the content of the document. At times I write by following a traditional outline of the sort that an author develops before writing a project. At other times I will sprint or speed write the composition, and when I edit, I try to *extract* an outline. In other words, I look for the outline *after* I write. It will be there; it has to be. If you really want to challenge yourself, you can do a quick edit by using what I call a "bone search" for the "outline" concealed in the paper. The process takes only a few minutes. Every writing project has a skeleton—or should have—holding the structure together. The bones should stand out. This trick is a tried and true "expression test" of your work. Simply read the first sentence of every paragraph in the project. *In theory, since the first sentence of each paragraph usually highlights the intent of the paragraph, you should be able to read all the first sentences and see a miniature version of the document—or an outline of your work.* Try it. If only minor discrepancies show up, you are on the mark. If there are any faulty alignments, they must be corrected.

After you examine the topic sentences to see that there is structural and logical continuity, begin to read the text through to see if it actually "reads" as a logical and orderly text. You want straight-line logic to be apparent. As I mentioned before, I have already had to pull many pages out of this text. They did not fit the "flow" of thinking and would confuse the readers.

As you read your work examine each paragraph for structure also. Each paragraph should consist of a group of comments that rally in support of a controlling idea, which you can see in the paragraphs on the left (Sample 6A). This is a critical concern because paragraphs defend and prove whatever it is that you want to say (see Chapter 10). Examine each paragraph to see if it is a logical argument or a support that helps prove your ideas. Revise accordingly.

Edit for Readers

Now, read through the text again for content, but with the reader in mind this time. This is an important concern in technical and scientific documents. You want the text to read smoothly. The reader should not find blips. I usually check to make sure each sentence reads clearly without a second thought. *Any sentence you as the author have to reread is going to be awkward for the reader! Rewrite accordingly.*

Then, make sure you use vivid and clear transitions between paragraphs. Ask yourself what the readers need from you to make the movement clear and simple. Usually the first sentence of a new paragraph is a transition as well as a topic sentence. You can use introductory phrases such as "First we must examine . . ." or "After determining the causes of the breakdown, we. . . ." Simple words such as *next* or *then* will also do the job. Notice the first word or two in the second, third, and fourth paragraphs in the text on the left (Sample 6A). Think of the paper in the reader's terms. The transitions you add should help the content flow.

Edit for Grammar, Punctuation, and Spelling

Next, move on to the final edits. Read the text for punctuation. Move on to another edit and correct spelling with a spell checker. Perhaps by now you have also corrected other errors as well, and so you are near the end. But one more edit should always be part of your writing tasks. You should develop a list of your most frequent errors, a "laundry list." Just as handwriting differs from person to person, so does writing style. Each of us has a unique way of writing. The style represents us as uniquely as our handwriting. Similarly, every writer creates predictable errors repeatedly. In a way, the errors you make are simply part of your style.

To Run the Grammar and Spelling Checker for the Entire Document

1. Click on **Tools.**
2. Click on **Spelling and Grammar.**

As you type, a green line will appear under sentences that are considered to have grammatical errors, and a red line will appear under misspelled words. If you are using a PC, you can find out about your grammar or spelling errors by right-clicking on the green or red line.

Working writers usually fail to realize that a high percentage of their errors are the same ones repeated over and over. If you always confuse *to* and *too,* put it on your list. Such errors may be your style, but your company will not accept excuses. Your computer cannot correct these errors because both are spelled correctly, although they serve different purposes. Your search feature will locate these possible culprits. If you spell *their* as *there,* put it in on the list. If you misspell technical terms you use everyday, add them to the list. Be sure your list includes the errors your software has not detected in the past. Spell checkers often overlook the most frequent spelling errors because nobody would misspell *to,* or *too,* or *two.* Instead, writers use the *wrong* one.

To Find or Replace Text

1. Click on **Edit** on the top toolbar.
2. To find text, click on **Find,** and then type in the text (or word) you are looking for in your document.
3. To replace text in your document, click on **Replace.**
4. Type in the word or words you are looking for, and then type in the text you want to replace it with.
5. You can replace each word individually by clicking on **Find Next** and then clicking on **Replace,** or you can replace the text for the entire document by clicking on **Replace All.**

I overlook some errors in my writing because they are elements of the way I *want* to write. I occasionally begin a sentence with *And* or *But,* even though older grammar books disapprove of this practice. And I keep on doing it. But I have my reasons, which have to do with style and tone and flow. You would avoid the practice if you want your style and tone to remain very formal. In a book of this sort, I will usually avoid most casual writing elements, but not all of them.

If you develop a list of your most common errors, you can hunt them down during your last edit every time you write a project. You may be surprised at the success of this tactic. Software can also be adapted to include your technical vocabulary in the spell checker system.

To Add Technical Vocabulary to the Spell Checker

1. Include appropriate technical vocabulary in your document as you are creating it.
2. Click on **Tools** from the top toolbar.
3. Click on **Spelling and Grammar.**
4. When the spell checker indicates that a technical term is spelled incorrectly although the word is correctly spelled, click on **Add** to add the term to the dictionary.

Finally, the edits for content are complete, and the edits for form and for the cosmetic touches are done. Give the project a quick read on Sunday to see what you think.

Stage Eleven

Finished Copy

All these sample stages—one through eleven—are options; they may not be needed. A business letter will jump from draft to finished product in an hour. A month-long project may include every stage we have looked at from one through eleven!

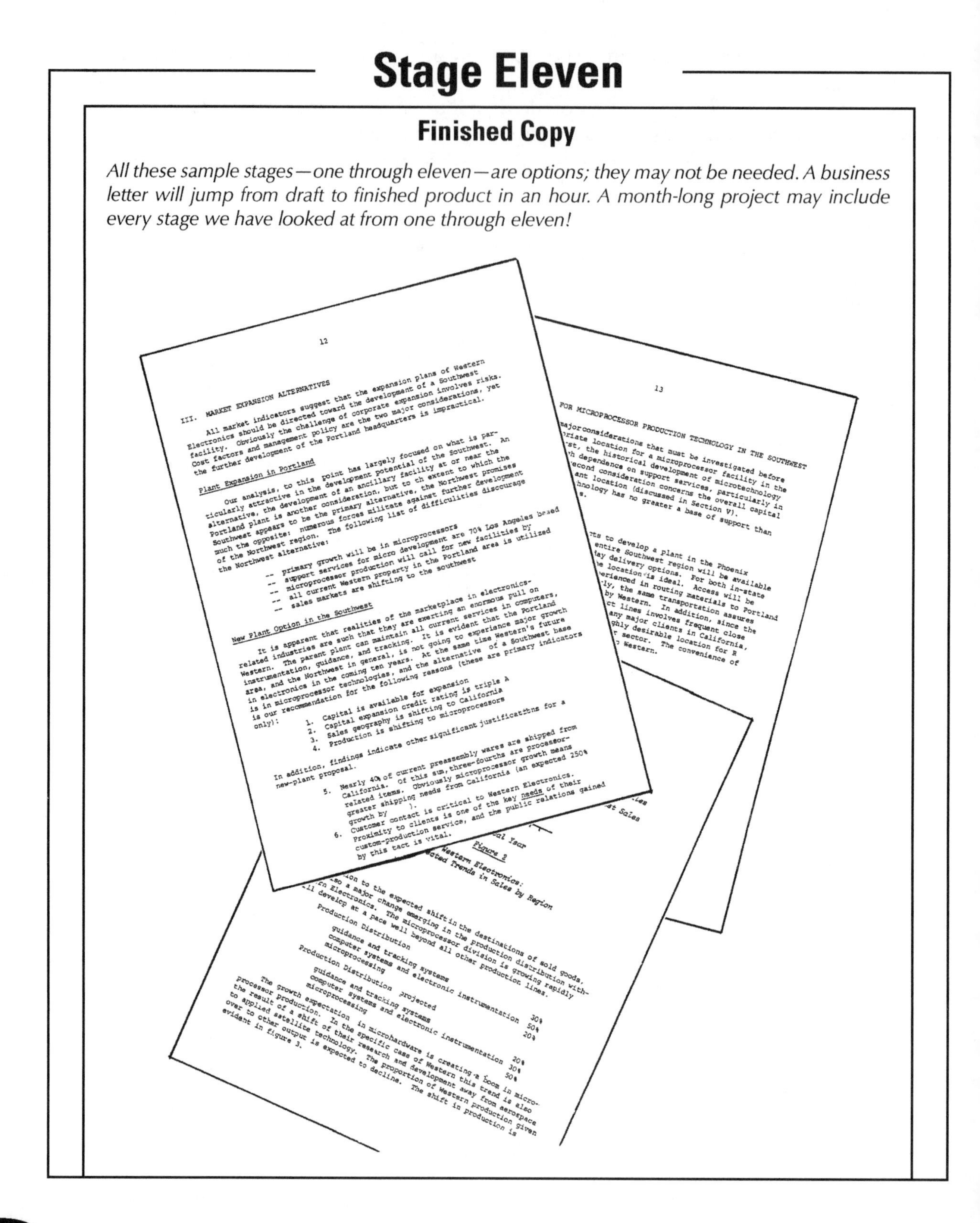

12

III. MARKET EXPANSION ALTERNATIVES

All market indicators suggest that the expansion plans of Western Electronics should be directed toward the development of a Southwest facility. Obviously the challenge of corporate expansion involves risks. Cost factors and management policy are the two major considerations, yet the further development of the Portland headquarters is impractical.

Plant Expansion in Portland

Our analysis, to this point has largely focused on what is particularly attractive in the development potential of the Southwest. An alternative, the development of an ancillary facility at or near the Portland plant is another consideration, but to th extent to which the Southwest appears to be the primary alternative, the Northwest promises much the opposite: numerous forces militate against further development of the Northwest region. The following list of difficulties discourage the Northwest alternative:

-- primary growth will be in microprocessors
-- support services for micro development are 70% Los Angeles based
-- microprocessor production will call for new facilities by
-- all current Western property in the Portland area is utilized
-- sales markets are shifting to the southwest

New Plant Option in the Southwest

It is apparent that realities of the marketplace in electronics-related industries are such that they are exerting an enormous pull on Western. The parent plant can maintain all current services in computers, instrumentation, guidance, and tracking. It is evident that the Portland area, and the Northwest in general, is not going to experience major growth in electronics in the coming ten years. At the same time Western's future is in microprocessor technologies, and the alternative of a Southwest base is our recommendation for the following reasons (these are primary indicators only):

1. Capital is available for expansion
2. Capital expansion credit rating is triple A
3. Sales geography is shifting to California
4. Production is shifting to microprocessors

In addition, findings indicate other significant justificattns for a new-plant proposal.

5. Nearly 40% of current preassembly wares are shipped from California. Of this sum, three-fourths are processor-related items. Obviously microprocessor growth means greater shipping needs from California (an expected 250% growth by).
6. Customer contact is critical to Western Electronics. Proximity to clients is one of the key needs of their custom-production service, and the public relations gained by this tact is vital.

13

...FOR MICROPROCESSOR PRODUCTION TECHNOLOGY IN THE SOUTHWEST

...major considerations that must be investigated before ...riate location for a microprocessor facility in the ...st, the historical development of microtechnology ...h dependence on support services, particularly in ...econd consideration concerns the overall capital ...ant location (discussed in Section V). ...hnology has no greater a base of support than ...s.

...ts to develop a plant in the Phoenix ...entire Southwest region will be available ...ay delivery options. For both in-state ...he location is ideal. Access will be ...erienced in routing materials to Portland ...ly, the same transportation assures ...by Western. In addition, since the ...ct lines involves frequent close ...any major clients in California, ...ghly desirable location for R ...r sector. The convenience of ...o Western.

Figure 2

Western Electronics:
Projected Trends in Sales by Region

...on to the expected shift in the destinations of sold goods, ...so a major change emerging in the production distribution within Western Electronics. The microprocessor division is growing rapidly ...ll develop at a pace well beyond all other production lines.

Production Distribution	
guidance and tracking systems	30%
computer systems and electronic instrumentation	50%
microprocessing	20%

Production Distribution projected	
guidance and tracking systems	20%
computer systems and electronic instrumentation	30%
microprocessing	50%

The growth expectation in microhardware is creating a boom in microprocessor production. In the specific case of Western this trend is also the result of a shift of their research and development away from aerospace to applied satellite technology. The proportion of Western production given over to other output is expected to decline. The shift in production is evident in figure 3.

The Due Date

Monday. If the final form of your project needs to be a hard copy, you will have to print out the document. Whether the project is hard copy or electronic, if you are supposed to process your work through the secretarial staff, there are a few minor tips I would offer.

Before you send it off for editing by the clerical department, why not take a colored pencil and write a few notes to the secretary to save time. Do you want it single-spaced? Double-spaced? Should the secretary correct "errors" or render it "as is"? If you want the headings in caps or in boldface, indicate your needs. Add any other notes you think might be helpful. Double-space the text if you are expecting additional input from other people. They need the space. I write all secretarial instructions in red so that they are not overlooked and so that they are not confused with the text. I forward hard copy and a floppy.

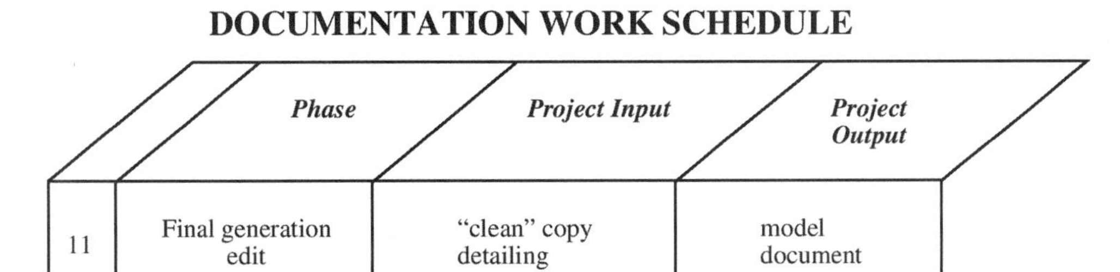

DOCUMENTATION WORK SCHEDULE

	Phase	*Project Input*	*Project Output*
11	Final generation edit	"clean" copy detailing	model document

Let's assume that you did not need secretarial support and you now have the final copy in your hands, but you do not want the supervisor to see it just yet. You must always proofread the final copy. This is an absolute must. The last thing you want is to have all this work crash because of any oversights. As you read the copy you might follow along with your finger or put a ruler under each line you read. The physical gestures help focus your eyes on each word so that you find errors. Sure enough, there are three typos, and one is a computer glitch amid the technical statistics, just where engineers and technicians often fail to look for errors. Even with secretarial support, you cannot be too careful—or expect someone else to catch all the errors—because secretaries, as well as authors, are human.

If a double-spaced typed page consists of approximately 250 words, that means that each page will contain about 1250 strikes of the keyboard (at five letters per word). If there is only one error in three pages, you or your secretary will have delivered nearly 4000 alphabetical symbols onto the pages—with only one error. It is miraculous, but we must not expect miracles. More than three *million* symbols make up the book you are reading. Imagine the number of weeds that show up in such a garden. (You have probably found at least one.) Proofread your final copy and make the necessary corrections. This last task is critical.

It is three o'clock on a fine Monday afternoon. The project goes into the supervisor's basket. Great job. What now? Reward yourself, of course! Take a copy home. Expect applause. With luck, maybe you will not have to write another one for awhile.

Technical Edits

Depending on the policy of your company, technical editing may be handled at a number of points in the process of developing a technical presentation. If you assume that you are going to present both the secondary research and the primary findings emerging from your company's research activities, you will need checkpoints. Since you are in the position of taking full responsibility for the technical data, you want to be sure of the accuracy and appropriateness of the material.

DOCUMENTATION WORK SCHEDULE

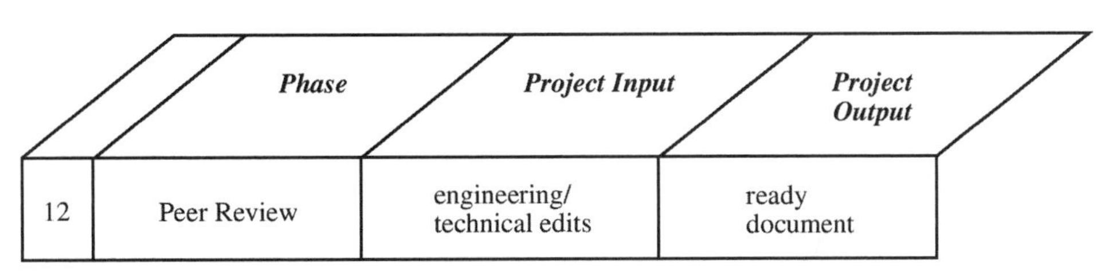

	Phase	*Project Input*	*Project Output*
12	Peer Review	engineering/ technical edits	ready document

Because teamwork is the standard practice in business and industry, it is important to have a peer review to debug any flaws in the technical areas of your presentation. The technical data are the most important editing consideration in your project, but also the least likely area where clerical assistants will be able to identify problems. Technical writers and editors are often not qualified to help you either, particularly in areas of higher mathematics. For content editing, there is no replacement for coworker evaluations.

In the course of these imaginary project activities, I have placed the technical edit last. You can assume that the edit might take place before or after supervisor approval, since this sort of editorial correction consists of checks for technical oversights. A document with serious theoretical problems would be a different matter and would call for a complex review process. Here, though, you are simply looking for that error in a schematic, that incorrect symbol in a calculation, that slip of a zero on the bottom line of a cost analysis.

Technical editing takes knowledge and patience. Engineers and technicians who know their disciplines are the readers of choice for this edit. Perhaps one reader is sufficient. More than one may be appropriate. The challenge for the author is to locate a reader with an attentive eye. Editing calls for scrutiny, and this sort of careful and methodical reading is not a skill that you can always count on among engineers and technicians. A good reader, however, will usually surface in every lab setting. That is the person you want as your "contact" for a technical edit.

Writers at Work

We have taken a very close look at the many phases of a writing project as they might transpire in a workplace setting. By way of a summary it would be appropriate to take a guided tour of a work environment where you can see the engineering technician at work. To show you the typical day-to-day processes involved in company writing, I would like you to meet Todd and Kayla. Both of these technicians are involved in industries where writing is a part of the daily work activity. They will walk you through brief projects they developed for the companies where they work.

In the list below you will find a brief summary of the typical writer's activities that are involved in developing a document. The summary is a review of the work schedule on page 85. As you read about the activities of Todd and Kayla, you will notice that writing projects vary greatly in substance, design and length, but the phases of writing projects will consistently involve activities from the following categories.

1) **The project is requested** and defined by a supervisor, often by memo.
2) **The author must plan an agenda** of activities to get from day one to the finished project.
3) **The author must organize key ideas** and think out the basic approach to the project.
4) **Research must be conducted.**
5) **Coworker responses and administrative recommendations are important** aspects of research in company projects.
6) **The author often constructs an outline** for the project.
7) **Pieces of the project are composed.**
8) **The rough draft is assembled.**
9) **A second draft is developed** with modifications that evolve from ongoing research and consultation.
10) **A third draft is developed** to reflect necessary changes and refine the writing.
11) **The finished project is tightened** for precision, grammar and punctuation.
12) **The ready document is submitted for review,** and the reviews involve both administrative and technical considerations.

Model 6.A (1)

ZETRON®

XYZ-Page

MARKETING REQUIREMENTS SPECIFICATION

APPROVALS		
Title	**Signature**	**Date**
Division Manager	CS	5-9-00
Sales Manager	Susan Stanton	5-1-00
Engineering Manager	John Abbott	May 10, 2000

Revision History			
Rev.	**Summary of Change**	**Originator**	**Date**
A	First version	TJR	04/08/99

Todd's Work Day

I work for Zetron Incorporated located in Redmond, Washington. Zetron is a high tech company that manufactures a variety of equipment used in many different wireless communications systems throughout the world. Some examples include, paging encoders, trunked radio systems, radio dispatch consoles and alarm monitoring equipment. I have worked for Zetron for almost two years. My job title is Marketing Research Analyst. I work closely with product managers and engineers to ensure our documentation at various levels stays up to date and is complete and accurate. I also spend much of my day doing company research and reporting on target markets, competition and market trends.

Purpose

The purpose of this brief explanation is to show you some of the work I did on a writing assignment that was given to me. In this case, the project involved a Marketing Requirement Specification. A Marketing Requirement Specification is a document used at my company that helps outline the requirements and issues concerning a new project the company is considering taking on. It serves as a useful tool to upper level management to help them decide if it is wise to allocate the necessary resources for the project. Most of the time these new projects are pieces or hardware or software designed by our engineering group. This particular example, however, deals with a piece of software that another company created that Zetron was considering reselling. I will tell you about some of the steps I took to create this document and show you the drafts (the smaller illustrations) as well as the finished product (Model 6.A).

Market Requirements Specification

1. PROJECT DESCRIPTION
2. MARKET OPPORTUNITIES
3. COMPETITIVE PRODUCTS
4. PRODUCT SPECIFICATIONS
5. RISK/ISSUES
6. PRODUCT PRICING

Item	Description

7. SALES FORECAST ESTIMATE

1st year:
2nd year:
3rd year:

Zetron C

Market Requirements Specification

ZETRON®

MARKETING REQUIREMENTS SPECIFICATION

APPROVALS		
Title	Signature	Date
Division Manager		
Sales Manager		
Engineering Manager		

Revision History			
Rev.	Summary of Change	Originator	Date
			xx/xx/xx

Zetron Confidential Page 1

Model 6.A (2)

1. PROJECT DESCRIPTION

XYZ-Page is a software product developed by StarterComp Incorporated. The software's main function is to forward e-mail to your alphanumeric pager. It can also be used to send alphanumeric pages to others and it can send you reminders of important meetings and appointments by monitoring the scheduling software on your computer or its own built-in scheduler. The software is highly configurable and allows the user to filter and manipulate the information that is being sent. For instance, the user can set options to have the software look for keywords in the subject or body of the e-mail or only forward e-mail from certain people. It can also break up longer messages or send them in shorthand to save characters. This product adds a great deal of functionality to alphanumeric pagers and would be a nice package for Zetron's commercial paging operators to offer their customers.

2. MARKET OPPORTUNITIES

XYZ-Page would be resold primarily to Zetron commercial paging operators who would then offer it as a service to attract or keep customers. The network version of this software could appeal to hospitals or other private in-plant operators who desire to add functionality to their alphanumeric pagers. In this case, the software would be sold direct.

Paging customers have been demanding more services for less money and increased functionality out of smaller devices. One such device on the market can receive, store and send e-mail; another has organizer features built in. There has also been a push to develop expansion cards for the increasingly popular personal digital assistants that give the device the ability to receive alphanumeric pages. The direction of the industry is to give customers access to multiple sources of information from one device. Currently, the solution on the hardware side is very expensive. XYZ-Page is a cost effective software solution. It doesn't require users to upgrade their alphanumeric devices. They do not need a separate e-mail address and they don't have to input appointments and calendar events at two different locations.

Project Assignment

I was given this task by my boss one Monday morning after he had returned from a trade show. He wanted me to evaluate a piece of software he had come across. He believed it might be something that we could resell. Zetron is well known in the paging industry and we are always looking for ways to help our paging operators increase their customer base. The piece of software was called XYZ-Page and it claimed it could forward e-mail to any alphanumeric pager. This functionality is definitely desired in the marketplace, so it was well worth taking a look at.

The first thing I needed to do was to test and evaluate the product and make sure it did what it claimed. Then, the main task would be to create a Marketing Requirement Specification (MRS) so that the right people in the company could decide if it was feasible to resell the software to our paging operators. An MRS is a "controlled document" at Zetron, meaning that it has its own template. (See the blank forms on the bottom of page 141.) This makes a document easier to write and provides some consistency company wide. With the template, readers can search the document efficiently for the information they need because they know the documents are all designed in a similar fashion. Different people at Zetron are interested in different parts of the document. For example, Engineering would be interested in the product specifications, while Marketing would be more interested in the market opportunities.

Evaluation

Before I installed the software on my computer, I wanted to get a general idea of how the product worked. Luckily, the software came with a pretty good manual. I read through it and I also took a look at the company's Web site. Next I needed to make sure I had all the necessary equipment. I had an alphanumeric pager I could use and the manual said my e-mail software would work but I didn't have a modem in my computer. I filled out a purchase requisition, had it signed by my manager and submitted it to the purchasing department. I received a new modem about a week later.

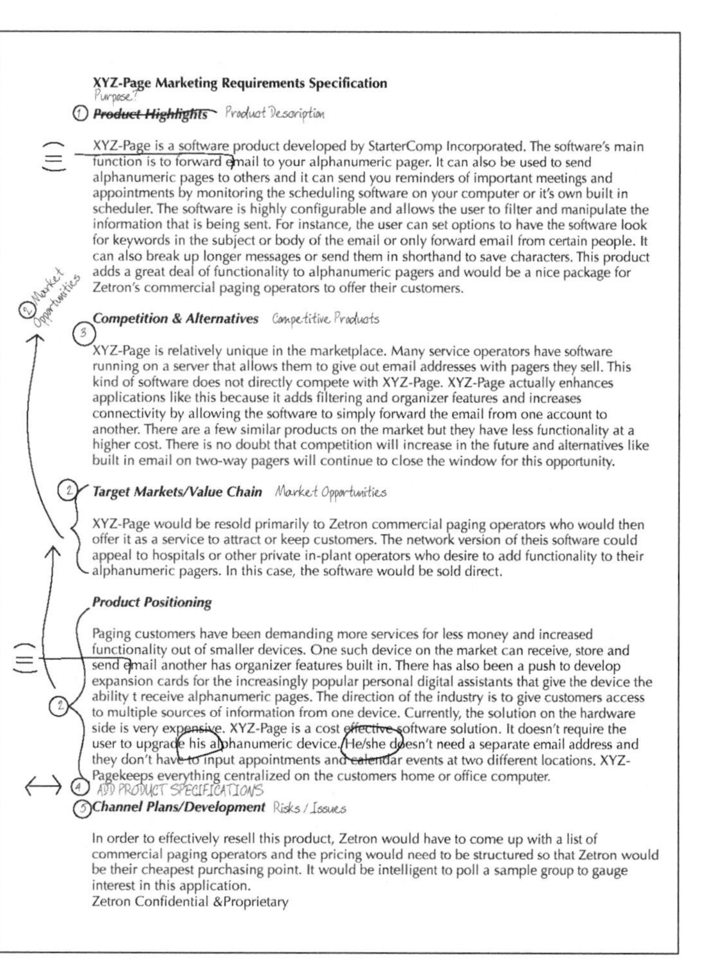

XYZ-Page Marketing Requirements Specification

Product Highlights

XYZ-Page is a software product developed by StarterComp Incorporated. The software's main function is to forward email to your alphanumeric pager. It can also be used to send alphanumeric pages to others and it can send you reminders of important meetings and appointments by monitoring the scheduling software on your computer or it's own built in scheduler. The software is highly configurable and allows the user to filter and manipulate the information that is being sent. For instance, the user can set options to have the software look for keywords in the subject or body of the email or only forward email from certain people. It can also break up longer messages or send them in shorthand to save characters. This product adds a great deal of functionality to alphanumeric pagers and would be a nice package for Zetron's commercial paging operators to offer their customers.

Competition & Alternatives

XYZ-Page is relatively unique in the marketplace. Many service operators have software running on a server that allows them to give out email addresses with pagers they sell. This kind of software does not directly compete with XYZ-Page. XYZ-Page actually enhances applications like this because it adds filtering and organizer features and increases connectivity by allowing the software to simply forward the email from one account to another. There are a few similar products on the market but they have less functionality at a higher cost. There is no doubt that competition will increase in the future and alternatives like built in email on two-way pagers will continue to close the window for this opportunity.

Target Markets/Value Chain

XYZ-Page would be resold primarily to Zetron commercial paging operators who would then offer it as a service to attract or keep customers. The network version of theis software could appeal to hospitals or other private in-plant operators who desire to add functionality to their alphanumeric pagers. In this case, the software would be sold direct.

Product Positioning

Paging customers have been demanding more services for less money and increased functionality out of smaller devices. One such device on the market can receive, store and send email another has organizer features built in. There has also been a push to develop expansion cards for the increasingly popular personal digital assistants that give the device the ability t receive alphanumeric pages. The direction of the industry is to give customers access to multiple sources of information from one device. Currently, the solution on the hardware side is very expensive. XYZ-Page is a cost effective software solution. It doesn't require the user to upgrade his alphanumeric device. He/she doesn't need a separate email address and they don't have to input appointments and calendar events at two different locations. XYZ-Pagekeeps everything centralized on the customers home or office computer.

Channel Plans/Development

In order to effectively resell this product, Zetron would have to come up with a list of commercial paging operators and the pricing would need to be structured so that Zetron would be their cheapest purchasing point. It would be intelligent to poll a sample group to gauge interest in this application.

Zetron Confidential &Proprietary

Model 6.A (3)

4. PRODUCT SPECIFICATIONS

StarterComp handles all the engineering issues including upgrades and bug fixes. The software works on almost any computer with 16MB of RAM and 6MB of free disk space. The software is compatible with most e-mail programs and scheduling software.

5. RISKS/ISSUES

- In order to effectively resell this product, Zetron would have to come up with a list of commercial paging operators and the pricing would need to be structured so that Zetron would be their cheapest purchasing point. It would be intelligent to poll a sample group to gauge interest in this application.
- StarterComp handles all quality and testing issues. The product is in its third version and is very solid. It is also noteworthy that they offer free technical support. Paging customers hate to lose messages, but if this happens, it will be the fault of the paging equipment or coverage area, not the software.
- Lead-time is minimal. StarterComp can ship several copies at a moment's notice.
- Zetron may require some order forms and marketing literature.
- The sales force should be educated on the product. StarterComp handles all other support.
- The reseller agreement would cover ordering and invoicing requirements.
- StarterComp is overdue to release the network version of the software. The latest projection is by the end of 2000. A mailing to the prospective commercial operators should be made about the time the reseller agreement is signed.
- This product requires a minimal amount of effort to support. The only question is how much value this product adds to our paging operators.

6. PRODUCT PRICING

Item	Description	Industrial Net	US Dealer
050-1222	StarterComp XYZ-Page Software	$70.00	$49.00

7. SALES FORECAST ESTIMATE

1st year:	$50,000
2nd year:	$100,000
3rd year:	$250,000

I then immediately began tinkering with all the options and preferences and ran some tests and took notes on all my findings. I was pretty impressed with the software and I met with my manager to tell him how the test runs were going. We talked about the product a little bit and he cleared me to start working on the MRS. He believed this software represented a good opportunity for Zetron and others in the company would want to know about it.

First Draft

When I worked on the first draft I put all my notes together and wrote down everything I could think of regarding the product. I kept the MRS template in the back of my mind, but I did not let this limit me. I wrote down everything I thought was relevant. Then I created a first copy that I marked up so it would be easy for me to figure out how to move the information into the template. (See the sample below on the right and the sample on page 143 and on the right.)

Final Draft

With some cutting and pasting, I moved the information into the template and made some final adjustments to the text. Now I was ready to submit the document for approval. Before I officially submitted the document, I let my manager look at it first so he could catch any mistakes I might have made or information I might have left out. With his OK I officially submitted the document and waited to hear what management would say.

Typically, the forms we use are a lot more complex that this one since they usually deal with the sophisticated computer equipment that Zetron manufactures rather than software that is to be resold. With more experience and training I will be able to create these documents as well.

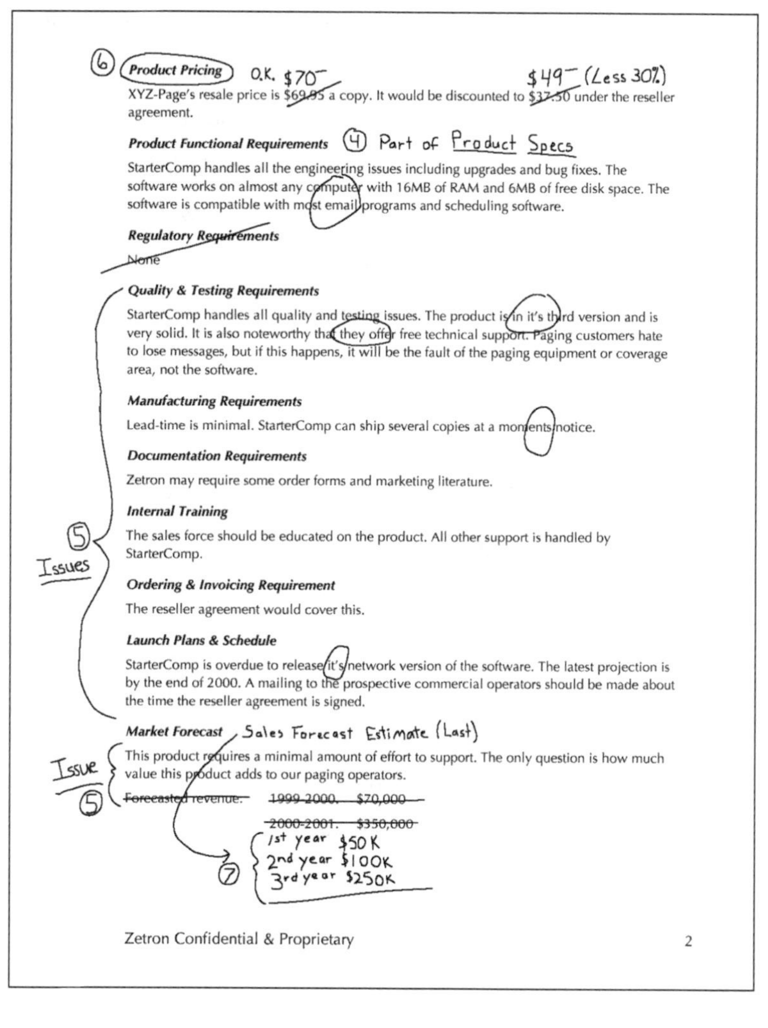

Product Pricing

XYZ-Page's resale price is $69.95 a copy. It would be discounted to $37.50 under the reseller agreement.

Product Functional Requirements

StarterComp handles all the engineering issues including upgrades and bug fixes. The software works on almost any computer with 16MB of RAM and 6MB of free disk space. The software is compatible with most email programs and scheduling software.

Regulatory Requirements

None

Quality & Testing Requirements

StarterComp handles all quality and testing issues. The product is in it's third version and is very solid. It is also noteworthy that they offer free technical support. Paging customers hate to lose messages, but if this happens, it will be the fault of the paging equipment or coverage area, not the software.

Manufacturing Requirements

Lead-time is minimal. StarterComp can ship several copies at a moments notice.

Documentation Requirements

Zetron may require some order forms and marketing literature.

Internal Training

The sales force should be educated on the product. All other support is handled by StarterComp.

Ordering & Invoicing Requirement

The reseller agreement would cover this.

Launch Plans & Schedule

StarterComp is overdue to release it's network version of the software. The latest projection is by the end of 2000. A mailing to the prospective commercial operators should be made about the time the reseller agreement is signed.

Market Forecast

This product requires a minimal amount of effort to support. The only question is how much value this product adds to our paging operators.

Forecasted revenue: 1999-2000. $70,000
2000-2001. $350,000

Zetron Confidential & Proprietary 2

Model 6.B (1)

SENDING AND OPENING ATTACHMENTS

- Click on the **Paper Clip** button. If it isn't on your toolbar, select **View** → **Toolbars.** Click on **Standard Buttons.** This toolbar gives you immediate access to the most commonly used options when working with your e-mail.

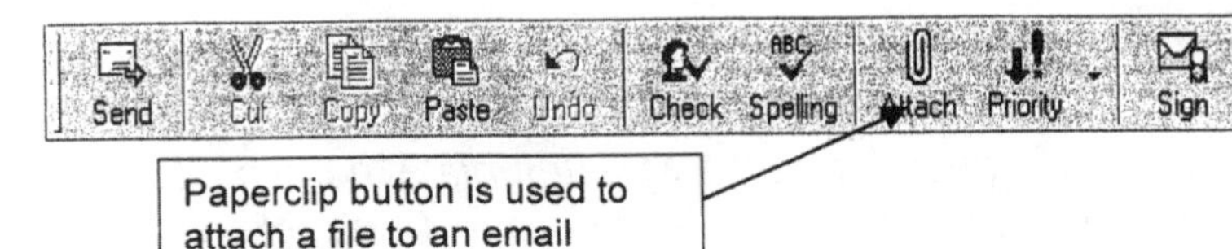

- The Attach File dialog box will appear.
- An icon will appear in the message that represents the attachment. The icon will be specific to the Application that is used to create the Attachment.
- To **open** an Attachment, double-click on the icon.
- You might prefer to InsertText from a File instead of attaching a complete file. To accomplish this, click **Insert** → **Text from File.** Browse to the location of the stored file, double-click to open the file, and the contents of the file will be placed in the body of the e-mail message at the location of the insertion point.
- You can indicate the importance of an e-mail message to the recipient by changing the **Priority button**.

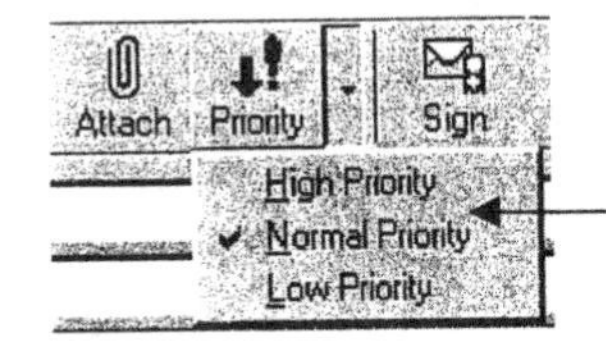

USING FOLDERS

- Outlook Express has several folders for storing your e-mail. All incoming e-mail messages are held in the Inbox. Outbound e-mail messages to be sent at another time are stored in the **Outbox.** Other folders in Outlook Express are **Sent Mail, Deleted Mail, Drafts, Calendar, Journal, Contacts, Notes,** and **Tasks.** Each of these folders is a storage location to assist to organizing your e-mail etc.
- To move between folders, double-click on the folder whose contents you would like to view. To view the folder list, click on **View** → **Folders.** Windows does allow more than one way to accomplish a task.

Kayla's Work Day

My name is Kayla and I freelance contracts in computer networks and systems training. A client of mine recently upgraded a company network. I was retained to develop reference documentation and provide software training. The majority of the users were not familiar with Windows, and were still solidly entrenched in a DOS command line environment.

The firm, Morrison Properities Insurance, was composed of two operations—property management and an insurance brokerage. Prior to upgrading their network, the firm was running standalone 386s. Most of the empolyees didn't have a computer at their desk and were sharing workstations between several people. Suddenly empolyees were faced with familiarizing themselves with both drastically different software applications and new hardware at the same time—while still having to be productive. E-mail, GUIs, buttons and icons, the Internet—to name a few newere features—were almost totally unknown to these people. Most of the empolyees didn't have a computer at home.

Prior to installation of the network, I set up four pc's for running Windows 95 in the office. This allowed small groups to begin training on the new software. We started with basics—mouse skills, learning to use Windows Explorer and components of the Microsoft Office suite. We focused on email and Outlook at the email client.

My task involved presenting each employee with the basic skills they needed to effectively use

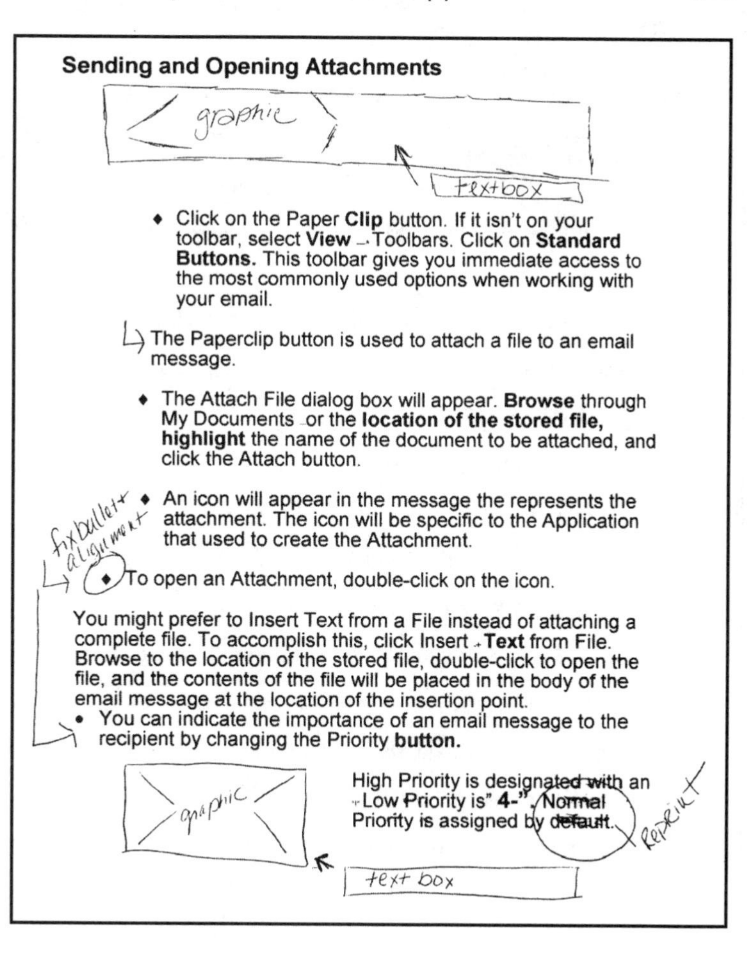

Sending and Opening Attachments

- Click on the Paper **Clip** button. If it isn't on your toolbar, select **View** → Toolbars. Click on **Standard Buttons.** This toolbar gives you immediate access to the most commonly used options when working with your email.

The Paperclip button is used to attach a file to an email message.

- The Attach File dialog box will appear. **Browse** through My Documents or the **location of the stored file, highlight** the name of the document to be attached, and click the Attach button.

- An icon will appear in the message the represents the attachment. The icon will be specific to the Application that used to create the Attachment.

- To open an Attachment, double-click on the icon.

You might prefer to Insert Text from a File instead of attaching a complete file. To accomplish this, click Insert → **Text** from File. Browse to the location of the stored file, double-click to open the file, and the contents of the file will be placed in the body of the email message at the location of the insertion point.

- You can indicate the importance of an email message to the recipient by changing the Priority **button.**

High Priority is designated with an → Low Priority is" **4-"**. Normal Priority is assigned by default.

Model 6.B (2)

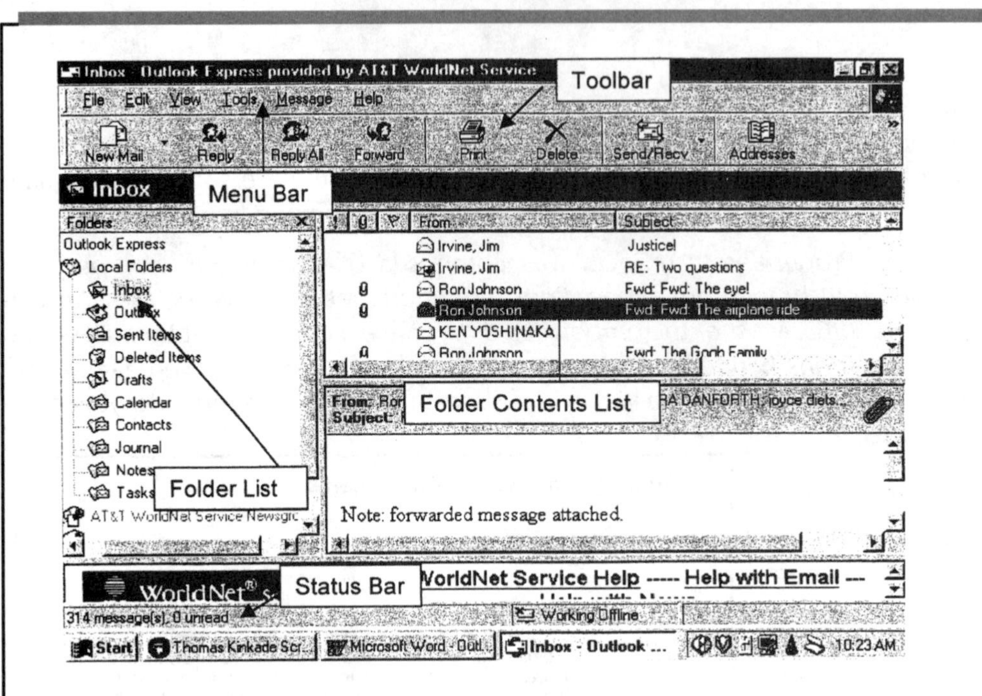

- The Status Bar is at the bottom of the window. It tells you how many messages are in each folder.

READING A MESSAGE

- To read a message, double-click on it.
- A window will open with the message in it.
- The Toolbar will be the Read Mail toolbar. To add other toolbars (from many other options) go to **View →Toolbars.**
- You can Reply, Reply to All, Forward, Print, Move, Delete, go to Previous Message, go to Next Message, or go to Next. Unread using the options in this toolbar. When you forward a message, the original message is automatically attached.

COMPOSING A NEW MESSAGE

- To compose a new message, click the **New Mail** icon on the toolbar.

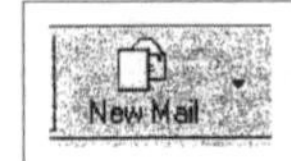

the power of email. I started this process by working with employees individually and I moved to small groups as their knowledge grew. I also wanted to provide all employees with reference material to guide them step-by-step through the completion of a task.

The manual was a complicated task that involved screen scans such as the ones you see in the sample on the left. In order to write the manual I had to blend text and graphic features on every page. Then I had to test the manual by seeing if the employees could follow the instructions. This process took weeks but I enjoyed the challenge and the company was impressed with the bonus; I was hired to retrain employees and not to provide them with a manual.

Two sets of documentation were given to each employee. The first was basic information: step-by-step instructions of basic email functions in Outlook. As each individual's knowledge grew more detailed documentation was added to enhance his or her newfound skills and functionality.

I looked critically at the tasks I routinely perform when I am using email. I broke each task into the essential steps. A basic instruction sheet was built upon that information. Then, the same analysis was made of higher-level tasks and more detailed tutorial documentation was developed.

Upon completion of each difficulty level section of the Outlook tutorial, I followed, the written directions step-by-step while looking for omissions. Once I had stepped through each task, I asked someone else to follow the documentation and evaluate the accuracy of each step to each task. Essentially the process was handled in repetitive stages: write; test; edit. Each revision was another test when I used it on a group of employees. Two full pages of the manual will suggest to you what the finished product looked like and there is a sample of a rough draft page that was developed before I added the screen scans.

The task of building the small text boxes that you see in the illustrations is not difficult and they point to the important parts of the scan image. The real challenge is to be sure the text is correctly understood. Clarity and precision can be difficult to achieve. I write with regularity both at the company where I work and on freelance contracts. Since the freelance work is a training activity, the companies are always surprised when I build employee manuals for them. Writing has helped my business grow.

By the way, you might be intersested in knowing what tools I use. I build text by using Office 97 (old maybe, but familiar) and the graphics depend upon the application being used by trainees. I use a Lexmark color printer. The black and white copy you see here was presented to the company in color.

Summary

- Use a pressure-and-reward system to coach yourself through the early stages of writing.
- Write as quickly as possible without regard for careless errors.
- Use only one side of any sheet of paper so that you can create spreads or layouts of the project.
- Use pasteups to reconstruct and modify the first drafts.
- Use any of the popular techniques that you find effective:
    - ✓ front-to-back writing
    - ✓ outline method
    - ✓ sprint and revise
- If you write front to back, do the introduction last.
- If you write from an outline, write whatever parts come easily and progress to the other areas of interest. Do not write in sequence unless you prefer to do so.
- If you sprint, write without judgment and do not review the product until you are done with a session.
- Develop subsequent drafts with a focus on content.
- Develop final drafts with a focus on the reader and mechanical matters concerning grammar, spelling, and punctuation.
- At some point a technical edit might be an important step.

Activities Chapter 6

Select one or more of the following suggestions and develop an analysis of your writing practices.

If you recently composed a project by using the front-to-back method, explain the procedures and events that went into the writing.

If you recently composed a project by using the outline method, explain the procedures and events that went into the writing.

If you recently developed a project by using the sprint method of free writing or speed writing, explain the procedures and events that you used to write the project.

If you have used several of the writing methods identified in the previous activities, explain when and why you used them. Compare the advantages and disadvantages of the methods in terms of your skills and experiences.

If any of the documents in the preceding activities involved two or more drafts, explain the roles of the different drafts you developed. Did you use any kind of peer review when it was completed? (Your response will be kept confidential.)

These projects should be 250 to 500 words in length, typed in a memo format, and addressed to your instructor.

Share a Project

Collaborate with three other members of the class to develop a short report and practice team editing.

- *Briefly explain your current employment to the team and identify a specific work activity in writing.*
- *Develop a two- to four-paragraph description of the work-related activity.*
- *Circulate the document among the team members for revisions, suggestions, and discussion.*

PART III

Design Basics: Superstructure Logic and Infrastructure Logic

Work in Progress

The Meetings

It was over a month since I had handed in my proposal and I still had not heard anything positive or negative from the company president. I felt that I had enough information on hearing protection, so I moved on to the next issue regarding chemical controls and hazardous material. These risks go hand in hand with respiratory protection, so I was ready to attempt another meeting with Matthew. This time I would try a different tactic. I asked Matthew if he would be willing to listen to parts of my proposal whenever it was convenient to his schedule. Half expecting a flat refusal again, I was surprised that he accepted my request and set a time later in the week. The next item on my agenda was to speak to Woodrow, my immediate supervisor.

My meeting with Woodrow was mainly to ask advise on how to proceed with Mathew, but I thought that it couldn't hurt my cause to get his general input on the safety program I was proposing. He was a working supervisor of a number of avionics technicians but he was available in the afternoon for an impromptu meeting. We sat down and I discussed my proposal with him. He was willing to answer my questions and gave sage advice concerning Matthew. He had known Matthew for several years and knew his mind set. His advice was to go slow and he told me not to dump too much information on him all at once because he will shut down and get defensive and angry.

Since the meeting with Matthew was scheduled for Thursday late morning, I took some extra time in the early morning to gently remind him that we had a meeting scheduled; would he still be able to make it? He was willing to get it out of the way right there and then, so we sat down and I dove right into the meeting. Starting with my strongest observations—decibels levels in and around the window shop and warehouse—coupled with the general guidelines required from OSHA and WISHA, I suggested that warning signs and boxes of "foamies," disposable earplugs, be placed in key, target areas. I deliberately kept the meeting short. I explained that I recognized that PAT is a money making organization, and that it is to our advantage to have a safety program, that allows the persons involved to be appropriately trained on various safety issues. I mentioned that full employee and management cooperation must be present for the safety program to be successful. Overall, I viewed the meeting as a challenge, but it was rewarding.

Late spring and early summer are months during which the company president is often on business trips to see new customers and to visit old customers on our vendor/customer list, so she had the proposal quite a while before actually reading it. I was beginning to have second thoughts about the whole program—or at least my role. After putting so much time and effort into the proposal, I started to get frustrated when two long months came and went with no word from Karen, the president. I was at a standstill and couldn't do anything more except be patient and wait for the word.

My approval arrived silently in the form of an office memo from Karen that politely requested my presence at my convenience to discuss the total assimilation of my proposal into a working safety program for the PAT employees. I couldn't believe it at first. She accepted everything I proposed! At our meeting we discussed a loose time frame of six months from write-up to implementation, and, of course, meetings with Matthew and Woodrow to "sell" them on my ideas. I was excited to finally begin the OSHA procedure project with the President's administrative support. L.C.L.

Document Logic

The next four chapters will deal with the physical aspects of your document. There are two distinct issues to be managed in the task of writing. The first is the activity of writing: the *process* of getting it done. The second is the writing *product*. Throughout this discussion you will see quite a few helpful illustrations of writing products.

It is important to view writing as blocks of thought rather than as pages of words. A page of words is too difficult to see in terms of structure. You can get 500 words on a typed page if you single-space the printout. Imagine a "structure" for a formula or calculation with 500 symbols. The perception is too difficult to imagine, much less control. Patterns provide structure for writing projects and it is important to see writing in terms of logical design considerations so that you can organize your thinking.

Usually, formulas or calculations are small, discrete units that have an apparent function that can be identified rather quickly. For a reader, the problem with a page of 500 words is that the page looks like an endless parade of language that is only minimally differentiated by punctuation. Writers often have the same perception. If a page is to be written, the words that are supposed to fill the page seem like a maze, and writers often become confused. Why? Because a maze has no apparent system. There are no guidelines.

Logic Blocks

You must view writing as a very orderly structural process composed of discrete units—units of logic. Each of these units, or *logic blocks,* should be an explanation or a logical argument that defends some point you want to make, or that structures observations if you want to be strictly descriptive and avoid decision making. In other words, each paragraph or two should represent, describe, or explain a discrete point.

A writing project is written to fulfill some sort of task. The project is supposed to demonstrate some set of outcomes, often in support of an overall proposal, so it must have an orderly, logical pattern to get the job done. Notice that I mentioned a "set of outcomes." If you divide the writing task into discrete activities, you will build the finished product block by block. Each block is simply one of the desired outcomes; each one is a logic block in the set. An environmental impact statement (EIS), for example, is a set of arguments that propose an outcome. One block of material must explain the impact of, let's say, a housing development or a bird sanctuary. Another block must explain the impact of population on the development of local schools. Another must address county roads. Still others deal with water resources, zoning regulations, and so on. Each issue is another logic block. In any specific block—whether it is a paragraph or pages of paragraphs—the logic must support the case.

Earlier we observed that the difficulty of a writing project can be somewhat measured by the involvement you must make in terms of judgments. Many documents are purely descriptive; they involve no judgments. The degree of difficulty, however, increases when you are using the document to support a specific proposal. To describe is one task; to support a proposal is a distinctly different activity.

Can you remove yourself from any decision making in a document? Absolutely. The use of your logic structure in a presentation can be purely descriptive or highly judgmental, or perhaps a mixture of both. *The challenge of a descriptive document is the logical assembly of the data.* The material may be a standard procedural format, such as a lab report, or a situation based on a work request for specific information. The descriptive text *presents* points.

An evaluative document has the more difficult task of making and supporting points with evidence. The evidence becomes a *demonstration* of a position taken or identified by the author. If, for example, you present conflicting alternatives, the presentation is evaluative, but you can, if you wish, avoid the challenge of selecting the superior alternative.

As I noted in our initial discussion of the author's involvement in a document, there is a third level of commitment. *You can describe; you can describe and evaluate; you can describe, evaluate, and offer a recommendation.* When a project involves the third level of commitment, the document will usually develop a logical argument that supports the author's position. In this case the author wants to *prove* a point.

The proof or the "argument" is an analysis of those conditions that point to a specific recommendation from the writer. A great many documents operate at the third level of involvement, and many of these are designed to be persuasive without making direct recommendations. For example, an EIS will never conspicuously argue. The craft of argument is very subtle. The issue is *logic,* and tact.

In general, if you see a writing task as a goal of proving a point or points, perhaps it is easier to manage the overall organization, and the logical arguments in the blocks are easier to define. I often am trying to make a point of some sort when I write, so I look at many of my writing tasks, especially work-related chores, as documents that are written to prove points and make recommendations. The writing "proves" the need for the suggestion that I propose to the reader.

I could call the logic blocks "explanations," but the explanations are ordered by a logical need, and that need generates the logical outcome. *The outcome seems to be clearest to me as a writer if I see my job as a logical argument that I must make.* I could say that writing is an attempt at persuasion, but persuasion is either a *motive* or an *outcome,* or maybe both. The document is in the middle of this transaction and you want to concern yourself with its *structure,* because this is where the logical position is developed.

Document Patterns

If we now turn to the structure to examine the overall design, the usual flow of events should look familiar to you unless you have done almost no writing in the past. In the following illustration, notice the three stages indicated in the top row: introduction, body, and conclusion. These are the three largest units of any document you would write.

Familiar patterns in the parts of a paper

First stage		**Second stage**		**Third stage**
Introduction	→	**Body**	→	**Conclusion**
Descriptive intent	→	**Demonstration**	→	**Observations**
Evaluation	→	**Evidence**	→	**Explanation**
Recommendations	→	**Argument**	→	**Discussion**
(Proposal)	→	**(Proof)**	→	**(Analysis)**

Sample 7A

Network Proposal for the Walnut Avenue Eementary School
—Outline—

I. Introduction
   A. Define the problems with current network installation.
   B. Describe the objective of upgrading the present network installation.
II. Explain Networking terminology and describe device functionality.
   A. Define the common network terms.
   B. Define the network devices to be used.
      1. Explain router functionality.
      2. Describe the purpose of a switch.
      3. Discuss the concept of bandwidth and throughput.
   C. Explain what network topology is and its implications.
III. Give an in-depth description of the problems with the current network installation.
   A. Discuss the problems with current cabling.
   B. Explain the problems with current hardware including
      1. Slow switch and router performance
      2. Lack of extended (100Mps) bandwidth.
IV. Introduce proposed network upgrades.
   A. Give an in-depth description of the selected topology.
      1. Advantages of Ethernet
      2. Network speed requirements
   B. Describe the proposed cabling upgrades.
      1. Explain the reasons for replacing existing copper cabling with Cat. 5.
      2. Talk about replacing existing backbone cable with Multi-Mode fiber.
         a. Describe advantages of fiber-optic over copper wiring.
         b. Discuss cost factors.
   C. Discuss propsed hardware implementations including
      1. Replacement of Routers with fiber-optic compatible models
      2. Upgrading of Switches to support 100Mps.
   D. Describe the advantages of the selected plan over existing configuration.
V. Conclusion
   A. Discussion of what a new installation will achieve upon implementation.
   B. Give final reasons as to why the upgrade is necessary.

The introduction is followed by the body, which is followed by the conclusion. Notice how these three components translate into logical activities in succeeding rows. The second row represents the tasks for developing a purely *descriptive* text. The third row represents the addition of the tasks for *evaluation*. The next row represents the addition of a request for the author's *recommendation*. Each activity calls for evidence to support the author's position. The challenge of the project is then to develop the logical points that verify that position. In other words, the "facts" are introduced and explained, or argued. If all goes well, we conclude that you are right and that your description or your proposal is correct: your position is true. (The final row represents the document as a process for logical argument.)

Recall the expression test and the reality check. In the world of paperwork, just about every significant document you will write is going to test your ability to structure accurate data and evidence and logic on paper. Readers will then accept or reject or qualify your success at being able to support and verify your position. If you keep the mathematical analogy in mind (the idea that writing is as logical as an equation), all your major documents will be, as you see in the rows of the illustration, attempts to use logic to demonstrate or prove your point. In scientific and technical documentation you should look at the project as a *demonstration* if it is descriptive, or as a *proof* if there is a need to develop an argument to defend a position.

Study the outline on the left (Sample 7.A). This outline was the backbone of a ten-page proposal. Because the structure was well-organized, the author followed it to the letter and developed a very successful document. Notice the subsections that become component parts of the introduction, the three major elements of the body, and the conclusion. A diagram of this very basic design concept follows, on page 160.

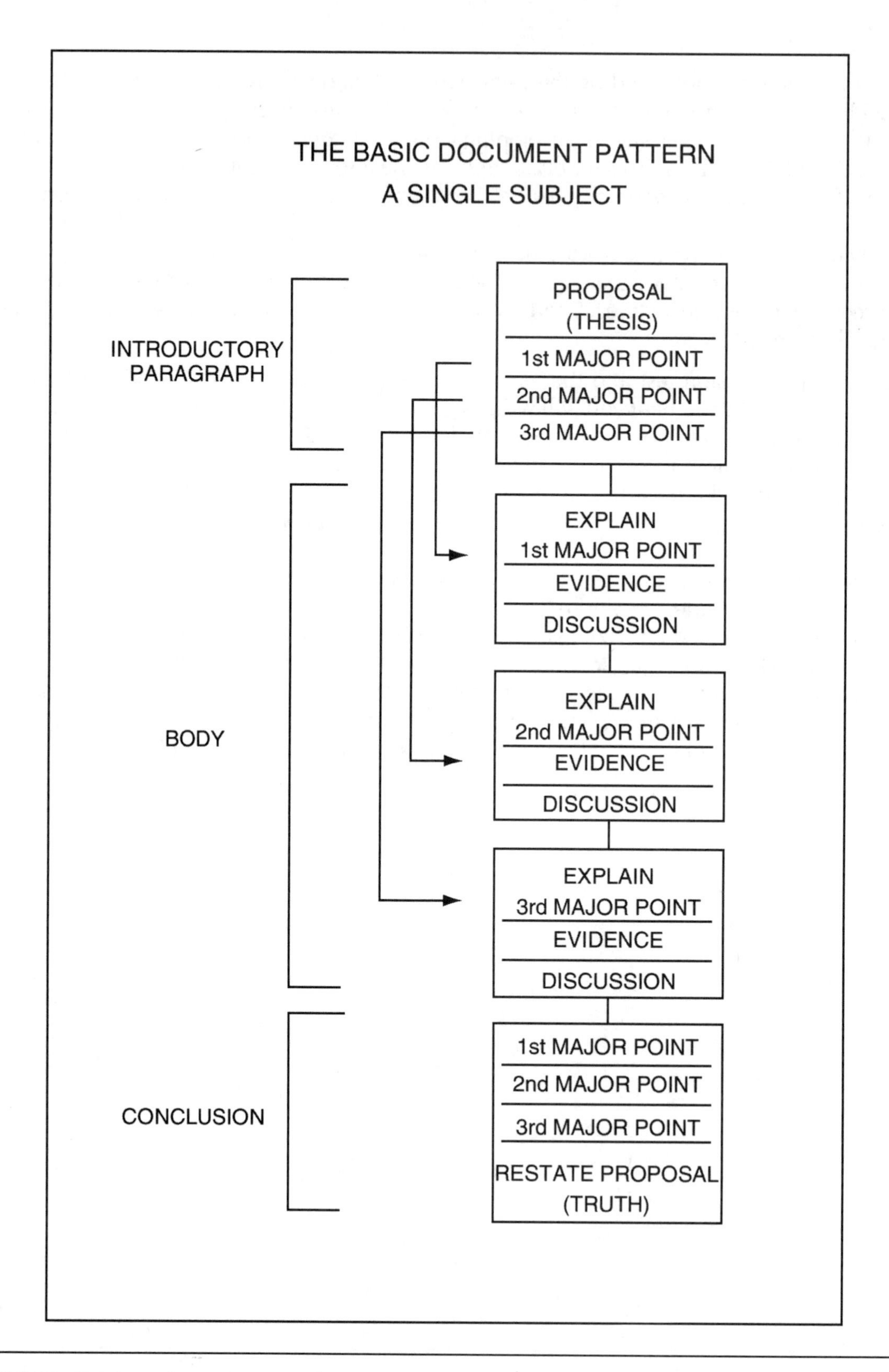
THE BASIC DOCUMENT PATTERN
A SINGLE SUBJECT
INTRODUCTORY PARAGRAPH
PROPOSAL (THESIS)
1st MAJOR POINT
2nd MAJOR POINT
3rd MAJOR POINT
BODY
EXPLAIN 1st MAJOR POINT
EVIDENCE
DISCUSSION
EXPLAIN 2nd MAJOR POINT
EVIDENCE
DISCUSSION
EXPLAIN 3rd MAJOR POINT
EVIDENCE
DISCUSSION
CONCLUSION
1st MAJOR POINT
2nd MAJOR POINT
3rd MAJOR POINT
RESTATE PROPOSAL (TRUTH)

The Basic Document Structure

On the left is the fundamental structure that is used for the basic logical progression of writing in many types of documents. The structure can be used for descriptive documents such as studies, laboratory narratives, or various kinds of reportage. It is also used for argumentation in any project that is evaluative and involves judgments. In our discussion here, we are somewhat more concerned with proving or demonstrating an author's proposal because argumentation is structurally important in many documents.

Much of your writing for any business will simply conform to the logic of what you observe in the model on the left. If you set out to write a business letter, for example, you can use the model—with considerable liberty, of course. If you set out to develop any proposal (thesis), the structure can be used in very nearly the form you see here.

None of the block diagrams you will be looking at in the following pages are totally applicable in all situations. They are not paint-by-the-number kits, but they represent the underlying logic structure of many basic writing tasks. Logic blocks and logical organization *will* be necessary in your work. "Seeing" writing tasks in logical patterns should greatly increase your organizational efficiency and speed your writing along.

The Introduction

The illustration to the left lists an introduction as the first order of business. An introduction is not difficult to construct if you follow a few simple guidelines. First, remember to write the introduction last. You have to write the document in order to introduce it. I write the body and the conclusion and then the introduction. A writer does not generally write in the order in which you see a finished document.

The easiest introduction is a brief outline of your project. Explain the key point of the project and highlight the main arguments. The *outline introduction* is my favorite for scientific and engineering documentation because the opening gets right to the point and immediately presents the logic structure to a reader. Scientific and engineering papers can be very difficult to read, and the most helpful introduction is one that explains the plan of the project. Besides, in business and industry we are very "up front" in our writing. Corporate documentation, as often as not, opens with *outcomes;* this means that conclusions are the starter for the project. This device—known as the *executive summary*—indicates the matter-of-fact style and functional approach we usually take in our writing.

If you have prepared an outline (discussed in Chapter 9), state the project proposal and introduce the central issues of the outline in your introduction. If you develop a project without an outline, then underline the topic sentence of each of the paragraphs when you are finished. These sentences are the key points to identify in the introduction. The outline introduction is *fast,* and it is a clear map that will help a reader through a dense subject.

The Body

The body of the document focuses on the highlights identified in the introduction. The body of the text presents the supporting evidence. The issues under discussion should follow the order promised by the introduction. Identify a main point, and provide your evidence. The evidence can be statistical, but numerical data is only one kind of evidence. There are many kinds. You might use a host of sources, from authoritative quotations (in the liberal arts area) or statistics (in business and accounting) to laboratory findings (in the scientific and engineering areas). You may have tables, charts, graphs, tabulations, photographs, readouts, electron micrographs, spreadsheets, and so on. All this material supports the logic that supports the proposal.

Often, however, the logic structure of the argument *is* the evidence! The logical argument can stand alone. You can argue your case to prove it, or you can demonstrate your case to prove it. More commonly, you will do both. In discussions with your friends you may notice how much generalized argument seems to enter the discussion. That is because you largely argue in the absence of facts. Most of us rely on the logic as a stand-alone approach to doing battle in conversation.

The logic blocks for the body of your project will pull on both your external resources and your logic resources. *Both* are valid. In fact, you do not even have to research a project before you write. I often write first and research second after I see which logical arguments I have to prove. To train writers in these respective techniques I usually encourage them to write without resources for ten or sixteen weeks and only then do they subsequently write with resources, for another ten or sixteen weeks. The first projects are highly deductive (general truths seen in sample cases); the second projects are highly inductive (data-based outcomes derived from evidence).

You must be sure that you explain the evidence. Do not assume the reader will understand the material. If you use a quotation, for example, your explanation of its significance should be as long as or longer than the quote itself. You need to demonstrate the correctness of your supporting materials.

The Conclusion

When you are done with the text, you *may* need a conclusion. Conclusions can be difficult to write if you do not have a clear idea of what you just tried to demonstrate or prove in the body of the paper. This is because most conclusions in technical documentation simply sum up the discussion, and if you do not understand the key ideas with clarity, you will not sum up with clarity. Lawyers use the technique of the summary extensively. For any complex matter, a summary allows a reader to have one final chance to understand the specific intentions of a document. You often need to summarize technical documentation for this reason.

If you notice a redundancy in the conclusion, you are correct in your observation. The conclusion is redundant. The introduction and conclusion can more or less say the same thing. Repetition adds clarity to complex documentation. A saying that was originally a tip for public speaking, "First you tell them what you are going to tell them, then you tell them, then you tell them what you told them," is quite useful in hard-to-read and hard-to-follow projects of the sort found in scientific and technical documentation. Your mission is clarity. Repetition is one of your tools. I have already made use of repetition to emphasize key points in several chapters you have read.

Simple Architecture

A "paper" is often seen as the most challenging task among the common documents developed in business and industry, probably because it can be quite long, but also because the structure of a paper is less immediately apparent than the structure of, for example, a business letter. As you can tell by examining the first block diagram (p. 160), papers do indeed follow structures or architectures, and the simplest one is the first diagram.

A project that focuses on a discussion of a single topic from the author's viewpoint is the most rudimentary and most common project design: the single topic/single viewpoint model. There are two important variations of the document depending on the author's role in the project. A single-topic project that is purely descriptive is the most fundamental design. A more complex project includes an evaluation and conclusions or recommendations. The architecture is the same but the author must now explain his or her proposal. The architecture does not change; the perspective changes.

The following two presentations (Models 7.A and 7.B) are documents that clearly show the organizational pattern of the basic block diagram structure. They are variations of what college educators call *essays;* business executives and engineers throughout industry call these same documents *papers*. We will discuss these presentations so that you can see how they were built.

Men and women engaged in college-level engineering and engineering technical programs composed almost all the documents you will see in the following chapters. There is a mix of documents, composed for entrepreneurial or small business practices, for a number of corporate purposes, and for college programs.

Samples are single pages with an illustration of material under discussion.

Models are illustrations that are composed of several pages, many of which are accompanied by commentary boxes on the facing page.

Model 7.A (1)

The Electromagnetic Spectrum

Sound, light, X-ray, radio, and television waves are forms of radiant energy that are considered to be electromagnetic oscillations. The frequency of oscillation determines whether the particular signal is observed as sound, light, X-rays, or radio waves. These forms of energy are referred to as *electromagnetic waves,* and the overall range of frequencies of these waves is commonly referred to as the *electromagnetic spectrum.* The spectrum extends from the longest waves of sound to the fastest waves of the speed of light. Between these extremes are the many emissions we use daily: infrared waves, ultraviolet waves, gamma waves, and others.

1

Sound waves are at the lower end of the electromagnetic spectrum and range from 20 cycles to 20 kilocycles (kc). The speed of sound varies with the kind of material in which it is traveling. For air, which is the most common medium, the speed is 1130 feet per second. To us, sound is the sensation produced in the brain by sound waves that strike the eardrum and cause it to vibrate in a manner similar to the original sound waves.

2

Our light response is the sensation produced in the brain by light rays. The frequency range of visible light is in the area of 3.75×10^8 megacycles (mc) and may be derived from mechanical, electrical, chemical, heat, or light energy. Objects are visible when the light rays from them reach the eyes and stimulate the optic nerve, which transmits this sensation to the brain. If the source of light is emitted from the visible object, it is said to be *luminous.* Most objects are not luminous, and the light that reaches the eye is actually the reflected light, usually from another luminous body. The speed of light is approximately 186,000 miles per second.

Radio and television transmitting stations convert sound waves and light waves to electrical impulses. These electrical impulses, which represent the original sound and light waves, are transmitted by the use of high-frequency alternating currents. These currents produce magnetic and electric fields that radiate in all directions as radio waves. Receiving instruments then convert the signals back into the sound and light images we experience on television. Radio waves also happen to travel at the same speed as light, or 186,000 miles per second.

A Descriptive Project Concerning a Single Topic

The simplest writing strategy is to approach the project with one topic of discussion and the simple goal of describing the topic in appropriate detail. On the left is a paper concerning the electromagnetic spectrum. This is a single-topic project. The details are developed in paragraphs that concern major frequency groupings in the electromagnetic spectrum, such as visible light and X-rays, but these are divisions of a single subject. The basic architecture for this simple tactic is illustrated in the diagram on the right. Any given topic will suggest a method of organization or architecture. Here the author used wavelengths as an ordering device and discussed the various aspects of the subject.

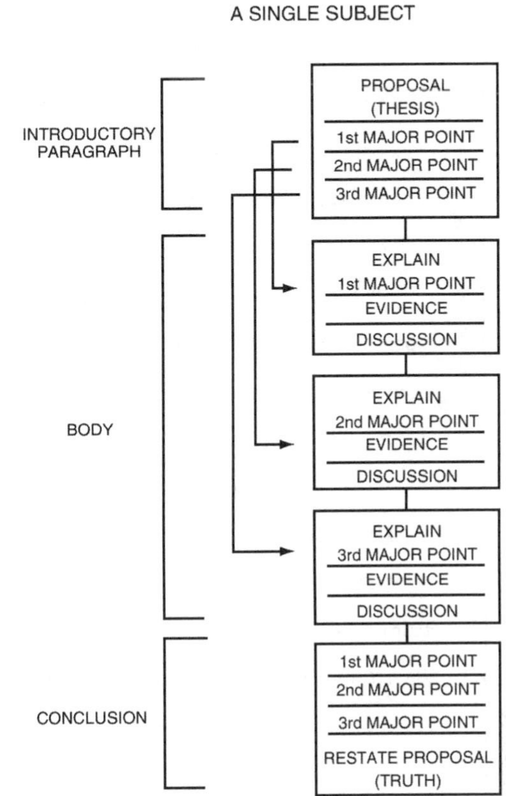

1

One of the most practical introductions for technical work is the outline. The introduction you see here is cleverly designed to avoid being too obvious. Read the first and last sentences of the opening paragraph back to back and you will notice that the author has constructed an outline by simply listing a selected group of wavelength families.

2

There is no need for the paragraphs to develop complicated transition sentences to interrelate the components of this text. The wavelength groups have a logical connection to one another. Each paragraph moves from group to group.

Model 7.A (2)

-2-

Infrared energy is invisible radiation ranging roughly from the lowest wavelength of red light identified by the human eye to the longest waves detectable on thermal detectors. This energy is most simply detected by its heating effects. It has many uses, which include the chemical analysis of different substances by infrared spectroscopy, and satellite infrared photography that uses special film for detecting "hot" but not luminous objects.

Ultraviolet radiation, sometimes called "black light," ranges from the violet limit of the visible spectrum to the X-ray region. The sun is the chief natural source of ultraviolet radiation. Artificial sources include arcs from such elements as mercury, carbon, and hydrogen. Our eyes are susceptible to ultraviolet damage, and this is the main reason for warnings against looking at the brilliant blue-white arc used in arc welding. The damage to the ozone layer has increased the ultraviolet risk in skin cancers.

Gamma rays are the most penetrating of the radiations composing the electromagnetic spectrum. They were discovered as one of the radiations emitted by radioactive substances. Gamma rays produced in the radioactive decay of radioisotopes, such as cobalt-60, are used in medicine for the treatment of tumors and in metallurgy for the detection of flaws in heavy-metal castings.

X-rays are produced by striking a target in an X-ray tube with electrons that are traveling very fast. The result of the collision with the target's atoms disturbs various parts of the atoms. X-rays are the result. X-rays are used extensively in the medical field for the detection of various disorders of the body.

These are a few of the many practical varieties of electromagnetic radiation that make up the electromagnetic spectrum. Many of the distinct regions of the spectrum have become part of everyday life. The entire range serves an endless variety of purposes in the tools we now design. From the laser levels at a construction site to the ultrasound labs of today's hospitals, the various radiations of the electromagnetic spectrum find practical and convenient applications that are constantly expanding as we discover endlessly promising new uses for the many wave groups.

1

6

2

3

4

5

1 The device that is used to establish continuity is the wave group under discussion—infrared radiation, for example. Notice that each body paragraph opens with the topic of the paragraph, which is immediately identified in the first few words of the first sentence. This is a consistent pattern.

2 Each paragraph of the second page continues to follow the subject matter in the order outlined in the last sentence of the introduction.

3 The paragraphs of this discussion are uniformly short. Since the text is pure description, there is no point or proposal that has to be proved with the evidence. Reportage material is often clipped because of the mission of describing. Little or no proof is required.

4 As you read through this project notice that the author is very objective in style. The text is quite matter of fact. In addition, there is a noticeable absence of any sort of evaluation. This text is designed to report and not to evaluate or judge or prove.

5 X-rays are mentioned in the introduction, but you will notice that the order in which the author identifies the wave groups does not position X-rays last. He could revise the introduction to reflect this detail. Always check to see that the essay conforms to the introduction. You can change either the essay or, in this case, the introduction to be certain they match.

6 As you reflect on the introductory paragraph, it is obvious that the text does not present a proposal or a thesis. Reporting is highly neutral, and this is often demanded of us in business and industry. This type of presentation or document is typically a gathering process in which material is identified or studied without further comment.

Model 7.B (1)

The Superconducting Supercollider: The Grand Experiment

In October 1994, funding for the proposal of the world's largest particle accelerator was denied by the Congress of the United States. The scientists and supporters of the superconducting supercollider (SCC) argued against this move and called it short-sighted. On the other hand, there were specialists in other fields of science and many concerned legislators who considered the project an unnecessary expenditure. They argued, quite justifiably, that the vast amount of money required to build the accelerator could not be justified by the predicted gains in knowledge it might provide.

With $2 billion spent and 20% of the project completed, the funding was stopped for what has been called "the grandest experiment of all time." The superconducting supercollider, when finished, would have been the world's largest and most scientifically advanced particle accelerator, a device designed to probe the innermost secrets of the universe. Since the 1930s, physicists have been using accelerators to smash atoms together, to analyze the debris, and then to gain the impressive results we see in theoretical nuclear physics. Matter, in all its complex forms, seems to be made up of just a few simple particles operating under a few basic forces, and the accelerators have helped us understand these forces.

An experiment of the magnitude of the SCC would have resulted in extensive spin-offs that would have created newer technologies. Improved superconducting materials used in computers and other electronics fields and advances in computer software could have emerged from these new technologies, which would have created new jobs in design, theory, and manufacturing.

However, the SCC had an original budget estimate of $5 billion, and the last predicted estimate was in excess of $11 billion! At a time when the entire country was concerned with the increasing national deficit, many congressmen and congresswomen and their constituents questioned the cost of such a venture. Admittedly, the $2 billion already spent was just a small percentage of the national deficit, but this was a project that this country could not afford, and it provides one more piece in the puzzle of reducing the national debt.

An Evaluative Project SingleTopic

A somewhat more challenging task for a writer is any document that involves the author's evaluations and conclusions. In the particle accelerator discussion, there is a single topic, but the document is clearly evaluative. It makes a judgment and defends the opinion. The architecture remains the same as the structure that was used in the first sample we examined; the shift to an evaluative stance has no effect on the basic structure. What changes is the function of the paragraph logic. In an evaluative project, the paragraphs are designed to demonstrate the author's viewpoint. This discussion definitely takes a position. The proposal is quite unmistakable when the author supports the critics of the SCC project.

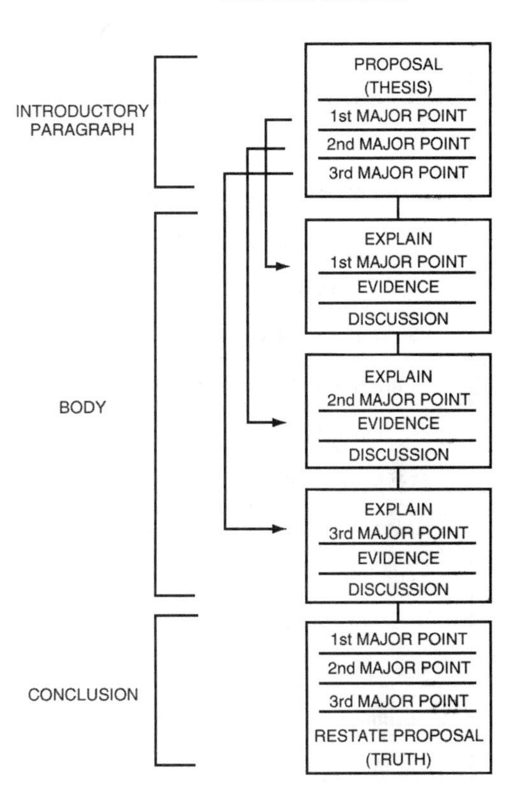

1 The introduction to the project concerning the supercollider focuses on the history of the funding conflict. It does somewhat suggest the highlights of the issue, but the author realizes that the reader must first have some historical background of the issue.

2 We quickly see that the SCC discussion definitely takes a position. There is a proposal, which is unmistakable when the author introduces the expression "quite justifiably."

3 The second paragraph provides more substance to the issue by explaining the history of particle acceleration. The third paragraph explains that there were economic perks to these massive projects.

4 The document then turns on a dime. It initiates the opposing view with one word: "however." This is a major transition, but the specific one I call a "reversal." In other words, there is a pivotal point at which the paper turns around and focuses on the opposite position.

Model 7.B (2)

-2-

Specialists in other fields of science and even different areas of physics resented the amount of money being heaped on a relatively small number of researchers at a time when research budgets were being reduced collectively. These specialists argued that the knowledge gained would not be of value in many of the fields they represent and that the principles of their branches of science are not based on particle theory in any case.

The original SCC budget submitted to Congress turned out to have omitted several crucial items that surfaced only after Congress had approved the project. The first superconducting magnets had to be redesigned after their first test run failed. In June 1993, Energy Department investigators reported that employees at SCC's Dallas headquarters were freely spending the taxpayers' money on lavish parties, liquor, and office decor ($54,000 for house plants, for example). The Secretary of Energy, Hazel O'Leary, fired the University Research Association, the nonprofit consortium running the project, for failure to track cost schedules. Apart from this abuse, the budget did not account for scientific failures or the basic reality of inflation.

At the European Laboratory for Particle Physics (CERN) in Geneva, Switzerland, a consortium of eighteen countries have pooled their resources to build an accelerator called the Large Hadron Collider (LHC). Whereas the particles in the SCC would have raced along at nearly the speed of light, the LHC will be able to provide only 40% of that speed, but the results are predicted to be comparable to that of the superconducting supercollider. A group of countries splitting costs and sharing the results makes good sense in these times of huge national deficits and increased government spending.

With budget estimates continually rising, the possible gains cannot justify such a large expenditure regardless of the gross mismanagement of the project. Congress had no choice but to make the decision to stop the funding to the SCC. The decision was a wise one. A consortium such as the one funding the particle accelerator at CERN presents a viable way of researching this particular field of physics at a cost that taxpayers can afford, particularly when the taxpayers in question know the bill is divided among eighteen nations.

1
2
3
4
5
6
7

1 After the reversal the author takes up the primary task of this project: he must now argue that the research is a super waste of money, rather than a valuable super accelerator.

2 Having taken the time to explain both the history and arguments in support of the SCC program, the author presents his own view, and the rest of the text stands in opposition to accelerator funding. If you take an overview and look at the way in which the project is developed, you will realize that the balance of the text is designed to oppose the expenditure involved in supporting the SCC.

3 One argument is that the project put too much money in the hands of a few, meaning that the same funds could potentially do a great deal more in the hands of many more scientists involved in a wider variety of research.

4 The misuse of funds is a reasonable argument, although it primarily appeals to our sentiments as tax-payers. It is a predictable tactic, although appropriate surveillance is probably the real issue here.

5 The European approach to the huge costs of an accelerator on this scale is to spread the costs among interested parties, and the accelerator is downsized and taken out of Texan dimensions. These are winning arguments.

6 Introductions often initiate a document by outlining the basic intention of the author. Conclusions often sum up by reviewing the key points that emerged. However, since the conclusion *follows* the discussion, it seldom repeats the exact structure or logic of an introduction. The introduction of this project lacked an alternative to the SCC project. An alternative was developed in the paper and appears in the conclusion as an option that, according to the paper, deserves our attention.

7 Notice that the text is evaluative. It asserts a position. You can see that these body paragraphs are longer than those in the reportage or descriptive work of the first paper.

Model 7.C (1)

BACTERIAL IDENTIFICATION

USING THE HEWLETT-PACKARD GAS CHROMATOGRAPH SYSTEM

Sheryl B.

October 16, 1999

Project 2

ENG 105-Applied Writing

Instructor: David R.

via:

Biomedical Electronics Program

Link Project

SEATTLE COMMUNITY COLLEGE

Layout Fundamentals

"Packaging" the material for an extended presentation such as a report is really quite easy. The simplest variation, the one beginning on the facing page, includes a cover page and the document. There are standards for these presentations, but in their basic version the formats are actually easier to construct than a common business letter. There are a great many types of papers, including engineering reports, project proposals, fieldwork analyses, variance studies, and status briefings. In their most basic layouts, they conform to standards that are established on the initial pages of the text.

To your left on the next several pages are two samples (Models 7.C and 7.D) for you to examine as prototypes while you read the following format specifications. You will notice that both samples conform to *all* the specifications. This format, with slight variations, is standard for anyone who writes reports with any frequency. I have included only the title pages and the first page of each text. All subsequent pages conform to the first-page format.

Use heavy bond 8½ × 11-inch paper (no holes). Double-space the text if the project is going to be reviewed and revised by supervisors or committees (or college instructors). Use single spacing or 1½ spacing if the document is being generated as a finished copy. I seldom use a single-space format for projects of this type because it is very compressed and not as readable as somewhat wider spacing. The book you are reading has a slightly wider spacing than is usual for textbooks because I requested this format.

First in a formal document of this type is the title page. It will usually look similar to the samples you see here. The layout may seem obvious enough, but it is often abused in different ways. There is both order and tradition in the layout. The conformation you find on the title page is quite common and includes the following material.

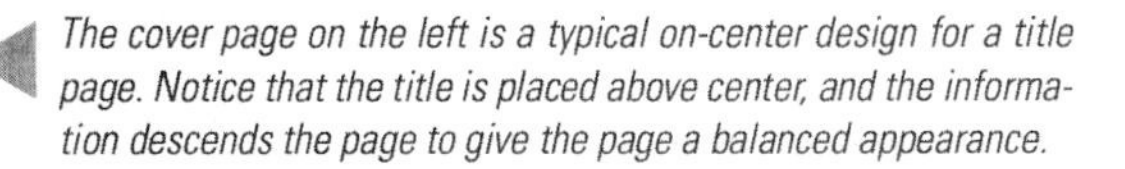

The cover page on the left is a typical on-center design for a title page. Notice that the title is placed above center, and the information descends the page to give the page a balanced appearance.

Model 7.C (2)

The Hewlett-Packard gas chromatography system consists of separate modules for sampling, inert gas pressure control, valve control, capillary columns, temperature control, oven, detector, cathode ray monitor, and printers. These modules are all controlled by the master computer, which can be ordered with various hardware, software, and data library options. We will first take a sample through a typical system, explain the operation of the instrument in a general way, and then look at each module and its options. Finally, we will look in detail at microbial sample preparation techniques and the specific modules that are used in bacterial identifications.

The sample, whether it is a calibrating standard, a blood or urine sample, a tagged enzyme, a prepared culture, or an unknown compound, is introduced into the instrument through the sampler. A valve, under control of the computer, opens, and pressurized inert gas forces the sample out of the sampler into the system. where the gases, controlled and monitored by the computer, position the sample in the appropriate module at the proper time. The next step in the analytical routine is usually to mix a reagent into the sample.

The sample containing the reagent is then vaporized and moved into a column, where it can be heated in an oven, have light passed through it, or be treated in some other way. A detector will sense light, heat, electrical activity, or other physical changes that occur and convert the measurement to digital form. This information is fed to a computer, which analyzes the results, compares these findings with a library of standards, and displays values, charts, waveforms, or other data that identify the sample. This same information may also be printed out or sent through an interface to other computers.

The master computer monitors and controls sequencing, valve and pressure settings, flow rate, temperature, and the electrical current at any point, in each module. The logistics of keeping track of an individual patient's sample is made easy by computer-monitored bar-graph coding. The computer

Title

Author

Date

Project type (or number)

Your division or group title (or course number)

Supervisor (or instructor)

Other data

Corporation name (or college)

These points of information are useful in any college or business setting.

I suggest that you want to be very thorough about the material on the cover page for several reasons. First, this is the wrapper on your product. You want it to look sharp. Second, a large corporation may stretch from coast to coast, and documents can get lost if they lack proper identification. The same is true of colleges and universities. I work in an environment that includes four central locations spread throughout a major city, additional service facilities elsewhere in town, dozens of departments, and close to 30,000 people. Your workplace may be similar; a paper might not be routed properly without very thorough detailing on the title page.

NOTE
The basic features discussed here are available in template format on your computer. See Appendix B, Templates and Tips. The appendix will explain how to locate the templates for the following software:

Word 6 for Macs
Word 98 or Office 98 for Macs
Word 7 for Windows
Office 97 for Windows
Office 2000 for Windows

The first page of the text does not repeat the title if there is a title page. The paragraphs are indented as a convenience to readers because the text contains many graphics. If new paragraphs are not indented, they are not apparent when placed below graphics.

Model 7.D (1)

LEAKAGE TESTING WITH THE HS-XXXX

Alan S.

February 2, 1999

Collaborative Project 1

Applied Writing

Instructor: David R.

Your Company Name Would Go Here

Observe that the material on the cover page not only follows a specific order but is usually positioned to begin above the center of the page. It can be liberally spaced. I put titles in caps. You are not supposed to underline the title, unless it is a company practice. The layout of this page is a matter of convention for the most part.

There are two frequent errors I see on cover pages. First, I often see a blank cover page with all the information single-spaced in the bottom right-hand corner like the health hazard warning on a pack of cigarettes. This is a popular practice in colleges. The second error is that the title page is overdone. Avoid desktop publishing, particularly on the title page. No two-inch letters, thank you. If you want pizzazz, you can boldface the information using 14-point type and alter the title so that it is about as wide as a pencil (18- or 24-point bold); no more extravagance is necessary. Also stick to one font. A company logo or perhaps a technical graphic may be used if a little bit of interest is desired on the page. Avoid clip art files; the graphics are too nonspecific.

Now let's look at the text. The usual margins of this sort of formal document run 1¼ inches on all sides. The left side is bumped to 1½ inches if you plan on binding the text in a folder. The extra quarter inch provides just enough room for the staples. Justify the right-hand side of the text if you can. Justification is a stretch-and-squeeze process that makes any given number of words fit in a line so that the right edge does not zigzag. If you do not justify, leave the right edge uneven rather than hyphenate too many words on the right.

For college projects, I usually use transparent plastic folders with a spine that slides onto the project. Staple *through* the plastic folder and the project and slide the spine on from the end. If the cover is not stapled to the text, the project will fall apart. Do not use brightly colored plastics for the folders. This is a very formal document. For a more professional look have the project bound at a copy shop.

The cover page is not numbered. Number the pages of the text with consecutive Arabic numerals, 1, 2, 3, and so on. I usually do not number the first page of the text. I start with the second—as 2 at the top center of the page. Some software prints the numbers at the bottom center. Because of the prevalence of numerical characters in engineering and

The cover page on the left is left-justified. A cover page looks particularly impressive if the company name, or perhaps the company logo, appears near the bottom.

Model 7.D (2)

Hospital personnel know that they work in a dangerous environment. They are aware of the many physical, medical, and biological hazards a modern hospital presents to both personnel and patients. They have also been trained to recognize the hazards of electrical shock in a modern hospital with all its electronic technology. The risk of electrical shock from an error or accident is high. But an unseen and equally dangerous hazard exists: leakage of current!

If a current of only 20 microamps, so small that you may not even be able to feel it, reaches the heart, it can cause fibrillation (fluttering) and death. The Underwriters Laboratories have determined that a maximum of 5 microamps of leakage current can be allowed to exist in a patient-connected electrical device.

Leakage current (the presence of current in an area where current of that value should not exist) is caused by a short or stray capacitance. Shorts can be caused by any number of conditions: component aging or failure, physical damage, operator error, faulty or worn insulation, line voltage spikes, electrical interference, design error, faulty or altered software, thermal expansion, dirt or foreign matter, or incorrect assembly.

Stray capacitance occurs when radio frequency energy builds up on wires or metal surfaces and causes them to act like small batteries, producing current where current should not be present.

Each patient-connected machine must be tested periodically for leakage, using a leakage tester. The following pages will introduce the Instrutek HS-XXXX Leakage Tester, outline the six leakage tests this machine makes, explain why each test is required, and give step-by-step instructions for each test.

Before performing any of the leakage tests, you must make one preliminary check.

scientific projects, the page number is a little more obvious at the top. Do not enclose the numbers with parentheses or slashes. Hyphens are used for this purpose. Do not write "Page" before the numbers. Add a blank sheet at the end of the project, called a *dust sheet,* to look professional. Do not number it. These projects are not signed with a signature at the end, and they do not say "The End."

Remember that any project you write can be impressive in more than one way. The physical presentation is important and functions as packaging. Writing always involves a little marketing. The appearance of the product is the initial impression your reader will have of the document. Granted, both of the samples here are very ordinary, nonetheless, they look professional. The professional look is what you want to achieve, and it is not a difficult task. You can build an attractive cover page in a matter of minutes.

The following additional hints may be helpful:

- If you are double spacing, you must double-double space between paragraphs. I prefer to indent five spaces or one tab.
- It is handy to recall, as we noted earlier, that double-spaced material runs about 250 words per page. Single-spaced material runs about 500.
- Most companies will prefer 12-point type and may use a standardized font (typescript design). This text is set in 12 point because the large type makes it easy to read.

Notice that the text body of this sample is justified, meaning that the software adjusts the spacing between words to produce a neat right edge with few hyphenated words.

- Projects that are developed for college use will not have a cover letter or letter of transmittal to explain the document. If you are drafting a project in a company environment, however, you may need to attach a letter that explains the intent of the project. Formal presentations are usually designed to respond to company directives or needs. When the finished product is forwarded, to your supervisor, for example, it is often appropriate to include a cover letter that explains the project. Do not use the first page of the text for this purpose (see p. 325).

NOTE
You will notice that, most of our writing models in this chapter and the following chapters are missing cover pages. Each project originally had the standard cover page, but it was not practical to include them in *Basic Composition Skills*. The practice of providing a title at the top of the first page of the text body is an acceptable, if less desirable, procedure.

Summary

- Look at any writing project as a logical challenge rather than a writing challenge.
- Look at every paragraph as a logic block.
- Look at any document you read or write as a structure. Try to see the structure in what you read; try to build the structure in what you write.
- Use the basic diagram (p. 160) whenever possible.
- Be sure you are addressing a single topic.
- Determine whether it is a report that is purely descriptive or an evaluation that involves your analysis of evidence and possibly your recommendations or a proposal.
- Build an outline by prioritizing major headings and key subheadings (see Chapter 9).
- Develop the body of the text first; write in any order or personal preference that will get you going quickly.
- Quickly write the rough first draft and do not judge the product.
- Use all your resources: advice, data, personal notes, and so on.
- Draw conclusions by summing up.
- Develop an outline introduction for the project to complete the rough draft.
- Pace yourself through revisions.
- Structure the final version by the standards described for papers and proposals.

Activities Chapter 7

Use the chapter summary as a guideline.

Develop a highly structured document that reflects the design shown on p. 160. Use a technical topic or interest, and develop a descriptive analysis. Do not evaluate or judge it. Simply describe the process or the device or the situation involved. You might describe how to use some feature of a Web browser, describe a new Macintosh or IBM computer, identify the computer upgrades needed at work, or explain some other interest.

Develop a highly structured evaluative document using the same design on p. 160. Again, use a technical interest, perhaps a recent discussion in one of your classes. In this case, go beyond description and evaluate the subject matter in some fashion. Include recommendations if desired. Try, however, to maintain a neutral tone and avoid the familiar and popular style of magazines.

For the preceding projects, refer to the section Layout Fundamentals *to develop the page layouts. Design a title page and double-space the text to conform to the standards presented in the discussion. Save the project in a file on your computer. Also, see Appendix B for instructions concerning templates.*

When the projects are returned to you:

At your instructor's request, resubmit *either or both of the preceding projects with corrections to the errors that are indicated on the original.* Boldface *all corrections so that the instructor sees the revisions. Use the* Writer's Handbook *as your reference.*

Share a Project: First Option

Work with two other members of the class to construct brief projects approximately two to three pages in length (750 words). It would be helpful if all members of the group are from the same engineering technology program.

Decide on a familiar technical topic that will allow you to develop all three papers in three distinct formats: (1) descriptive observations, (2) observations and evaluations, (3) observations, analysis, and recommendations.

- *Have each member of the team select one of the formats.*
- *Agree to the objectives of the document.*
- *Assign a due date and bring three copies of each project to the meeting.*
- Distribute the copies, *read the projects, and discuss the outcomes of the activity.*
- *Provide a set of the projects for the instructor.*

Share a Project: Second Option

Work with three other members of the class to discuss and edit the finished version of the first set of documents.

- *In preparation, read the introduction to Appendix A, which discusses peer reviewing.*
- *Before holding the meeting, review the first and second editing checklists in Appendix A for papers that involve a descriptive analysis and an evaluative analysis.*
- *On the day the projects are due, have each member of the editing group explain his or her project in terms of objectives and project development.*
- *Next, hand the papers around for a critical reading and editing.*
- *Have each member edit the texts for both writing errors and technical errors.*
- *Have each member write a one-paragraph critique at the end of the paper.*

Work In Progress

I Get the Go-ahead

Carrying my permission to proceed, which helped defend me against any skeptical employees, I gathered my research and began the complex task of assembling my final projects. One obstacle remained—assimilating all my notes into quick, understandable, and useful training tools that would be written for all Pacific Aero Tech employees to accept and follow.

I first chose to work on Hearing Conservation because noise affected all PAT employees in varying degrees of exposure, and I was confident that this topic was a good place to start. I had to keep in mind that this and the other safety documents had to allow PAT's employees "to read it, get it, and go." I kept that phrase in the forefront of my mind while writing these specifically tailored programs.

I separated my Hearing Conservation document into two parts for the sake of clarity and understanding. The first part introduced and described the individual responsibilities of the employer, supervisors, and employees of Pacific Aero Tech. All workers as well as their supervisors have a responsibility for safety, but the burden of the program rests on management to provide an environment free from safety and health hazards. The following is a sample of text from the first part of my Hearing Conservation recommendations.

> Management is responsible for budgeting funds to provide training and hearing tests, providing hearing examiners, coordinating testing schedules with supervisors, notifying employees of hearing problems and maintaining audiometric tests and related program records. This is accomplished by appointing a program administrator.
>
> Management must provide annual audiometric (hearing) tests at no cost to the employee. At the time of the testing the examiner will review the results with the employees and inform them of any problems or shifts in their hearing. Although employee participation in the hearing testing is voluntary under this program, it is strongly encouraged for every employee's protection. A baseline test is the employee's first hearing test conducted in this program. All future annual tests are compared to the baseline to identify changes in hearing levels.
>
> Supervisors are responsible for determining which of their employees are required to be included in the program, and for assuring that they will attend training sessions and are properly utilizing hearing protection.
>
> When an employee is exposed to excessive noise levels, employees will wear hearing protection with a noise reduction rating sufficient to reduce the level below 85dB for an eight-hour time-weighted average.

For the second portion of my document I determined that I needed an organizational device that would be the standard for all the documents that follow this one. I also wanted section headings to clearly indicate all the parts.

L.C.L

Double-Topic Architecture

Although there probably is no statistical calculation for determining what percentages of corporate documents are single-topic projects, the likelihood is that most of them fall into that category. Nonetheless, there is another type of writing project that is equally important, if less frequently utilized: the double topic. If, for example, you want to develop a document that is a comparison of two products, of two theories or two therapies, or two of anything you have to split the discussion in two. As a writing consideration the double-topic approach is somewhat more involved than the first models you examined, but the structural similarities are obvious. The structure doesn't exactly change; it doubles or duplicates the original single-topic technique.

In the following diagram you see a simple way of looking at the organization involved in a double-topic project. As a writing procedure, you must subdivide the subjects of the comparison and try to make the points of discussion identical or similar. Similar is close enough; the world of logic is not always symmetrical.

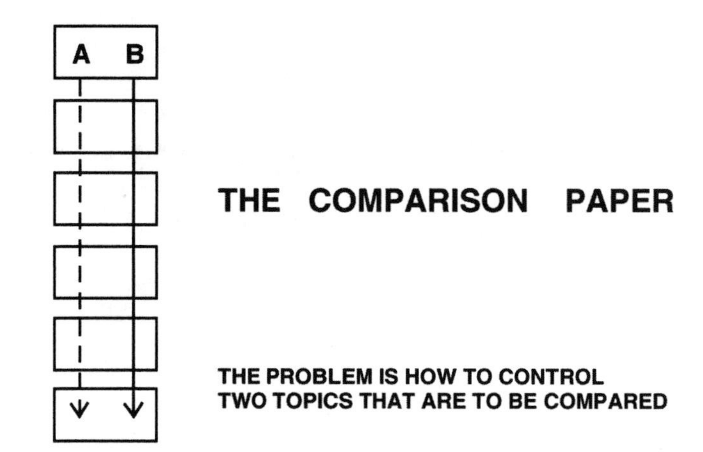

A B
THE COMPARISON PAPER
THE PROBLEM IS HOW TO CONTROL
TWO TOPICS THAT ARE TO BE COMPARED

For example, if you wanted to compare the allegiances of pro-life and pro-choice groups, you could find symmetry in certain issues. Consider the rights issue. Pro-life protects the "child's" rights. Pro-choice protects the mother's rights. The rights issue is a common ground. But religious perceptions are less symmetrical. They are dependent on belief systems that may not match. Identify the two units of comparison and then identify the elements that must be discussed, including those that do not match. This will be the basis of your logic structure.

An outline would be helpful for this project. The double-topic outline will appear puzzling at first because you cannot easily create a conventional outline (see Chapter 9). Divide a sheet of paper down the middle and proceed to develop *two* outlines, one for each "side" or subject. Try to get the issues to match if possible. A symmetrical design will usually take shape, although some issues will not cooperate. The world around us is not always in balance.

Since the comparison document has two topics, the two-sided outline will best reflect the issues involved. Try to keep the vertical sections balanced (equal), and try to keep the horizontal sections balanced (equal). If this concept sounds slightly complicated, study the two diagrams on the following pages. The concept of the first diagram is easily transferred to an outline as you can see in the concept of the second diagram. (For samples, see pp. 234 and 236.)

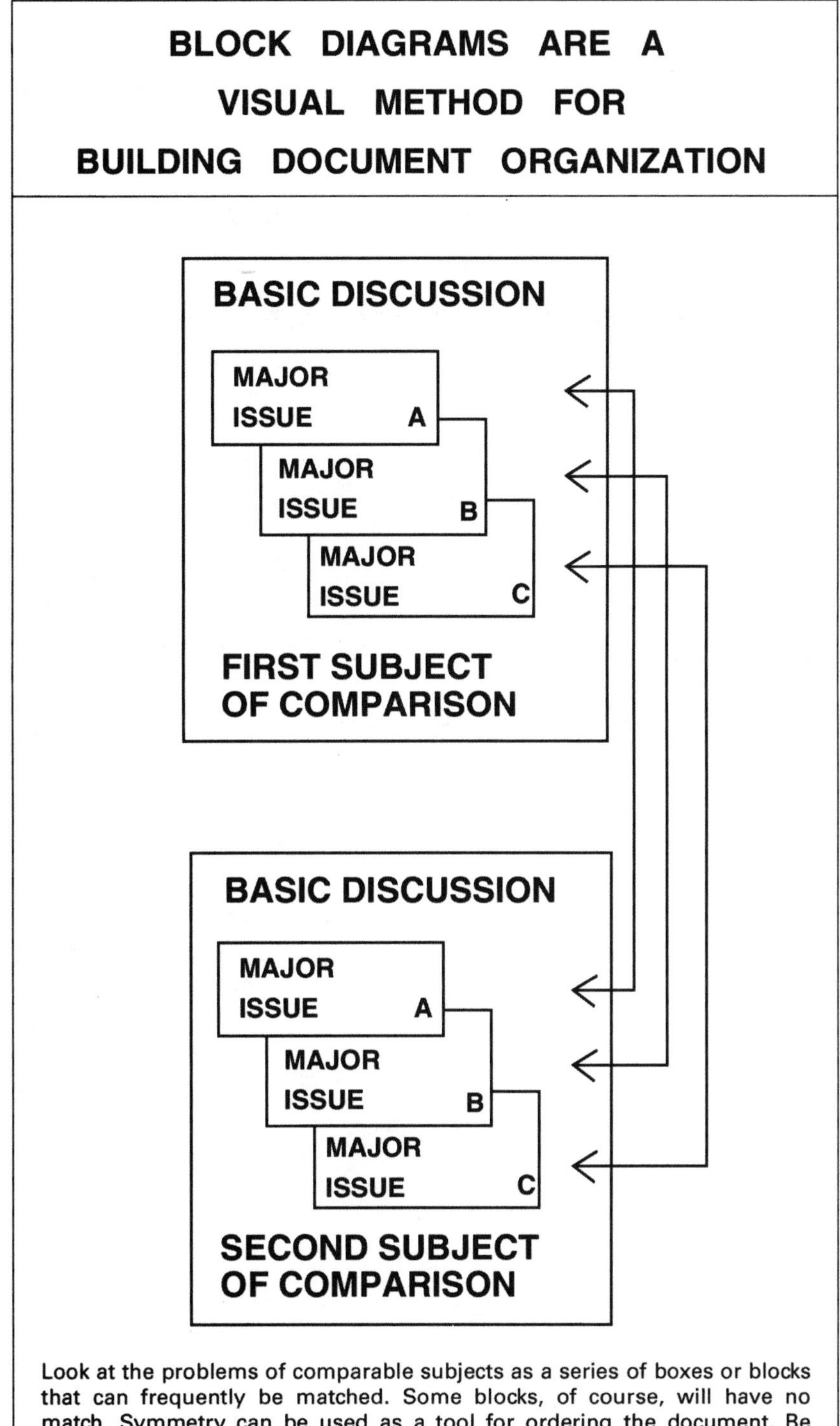

Look at the problems of comparable subjects as a series of boxes or blocks that can frequently be matched. Some blocks, of course, will have no match. Symmetry can be used as a tool for ordering the document. Be warned: the actual subject may be difficult to diagram at times.

Sample 8A

COST OF NETWORKING

Audience: This paper is for the networking student who wants to be familiar with the impact of cost for setting up a computer network.

Objective: The objective of this paper is to teach the networking student the importance of cost considerations and how the choice of wiring can affect network operation and set up.

Token Ring

Token Ring is more expensive to set up than Ethernet.

Token Ring is more expensive to wire than Ethernet because fiber optics are more difficult to work with.

Token Ring is much faster because it uses fiber optics cable.

Token Ring can carry more computers per network because it uses fiber optics.

Token Ring is less popular than Ethernet.

Ethernet

Ethernet is less expensive than Token Ring to set up.

Ethernet is easier to set up because it uses a much more flexible cabling.

Ethernet is much slower because it uses a much slower cabling.

Ethernet handles fewer computers per network than token ring.

Ethernet is more popular than Token Ring.

The diagram on page 189 suggests the general conceptual issue of a comparison, but it does not explain the organization of a writing project. The same material outlined in the two-column style suggested below provides two organizational plans for writing a double-topic presentation.

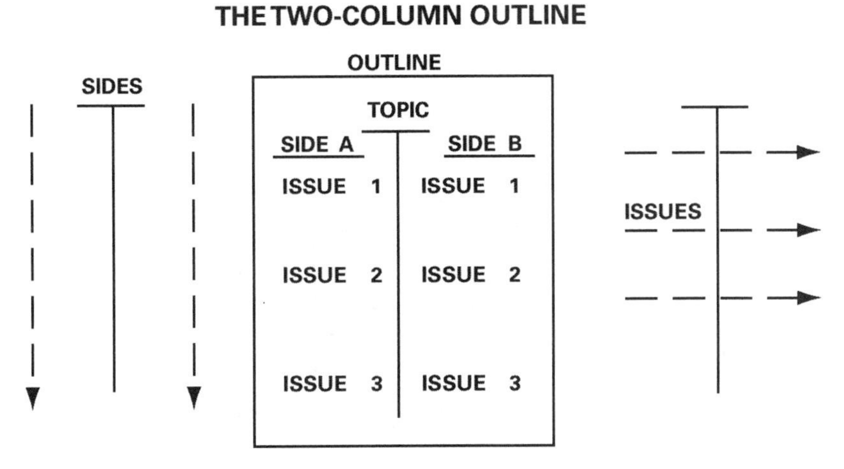

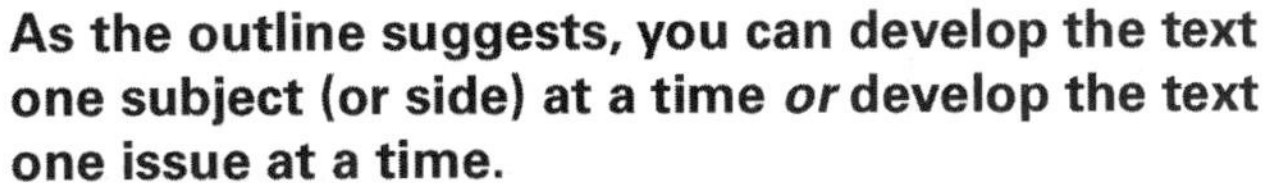
As the outline suggests, you can develop the text one subject (or side) at a time *or* develop the text one issue at a time.

There are two ways to look at the result. Looking at the outline vertically, you now see your perspectives regarding whatever two matters you outlined. Looking at the outline horizontally, you have also the criteria (the issues) that represent your effort to be systematic in your comparison.

As a result of the two ways in which you can view the outline, you have two options for writing the document: either you can discuss one viewpoint and then the other, or you can look at the project in terms of each relevant issue you have discovered.

Writers who have difficulty with outlines are not visualizing the logic structure of their ideas. They are losing ground before they start. The outline is the first acid test for the more complicated varieties of project architectures. If there is no structure on the drawing boards, the problem needs to be recognized early on (see Sample 8.A).

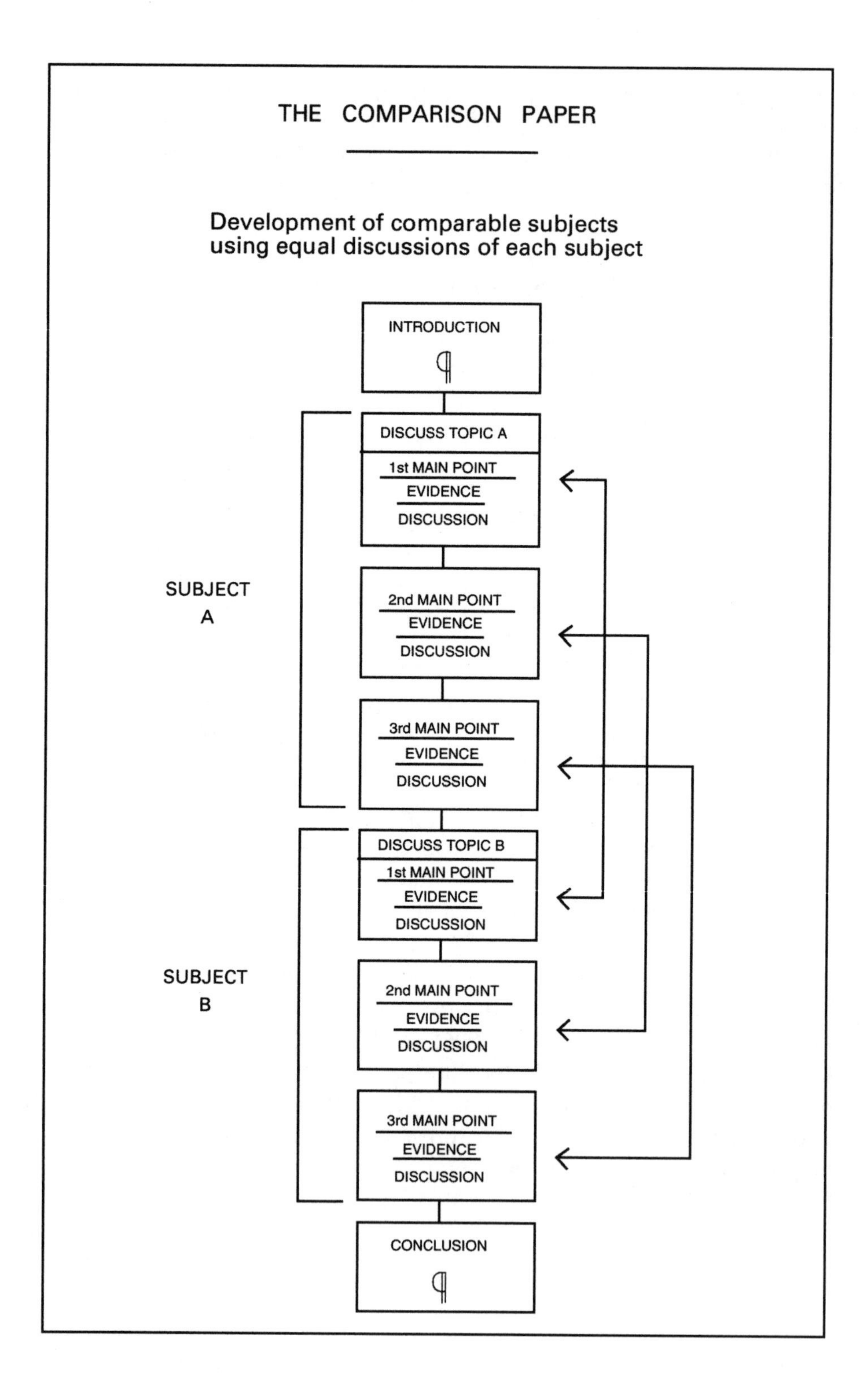
THE COMPARISON PAPER
Development of comparable subjects
using equal discussions of each subject
INTRODUCTION
¶
DISCUSS TOPIC A
1st MAIN POINT
EVIDENCE
DISCUSSION
SUBJECT
A
2nd MAIN POINT
EVIDENCE
DISCUSSION
3rd MAIN POINT
EVIDENCE
DISCUSSION
DISCUSS TOPIC B
1st MAIN POINT
EVIDENCE
DISCUSSION
SUBJECT
B
2nd MAIN POINT
EVIDENCE
DISCUSSION
3rd MAIN POINT
EVIDENCE
DISCUSSION
CONCLUSION
¶

Formatting by Topics

The organization of the double-topic project is not quite as simple as that of the single-topic technique, but the double viewpoint can be very important in one critical area of corporate concern: decision making. The either-or logic that is implied in the double-topic analysis calls for *choice*. A comparison document may or may not make the choice, but a company will use it as a tool for decisions.

There are several structures you can use to develop two topics. The simplest is usually the one you see on the left. Most writers favor this model at first, and they probably have selected the path of least resistance. Essentially, you develop two single-topic papers and then join them together. The introduction and the conclusion are used to explain and link the organization for the reader. The point-for-point similarity of the two discussions creates the organization.

When you develop the second topic in the second half of the paper, write with reference to the first half of the project. The second half has the role of making the reader *see* the comparisons. For example, I might open the second half of the discussion by saying, "Unlike the apple, the orange is a citrus fruit." Connecting comments pull the elements together and assert any judgments you plan to make. Let's examine Model 8.A.

The sample projects you will read (including those in the last chapter) are shorter than the originals. The original documents were two or three times as long as you see them here, and they often contained a number of figures. I do not have the space to present complete texts, but the models should provide you with the insights you need to build your own presentations. It is a credit to the writers that I could heavily abbreviate their work yet the integrity of the documents remained intact. It is exactly this logical structure that you want to examine.

Observe also that the projects are only moderately technical, so that you are sure to understand them. Most of the presentations I read are very technical, and only a specific group of technicians or engineers would fully understand them.

Model 8.A (1)

AM and FM Radio

The popularity of radio and radio programs began to decline in the 1950s as television began to grow in popularity. As music increasingly dominated the radio waves there was a slow but obvious shift away from traditional AM radio to the "new" FM radio stations. Classical music stations were among the earliest converts. Then came jazz, and rock arrived in the seventies and eighties.

The production of superior sound is the primary reason for using FM bands, but there are obvious problems with FM reception. AM radio was, in the early days, the clear choice for practical, day-to-day broadcasting over vast distances. Granted, AM never had the sound quality of today's music system, but AM radio was a workhorse. Today FM radios are the preferred medium for music, whatever our tastes.

1

The first component of an AM radio (see Figure 1) is an antenna to receive the incoming signal. Next is the RF amplifier. The RF amplifier is used to tune the radio for a station. It also amplifies the incoming signal to a usable level. The signal may be small due to the distance from the station. This signal consists of a carrier (station-assigned frequency) and the audio. Then, the signal is sent to a mixer.

2

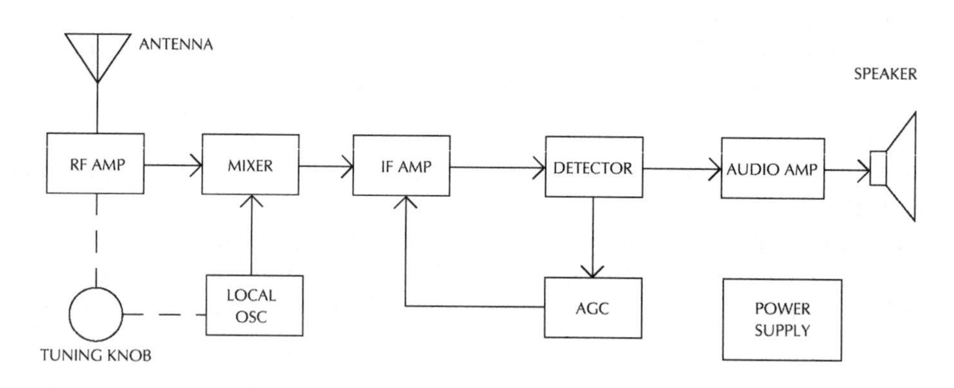

Figure 1. AM Radio Block Diagram

In the mixer, the incoming signal is mixed with a local oscillator signal to produce an intermediate frequency (IF). This IF consists of a carrier (455 kHz) and the audio. Then, the IF is sent to an IF amp. The IF amp amplifies this signal to a usable level and sends it on to a detector. The detector pulls out the audio and controls the amount of amplification the signal gets in the IF amp. From the detector, the audio is further amplified and sent to a speaker for the listener's pleasure.

A Descriptive Comparison Developed by Topics

The project on the left has been formatted by topic (subject). The AM and FM discussion is, of course, a comparison project, and you will quickly realize that it is designed to follow each topic separately in each half of the paper. The introduction is strictly historical in nature. You will find that historical background is frequently an appropriate springboard for technical documents. After the introduction, the project promptly begins with a discussion of basic AM and FM differences. You will notice that the project is objective and descriptive.

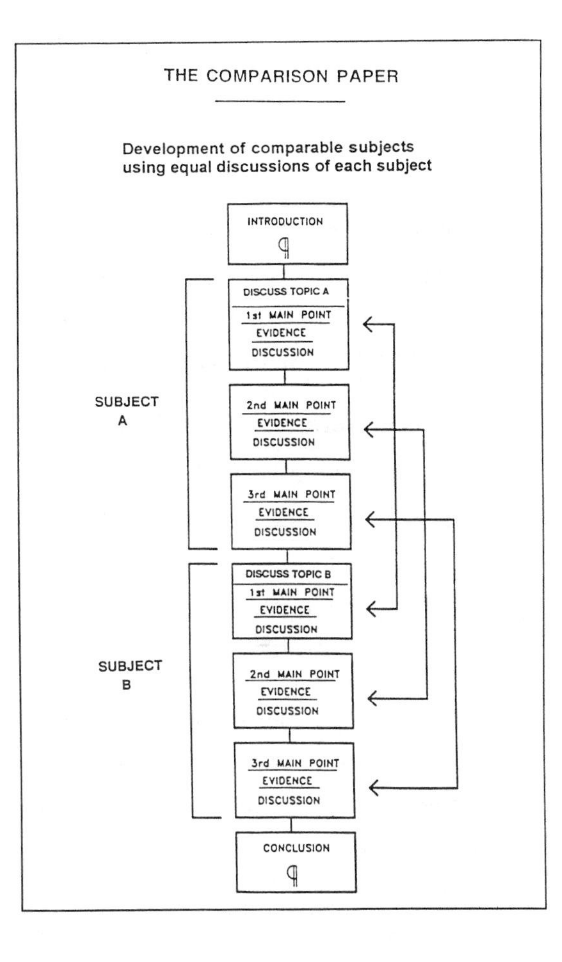

1 Since the first paragraph set the stage, the second paragraph outlines the conflict that is going to be developed. It is a brief survey that identifies the AM and FM differences. We will later observe that the outline introduction can follow one of a variety of openings such as this brief historical introduction.

2 With the third paragraph, you see a distinct shift into the body of the text. Your attention is turned to AM radio, and the author has begun a discussion of the topic that will continue for five paragraphs. Remember that this tactic is somewhat easier to manage than a text that develops by issue (requirements, standards, or some other criteria).

Model 8.A (2)

-2-

One advantage of an AM radio (seeTable 1) is that it is a very simple machine.This makes it relatively cheap to make and repair.The AM frequency band is another advantage. AM has a very long range, which means the station can be received at a long distance from the transmitter.

Table 1. AM Radio

Frequency	*Advantages*	*Disadvantages*
540 kHz | 1.6 MHz	1. Simplicity 2 Low cost 3. Long range	1. Atmospheric noise 2. Poor fidelity a. Limited audio range b. Crowded band

The AM radio does have some disadvantages. The first is atmospheric noise (static), which you can hear if you ever listen closely to an AM station. There is no practical way to remove the interference. The other major problem is poor fidelity. This fault is not due to the AM receiver. Poor AM fidelity is caused by the overcrowding permitted in the AM broadcast band. Each station is limited to the amount of space it is able to use in the broadcast frequency allocations, which does not allow the full range of audio frequencies needed for a quality broadcast. The audio range received is only 0 to 5 MHz. This means that listeners never receive the higher tones, and the result is an inferior quality signal and poor sound reproduction.

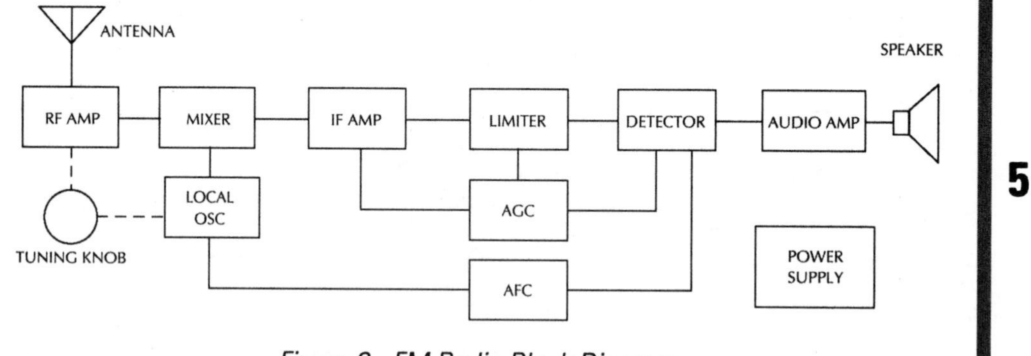

Figure 2. FM Radio Block Diagram

The FM radio was developed to eliminate these problems. Figure 2 shows that an FM radio is basically the same as an AM radio except for two added circuits: a limiter and an AFC circuit.The an-

1

2

3

4

5

6

1

If you read the first sentence of each of the body paragraphs you will notice that the author uses the key words "advantages" and "disadvantages" as sentence transitions to shift your attention as the text moves along. The author uses this simple tactic in the topic sentences to alert the reader to any change in the discussion.

2

The table reflects the discussion, which is easily reduced to a graphic. Again, note that if there is a table concerning AM radio there is likely to be a matching table in the second section of the proposal for FM radio. Everything is kept in balance.

3

Tables can conveniently render data or observations in a visual manner that is easy to understand. The table reinforces the discussion for a very simple reason: there is usually a redundancy involved. The table explains the text, and the text explains the table.

4

Two paragraphs focus on the criteria: advantages and disadvantages. These terms are usually part of any project that has a pro-and-con perspective. As a result, they make logical paragraph topic sentences.

5

Because this is a general-interest article, the illustrations are no more technical than the basic block diagram. Since the project has two distinct parts, here you see a matching FM block diagram that corresponds to the AM block diagram.

6

The shift to the second half of the comparison is easily managed in this presentation. The author moves to FM by saying that "FM radio was developed to eliminate these problems (the problems discussed in the previous paragraph). This transition moves the reader from the first half of the paper to the second half.

Model 8.A (3)

-3-

tenna, mixer, IF amp, detector, audio amp, and AGC circuits work in exactly the same way as the same configuration in the AM radio. The only difference is the frequency in the RF amp, the local oscillator, and the IF amp.

The limiter makes sure the signal to the detector is a constant size. In this way, the unwanted static picked up through the air is eliminated before the signal reaches the detector. Because the local oscillator in the FM radio has a tendency to drift off-frequency as it heats up, FM radios need an AFC (automatic frequency control). With the AFC, the listener does not have to retune the station as the oscillator drifts or as the station fades out.

The advantage of the FM radio results primarily from the limiter and the type of broadcast (FM). Atmospheric noise (static) is eliminated, so the listener receives better quality. Because of the frequency of the FM broadcast band, the music fan also receives more audio signals (20 Hz to 20 kHz). These signals provide more audio on the higher tones.

Table 2. FM Radio

Frequency	*Advantages*	*Disadvantages*
88 MHz	1. No atmospheric noise	1. Cost
	2. Good fidelity	2. Line of sight
	a. Full audio range	
108 MHz	b. Wider band	

FM does, however, have a few drawbacks. The first is the cost. Because of the added circuits, the FM radio costs more than an AM radio. Mass production has not changed this difference in cost. In fact, premium FM tuners can be fifty times more expensive than a basic AM radio you would take to the beach. Broadcast distance is the other problem. Because of the broadcast frequency, the antenna of the FM receiver is restricted to line-of-sight reception. FM reception range is much shorter than AM range. This means you have to be closer to the transmitter to receive the station. In truth, usually you must be in town to receive quality reception.

The AM radio, a simple receiver that is low in cost and long on range, often outweighs the disadvantages of static and poor fidelity. FM radio's advantages of static-free sound and good fidelity come with a substantial expense and shorter range. The choice will depend on the listener's needs.

1

2

3

4

1

> The target audience, by the way, is clearly expected to know the basics of radio. You will notice in the block diagram and the paragraph that follows it that the author assumes some fundamental familiarity with electronics.

2

> The structure of the second half is similar to that of the first half, particularly in that the advantages are discussed and then the disadvantages are identified, also.

3

> The tone of this document is neutral. We would be hard put to find any word phrase or sentence that favors AM or FM, even though we all know that FM produces the better product.

4

> In the conclusion, the author's objectivity is clear. The comments are evaluative, but only with respect to the problems of both transmission methods. Since readers realize the transmission powers of AM radio and the reception problems of FM radio, they understand the reason for his caution.

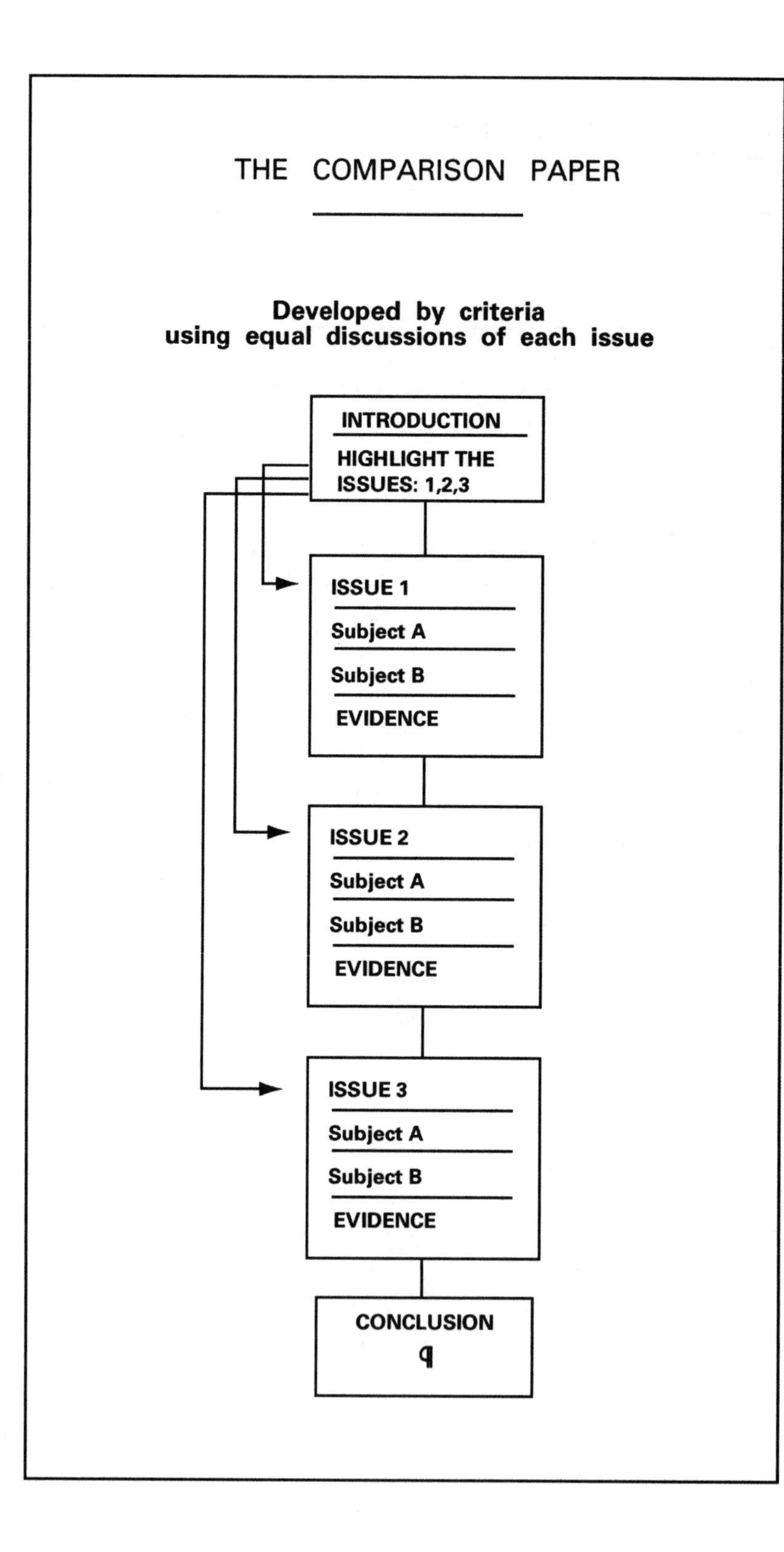
THE COMPARISON PAPER
Developed by criteria
using equal discussions of each issue
INTRODUCTION
HIGHLIGHT THE
ISSUES: 1,2,3
ISSUE 1
Subject A
Subject B
EVIDENCE
ISSUE 2
Subject A
Subject B
EVIDENCE
ISSUE 3
Subject A
Subject B
EVIDENCE
CONCLUSION
¶

Formatting by Criteria

The double-topic document can be composed in another way. Usually, the easier way to write it is to develop one side and then the other. By taking this approach, an author is essentially using the more common writing practice of developing a single topic but doing it twice. Developing the text by topic may be easy for the writer, but it may not be easy on the reader. If a reader has a long document in hand, the actual points of comparison may get lost in the length of the discussion. Issue C from one subject under discussion may be ten pages away from Issue C of the other side of the discussion.

The alternative is to develop the document by the issues involved—the criteria of the discussion. In this approach, the body paragraphs are based on the discussion of each issue. Suppose your supervisor wants a committee report readied on a cost analysis of a microcomputer under development. Two groups of engineers disagree on financial strategy. They represent the two sides or viewpoints. You could develop the document by viewpoint or, instead, you could divide the cost analysis into output moneys, input moneys, and aftermarket margins. You would then devote the three sections of the report to the *issues:* investment, short-term profit, long-term profit. You would discuss the argument between the engineers, but you would structure the essay around the issues and not the viewpoints. This technique is often the preferred model because of the point-for-point comparisons. This is also a popular strategy when a given set of criteria or guidelines are going to be used for a decision-making process.

The introduction to this type of comparison paper should identify the issues that will be under discussion or the criteria that you are going to use to evaluate the subjects of interest. The reader will need to know the basis for your analysis, and the criteria you will be using. In effect, you will be constructing an outline introduction because the issues or criteria shape the paper as it develops.

Like the first comparison (Model 8.A) the following project (Model 8.B) is simplified, but you will notice that both sides of each issue are discussed more or less equally.

Model 8.B (1)

Choosing A Network Operating System

1

Designing and planning a computer network can be a difficult task. The most important of the decisions that must be made is the choice of a network operating system. This paper discusses the factors that must be considered in choosing a network operating system and compares the two most prominent systems on the market today: Novell Netware and Microsoft Windows NT.

2

A comparison of network operating systems should focus on several issues. Among them are installation and configuration, management and administration, connectivity and compatibility, and performance. A combination of favorable ratings in all these factors makes for an efficient and powerful system that can easily handle most local area networks (LANs). Netware and Windows NT are comparable in features, especially with regard to supported clients, protocols, fault tolerance, and directory and file tasks. Windows NT does have some advantages in hardware compatibility, tutorials, and dial-in access, but it also requires more RAM (Random Access Memory) to function effectively. These features form the basis for the comparisons between the two products.

3

Installation and configuration of network operating systems can be extremely difficult. The ideal operating system can be installed in few steps and configured in an intuitive manner. Windows NT has the advantage in this category due to its hardware autodetection and configuration capability. This enables Windows NT to determine the hardware installed on the server and to automatically configure itself for the system. Netware does not have this capability and provides little on-line help. Thus, installing and configuring Netware can be very complicated and may require more technical support.

Windows NT uses the same graphical user interface as Windows. There is a program manager, which contains groups of icons for various programs. The program manager for Windows NT contains groups of program icons that run the system tasks and provide access to other nodes of the network. In contrast, Netware has no graphical interface, and programs are stored as files on the hard disk. However, each Netware utility runs with a graphic presentation that makes tasks easier. The command line capability is still available for greater control.

A Evaluative Comparison Developed by Issues

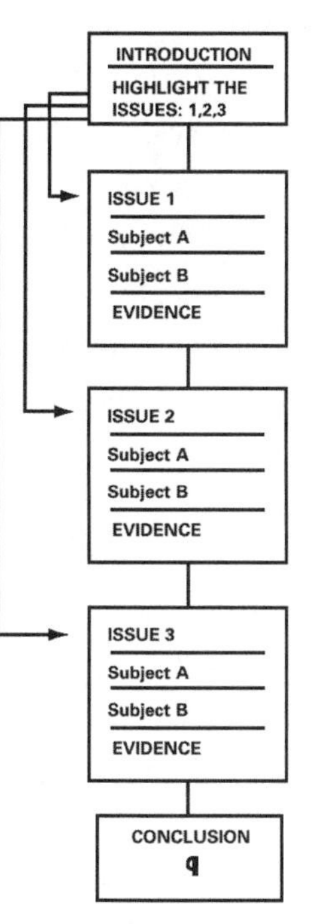

The presentation concerning network operating systems uses criteria to design the project. As noted earlier, an author can control documents in several ways. The AM-FM project had criteria for the discussion but looked at each system by turn. Here, the network discussion of each system looks at a number of installation requirements. The author's comment to me was clear: "In my business, the issues are bottom-line considerations such as efficiency and cost." Companies want point-by-point comparisons so that they can make cost-effective decisions. As a result, the approach you see here is quite popular when the issue is less concerned with products than with business needs.

The evaluation of the network systems is not highly judgmental but it clearly differs from the AM-FM discussion. Our first author described the given conditions of this subject matter. The discussion of the network operating systems establishes a set of criteria that become the yardstick of the analysis.

1 The introduction uses a strategy that is quite practical. The author points out the importance of the topic under discussion. He clearly explains his plan of action and identifies the two networks he will discuss: Netware and Microsoft Windows NT.

2 Since the first paragraph did not explain what requirements were going to drive the discussion, the second paragraph is used to establish these critical issues. The author also explains his selection of networks, since both are comparable in a number of respects and both are quite popular.

3 Notice that the third paragraph opens by referring to the "installation and configuration of the network." If we take a quick look back to the preceding paragraph, we realize that the author is right on target. It was the first of the issues he identified as a critical consideration for a potential buyer.

Model 8.B (2)

-2-

For both Windows NT and Netware, account administration is handled in a similar manner. Windows NT, however, has a better on-line help system and a better client interface. Netware has the advantage of requiring fewer steps to create user accounts and other tasks. Novell has added a new network administration tool, NWADMIN, that makes most administration tasks easier. Changes can be made in many accounts at once using a simple drag-and-drop method. Netware also offers a single network log-in for complex networks. This is possible in Windows NT only after network "trust" relationships are set up.

1

Support is roughly the same for both products, but there is a price difference. Although users can call for support 24 hours per day, seven days per week, they will be charged $200 per incident in the case of Netware, and $150 per incident in the case of Windows NT (as of this writing).

2

A network operating system in today's office should be able to administer a variety of desktop operating systems. With so many different types of workstations available, compatibility is a major concern. Fortunately, both Windows NT and Netware can support many different clients. Each can interface with DOS, Windows, OS/2, Macintosh, and UNIX. They also both support several different network protocols, including TCP/ IP, IPX, and Appletalk.

3

Most important to users, who do not see most of the network operating system, is performance. Netware and Windows NT have been compared on several performance issues. Netware surpasses Windows NT, except in the time required to log in to the system. In searching a large database and retrieving a file, Netware takes significantly less time. The other tests, which count the number of steps required for certain tasks, show that Windows NT can handle print jobs and password changes more easily but does not handle the task of adding a user as well.

4

Pricing structures of network operating systems are scaled according to the maximum number of users on the system. The number of simultaneous users logged on to a system is also taken into consideration. Thus, the best way to evaluate the cost of a system is to look at the cost per user. Windows NT is less expensive than Netware. This is primarily due to the difference in market share of these two software systems, and the difference is not necessarily a reflection of the quality of the products.

5

6

1

As the paragraphs develop you can see that the author's tactic is quite different from the topic-by-topic approach that you saw in the AM-FM discussion. Here the organization is based on an issue-by-issue strategy whereby both networks appear in the discussion in each paragraph.

2

If you read the first sentence of each of the body paragraphs, you should see an outline of the overall structure. This is a vivid test of the effectiveness of the topic sentences. Notice that support, variety, performance, and cost are signaled in each succeeding paragraph.

3

At times, a criterion of concern is not going to stand out as a "difference" between the topics under discussion. The author's paragraph concerning support services is a very important investment consideration. However, since the systems are equal in their support service management, only cost is mentioned.

4

The objective style is apparent throughout. The writing is tone-neutral and matter of fact. One system is not consistently superior, which indicates the difficulty of the analysis even though it is evaluative. Here, for example, Netware scores many points for performance, but, several paragraphs above, we noted a significant cost in the Netware support fees.

5

The cost of the systems is reserved for the last paragraph. Although cost will usually drive a discussion of this sort, power and flexibility are more important in an overall discussion of what the buyer gets for each dollar invested.

6

Notice that the paper lacks a concluding paragraph. Closure is not always important. Although this project is evaluative, the author was concerned about the point-for-point comparisons. He does not put forward a final judgment, and so he does not see a need for a conclusion.

Comparison Block Diagrams

Remember that the easiest method of comparison may be to develop a discussion by dealing with all the material in the largest units possible. For example, if you wanted to compare chemotherapy with radiation therapy, it might be convenient to discuss one therapy at a time. Do not move to the second therapy until you complete the entire discussion of the first one. Usually the criteria—or points of comparison—are the same for whatever you compare; you might select risk, discomfort, and length of treatment as points of comparison.

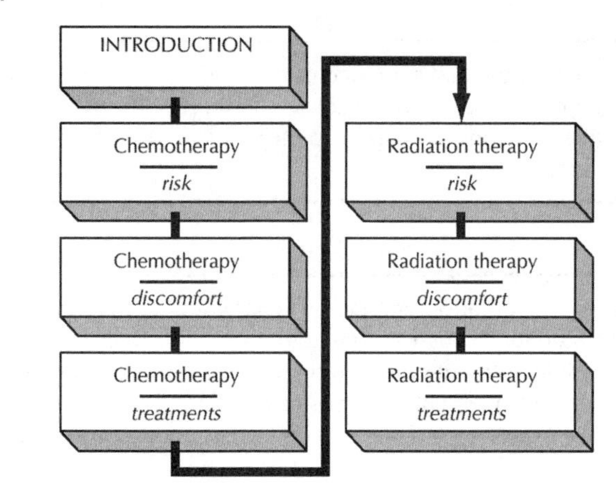

With a little sketching you can organize any kind of comparison of as many items as you want—theories, methods, policies, or software capabilities, for example. The following diagram illustrates the other method of comparison, a discussion based on criteria. Here we see a point-by-point comparison of desirable features of a group of products.

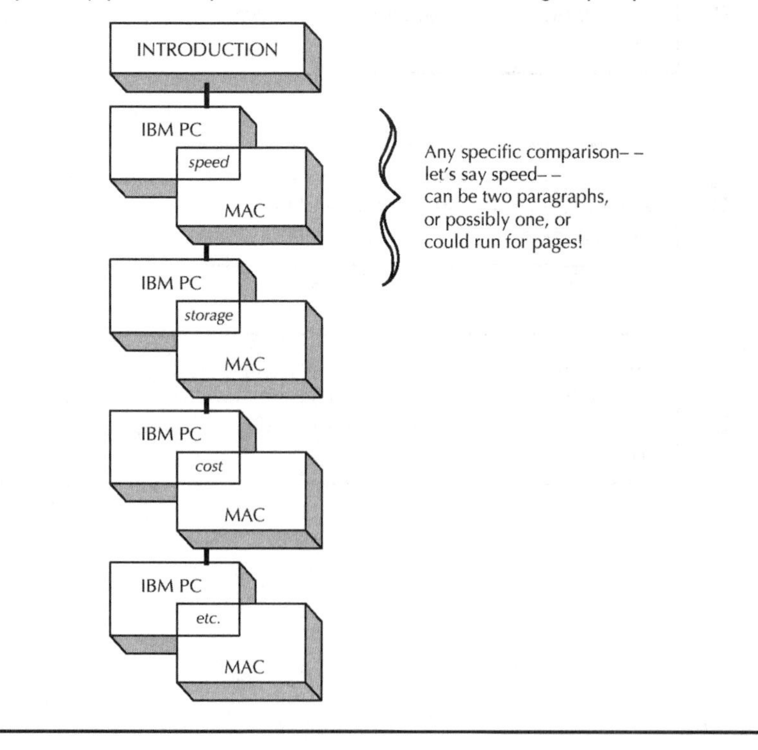

Formatting for Multiple Comparisons

The three architectures you have examined in the last two chapters will be valuable tools for your writing tasks, both at the office as well as on campus. The single-topic document and the two double-topic documents are daily fare. If you master them, they will give you quick logic controls for most of the material that goes into papers and essays, proposals, or any writing challenge that involves the methodical analysis of parts of a discussion. You need to be aware, however, of the need to adapt the basic structures to whatever needs arise. In particular, you may have occasion to expand on the basic architectures to meet the challenge of a project with, for example, ten topics.

The utility of comparisons is obvious enough: comparisons allow readers to measure and to judge. Everyone depends on comparisons; it would be poor cost accounting not to. Both in the popular press and in scientific and engineering journals the use of comparisons is regular fare. Whether you are deciding on an upgrade for a computer or the selection of a new car or the risks of a new job, you certainly must evaluate your options. You need to realize that comparing three, six or nine of something is essentially the same as comparing two.

If you were to photocopy a half dozen of either of the comparison diagrams, you could structure an enormous comparison project on your desk. The model is absolute. Additional comparisons are just a multiplication of the basic concept.

Certainly, a comparison of two is more easily managed than a comparison of ten, but you must recognize the need for multiple comparisons when they arise. They are not difficult to develop, and we conclude this chapter with a sample. Model 8.C was a four-topic analysis of heating fuels that was developed for an HVAC (Heating, Ventilation, and Air Conditioning) program in heating and air management engineering. It illustrates the utility and variety that can be gained by adapting the basic structures to an author's needs.

Model 8.C (1)

A COST ANALYSIS OF HEATING ENERGIES

The four most popular energy sources for heating in the local area are natural gas, fuel oil, LP gas (liquefied petroleum), and electricity. Because each of these energies is measured and sold in different ways, it is difficult to make a direct "apples-to-apples" cost comparison. One way to make an accurate comparison is to define each type of energy in terms of a common denominator. One such common denominator is the number of units needed to produce a certain number of Btus (British thermal units). A British thermal unit is a unit of measure that expresses the quantity of heat required to raise the temperature of one pound of water one degree Fahrenheit at a specified temperature. For purpose of easy comparison, we will use 1 million Btus.

1

In order to determine the cost of producing 1 million Btus, we will need to answer the following questions for each energy source.

1. In what unit is the energy sold?
2. What is the heating value of the energy?
3. How many units are needed to produce 1 million Btus with 100% efficiency?
4. What is the operating efficiency of the equipment?
5. How many units are needed to produce 1 million Btus with the efficiency factor?
6. What is the average cost per unit of energy?
7. What is the cost to produce an output of 1 million Btus?

2

Once these questions have been answered, we will be able to pinpoint the most cost-effective means of producing heat in and around the local area.

EFFICIENCY Before we begin analyzing each energy source, it is necessary to define *efficiency*. The efficiency of equipment is based on the ratio of Btu input to Btu output. In other words, we need to ask the question, "For every Btu of energy put into the system, how many Btu are returned in the form of heat energy?" For example, if a natural gas furnace has an input of fuel equal to 100,000 Btus and an output of only 80,000 Btus, the efficiency of the furnace is said to be 80% (80,000/100,000). Our next question undoubtedly is, Where did the other 20,000 Btus go? The answer is quite simple. During the process of combustion, some of the Btus are lost as a result of the chemical reaction, and some of them go up the chimney stack into the atmosphere instead of into the room being heated.

3

An Evaluative Comparison

A writer might be inclined to think that there are dozens of structures that could govern document design. In fact, the contents will often follow one of the patterns you have examined unless some external condition is driving the document, such as military specifications for proposals. The basic structures you have examined will serve a great many purposes. Writers enlarge the basic structures when they need to.

In this model the subject concerns a variety of heating possibilities. Cost is the bottom line for the analysis, but the comparison of heating fuels is not a simple matter.

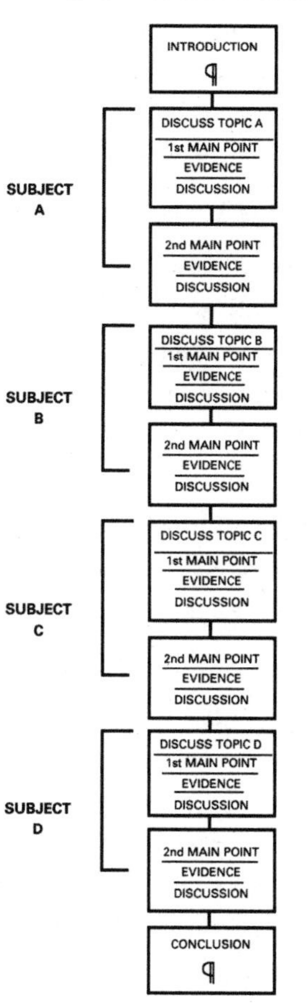

1 The introduction to this cost analysis indicates that the reader's ability to understand the calculations is going to depend on Btus and not on dollars. The focus shifts to technology. Readers can calculate costs but only after they understand the HVAC engineer's method of analysis.

2 A common tool, the question is handy as an occasional transition for writing. I discourage writers from asking many questions, but notice how effective the list of questions is in this case. The author is using the questions as a procedural plan for her analysis.

3 Technical documents often interrupt the development of a project in order to explain technical considerations. Here, the author must first explain the concept of fuel efficiency. This is clearly a detail that would not fit the project diagrams conveniently, but the diagrams are only concepts. Reality drives the actual project that develops.

Model 8.C (2)

-2-

From our previous example, if we desire an *output* of 100,000 Btus and our equipment is 80% efficient, we actually need an input greater than the original 100,000 Btus. We will need an input of 125,000 BTUs (100,000/0.8). We can look at each fuel with an understanding of efficiency, in precise terms.

1

NATURAL GAS Natural gas is a colorless, odorless fuel that is readily available in most areas of the city. The gas flows through pipes from utility companies directly to the equipment. Natural gas is sold in units called *therms.* One therm is equal to 100 cubic feet of gas. The heating value of 1 therm is equal to 100,000 Btus. The number of therms needed to produce 1 million Btus with an efficiency of 100% is 10 therms (1,000,000/100,000). The efficiency of natural gas furnaces ranges from 80 to 90%. The higher efficiency furnaces cost significantly more. If 10 therms are necessary to produce an output of 1 million Btus with an efficiency of 100%, then 12.5 therms are needed with an efficiency of 80%, and 11.1 therms are needed with an efficiency of 90% ($10 / 0.8 = 12.5$, and $10 / 0.9 = 11.1$). The average cost of one therm of natural gas is approximately 54¢ (as of this writing). Therefore, the cost to produce an output of 1 million Btus ranges between $6.75 and $5.99 ($12.5 \times 0.54 = 6.75$, and $11.1 \times 0.54 = 5.99$).

2

3

FUEL OIL Fuel oil is a petroleum-based product made from crude oil. It is classified into different grades based on its weight and its ability to resist flow, called *viscosity.* Grade 2 fuel oil is the most commonly used fuel in residential heating applications, and we will use this fuel for all our calculations. Fuel oil is delivered by truck to the consumer's own storage tank. It is sold by the gallon and has a heating value of 140,000 Btus per gallon. At 100% efficiency, 7.14 gallons are needed to produce 1 million Btus (1,000,000 / 140,000). The efficiency of a fuel oil–burning furnace is only 70%. At 70% efficiency, 10.2 gallons are required to produce an output of 1 million Btus (7.14/0.7). The average cost per gallon of fuel oil, including delivery charges, is 90¢. Therefore, the final cost to produce 1 million Btus is $9.18 ($10.2 \times 0.90$).

4

LP GAS LP gas is also petroleum based and is refined from crude oil or natural gas. LP gas is collected as a gas but is reduced to a liquid for transport. Because a liquid occupies less space than a gas, LP gas is less expensive and easier to transport and store. It is sold by the gallon like fuel oil and has a heating value of 90,775 Btus per gallon. A volume of 11.01 gallons

1

Once the analysis of the four fuel options begins, you can look back to the introduction to see whether there is consistency. Here the author is following the exact order she presented in the introduction: an analysis of natural gas first, followed by analyses of fuel oil, LP gas, and electricity.

2

No previous presentations had headings because I want to encourage you to learn to write your way from one sector of a project to the next by using transitions. Once you master the skill, however, you can begin to use headings. Here you see that if you removed the heading "Natural Gas" the first sentence would do an equally effective job of creating the transition. This is a good test.

3

You can take any one of the four fuels under discussion and check to see if the list of questions is being properly addressed. The first question asked for the unit of energy. The last question asked for the cost per million Btus. If you look at the natural gas paragraph as an example, you see that the second sentence identifies the unit of energy as the therm, and the last sentence calculates the Btu cost to the penny. This is tightly organized paragraph logic.

4

Just as the overall organization of this project is orderly, the makeup of each paragraph is also very orderly. The paragraph for the fuel oil analysis clearly repeats the internal paragraph pattern of the natural gas discussion. The four topics are dealt with in four paragraphs in an *identical* fashion.

Model 8.C (3)

-3-

is needed to produce 1 million Btus at 100% efficiency (1,000,000 /90.775). LP gas–fired furnaces are usually 80% efficient. Therefore, to produce an output of 1 million Btus with only 80% efficiency, 13.8 gallons are needed (11.01/0.8). The average cost of 1 gallon of LP gas is 82¢. Therefore, the final cost to produce 1 million Btus is \$11.32 (13.8 × 0.82).

ELECTRICITY Electricity is a common and familiar source of energy. Electrical "furnaces" create heat by energizing resistant heating elements. Electricity is sold by utility companies in units of kilowatts (kW) and has a heating value of 3416 Btus per kilowatt. At 100% efficiency, 292.74 kW are needed to produce an output of 1 million Btus (1,000.000 / 3416). Electrical furnaces have a 100% efficient rating, because no combustion takes place. Therefore, the final cost to produce 1 million Btus is \$13.17 (292.74 × 0.045).

1

Compilation and comparison of the data shows that natural gas is by far the most cost efficient fuel for heating in the local area (see the following chart). Switching to a 90% efficient natural gas furnace from fuel oil could save a consumer a substantial 35%. Consumers switching from LP gas would save 47% of the fuel costs. Consumers with electrical furnaces would enjoy the most significant cost savings, 54%.

2

A Cost Analysis of the Four Most Common Heating Energies

Fuel	*Selling Unit*	*Heating Value in Btus per Unit*	*Number of Units in 1 Million Btus at 100% Efficiency*	*Efficiency of Operation*	*Number of Units for Output of 1 Million Btus*	*Average cost per unit*	*Cost to produce 1 Million Btus*
Natural gas	Thera	100,000/ thera	10 theras	80%–90%	12.5–11.1 theras	0.5 4/ thera	\$6.75–\$5.99
Fuel oil	Gallon	140,000/gal	7.14 gal	70%	10.2 gal	0.90/gal	\$9.18
L/P gas	Gallon	90.775/gal	11.01 gal	80%	13.8 gal	0.82/gal	\$11.32
Electricity	Kilowatt	3,416/kw	292.74 kw	100%	292.74 kw	0.045/kw	\$13.17

3

1

The third page contains an evaluation of the electricity as a fuel. The analysis follows the patterns used with the other fuels, addressing each of the introductory questions in turn.

The project concludes with an obvious and, therefore, brief summary. Since the costs of a unit of each fuel are not directly comparable, the HVAC method of comparison gives the only reliable way to evaluate them. The author sums up her data by comparing percentages of cost savings due to switching from fuel oil, LP gas, and electricity to natural gas.

3

The final feature of this text is the table. A table is a convenient assist for a wide variety of projects. A paper that compares only two items can include a table if there are a wide variety of criteria or issues to consider. In this case, the author uses a table for four products as well as seven criteria (which match the seven questions on page one of her paper).

Summary

- Analyze complex writing chores to determine how to structure them.
  - ✓ Determine how many topics will be under discussion.
  - ✓ Decide on a descriptive or an evaluative position.
  - ✓ Decide whether recommendations will be an outcome of the evaluation.
- Develop an outline for *each* topic of discussion in your comparison. (See the outline summary at the end of Chapter 9.)
- Where possible, consistently apply the criteria (or issues) to all the items of the comparison in the same order.
- Prioritize the criteria used in the outline. Move from the most important feature to the least important feature.
- Use the outline subsections to evaluate the topics, for significant strengths and weaknesses for example.
- To develop each *subject* (or *topic*) one at a time, use a "vertical" perspective. Develop the text for each outline in turn. (See the illustration on p. 192).
- To develop by *criteria* (or *issues*), use a "horizontal" perspective. Follow the criteria (sideways so to speak), one by one, across the group of outlines. (See illustration on p. 200.)
- Write the project as quickly as possible and revise with care.
- Draw conclusions if desired or requested.
- Structure the document by the standards described for papers and proposals.

Activities Chapter 8

Use the chapter summary as your guideline. Write a comparison paper of 1500 words or more. Use one of the following options.

Develop a descriptive comparison of two items or concerns that interest you. The comparison can involve products, ideas, practices, laws, or any technical interest. Develop the text by fully discussing one item of interest before moving on to the other item.

Develop an evaluative comparison of two of something. Again, develop the text by completing the discussion of one concern before moving on to the other concern. Be sure the paper includes an evaluation in the conclusion.

Develop a descriptive comparison of two interests, but develop the text by criteria. Design the text to move from one criterion to the next: size, cost, energy consumption, and so on.

Develop an evaluative comparison of two interests. Again, use criteria as the organizational tool for structuring the paper.

Develop a comparison of three or more interests by developing one of the basic comparisons (of two) first, so that you become familiar with the challenge of balancing a discussion of this complexity. Then, continue by comparing additional items. The comparison can be descriptive or evaluative, and it can develop by subject or criteria.

When the projects are returned to you

At your instructor's request, resubmit *the project with the corrections to the errors that are indicated on the original.* Boldface *all corrections so that the instructor sees the revisions. Use the* Writer's Handbook *as your reference.*

Share a Project: First Option

Work with three other members of the class to discuss and edit the set of comparison documents. Use the same groups that you used previously, or develop new groups.

- *Consult Appendix A and reread the discussion of peer reviewing. Before holding the meeting, review the third editing checklist for papers that involve comparisons.*
- *On the day the projects are due, have each member of the editing group explain his or her project in terms of objectives and project development.*
- *Hand the papers around for critical reading and editing.*
- *Have each member edit the texts for both writing errors and technical errors.*
- *Have each member write a one-paragraph critique at the end of each paper.*

Share a Project: Second Option

Work with three other members of the class to construct brief projects of approximately three to four pages in length (750 to 1000 words). It will be helpful if all members of the group are from the same engineering technology program.

- *Decide on a familiar technical topic (or several) that will allow you to develop all four papers in four distinct variations of comparison:*
  - ✓ *descriptive comparison (of two) developed by subjects*
  - ✓ *evaluative comparison (of two) developed by subjects*
  - ✓ *descriptive comparison (of two) developed by criteria*
  - ✓ *evaluative comparison (of two) developed by criteria*

- *Include one paper in the form of a multiple comparison in any of the above methods if your group wishes to do so.*
- *Have each member select one of the formats.*
- *Agree to the objectives of the document.*
- *Assign a due date and bring four copies of each project to the meeting.*
- *Distribute the copies and discuss the outcomes of the activity.*
- *Provide a set of the projects for the instructor.*

Share the Preparation of a Document

Collaborate with three other members of the class to build a multiple comparison with a full text. This project will be developed over a period of several weeks and you will have opportunities to meet to discuss matters concerning design, research, production, and progress.

- *Select a familiar technical matter that calls for a comparison. You could compare energy, processes, policies, devices, parts, systems, and so on. (Omit motor vehicles because the systems are too complex.)*
- *Develop a text of 1500 words (six pages) that will include a table for the comparisons. The table is to be an original production of the team.*
- *Delegate parts of the task—including parts of the comparison and the table—to group members.*
- *Discuss the target audience, design, research, production needs, and progress at each meeting.*
- *Submit the final project to your instructor.*

Work in Progress

Outlining the Project

The next section of the document—Respiratory Safety—took days to compile and write. I stretched my "read it, get it and go" motto past my standard two page training guide. Since Pacific Aero Tech required several different stages of respiratory protection for their employees, I valued the importance of the information—quality and not quantity—and so I ended up with about thirteen pages of recommendations. In this case I consulted the president, Karen, and Matthew about the potential length of the document and my initial concerns. Their response was positive. I was to write all that was necessary and break it up into various areas of instruction—each a level of training in itself. I had to divide my final document into four separate parts with many subsections. This eventually allowed me to return to the original concept of two pages per topic. Here was the outline of my plan:

- *Part one: Responsibilities of program administrator, supervisors, and employees*
- *Part two: Identifying the nature of the hazard*
- *Part three: Worker activity and respirator use*
- *Part four: Approved Respiratory devices, maintenance and care.*

In the first part, OSHA dictates the rules and responsibilities of each individual exposed to harmful dust, fumes, sprays, mists, fogs, smoke, vapors, or gases. The following is an excerpt from part one of the respiratory document:

> ***PAT will provide*** respirators that are applicable and suitable for the purpose intended to protect the health of the employee. A suitably trained program administrator must administer the program.
>
> ***Supervisors will ensure*** the proper use, maintenance and care of all employees on the respirator program.
>
> ***The employee will use*** the provided respiratory protection in accordance with the instructions and training received. The employee will guard against damage to the respirator and report any malfunction of the respirator to his or her supervisor.

The second part identifies the nature of the hazard. The following factors concerning the nature of the hazard were subsections that were considered in the respirator selection: contaminants, physical properties/chemical properties, physiological effects on the body, actual toxic concentration of material, and warning properties of a material.

Out of all the documents that I had to write, this particular one, with all of its sections took me the longest, yet it was the easiest because I didn't have to scale down the information—I just broke it up into pieces.

L.C.L.

Structural Deviations

One writing difficulty that you must be keenly aware of is the very likely prospect of losing control of projects. I read hundreds of presentations every year, and I see vivid evidence of how difficult it is to be exacting in the structure of ideas. In order to have readers accept the outcomes of projects, you have to design your documents in a very orderly and obvious fashion. The structural diagrams I identified in the last two chapters may seem easy to follow, and I think they are, but you should look at the likelihood of error so that you see what can go wrong.

Problems emerge as a document begins to take shape. The evolution of a writing project can seriously distort an author's original intentions. If the structure has been compromised by the process of writing, the structure must be restored.

A student created the following diagram to illustrate the general problem of losing control of a document. The shaded areas of the logic blocks show the problem. Writers start a paper with the best of intentions (the shading) but then their inexperience or the difficulty of the project takes over, and in no time they start to lose their grip on the issue. The project progresses, but the actual writing says less and less about the original ideas that were proposed in the introduction. Any writing project is likely to encounter this problem. The writer speaks yet says nothing.

NAVIGATION ERRORS

This illustration is intended to suggest what generally happens when a writer has difficulty controlling the progress of a project. The reader is told what to expect in the discussion. However, the author needs organization and continuity to guide the project. Without control, both the author and the reader will drift away.

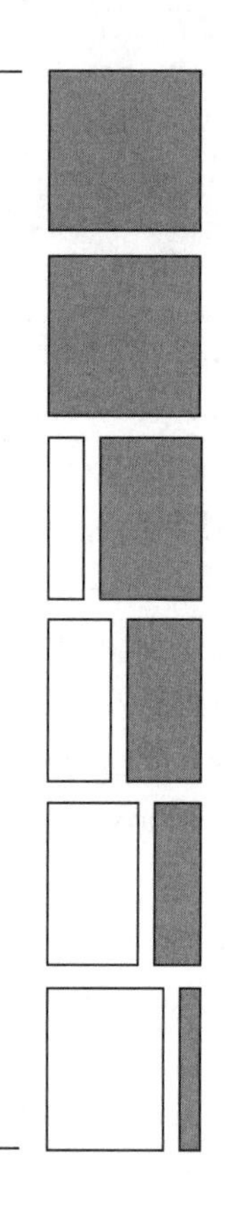

There are five classic causes for the problem of getting off track, namely, rambling, the double thesis, extraneous material, drift, and repetition. Let's look at each in turn.

A document that **rambles** is one that wanders aimlessly and is usually long-winded. The usual cause is the inexperience of the writer, and the best cure is plenty of time devoted to the rough draft and second draft of the paper. The novice is smart to add a third draft after having someone read the second draft. Just ask the reader one question. Do I lose you at any point when you read the paper?

The **double thesis** is a common result of any project that starts out on one issue and ends on another. If a document concerning drug abuse in America turns into a concern for

Breakdowns

You have to stick with it. If the structure starts to slide, the causes are often easy to identify. Five major problems are the usual villains. It is easy to visualize the way each lapse will disorganize a project because they simply interrupt the straight line of the logical discussion.

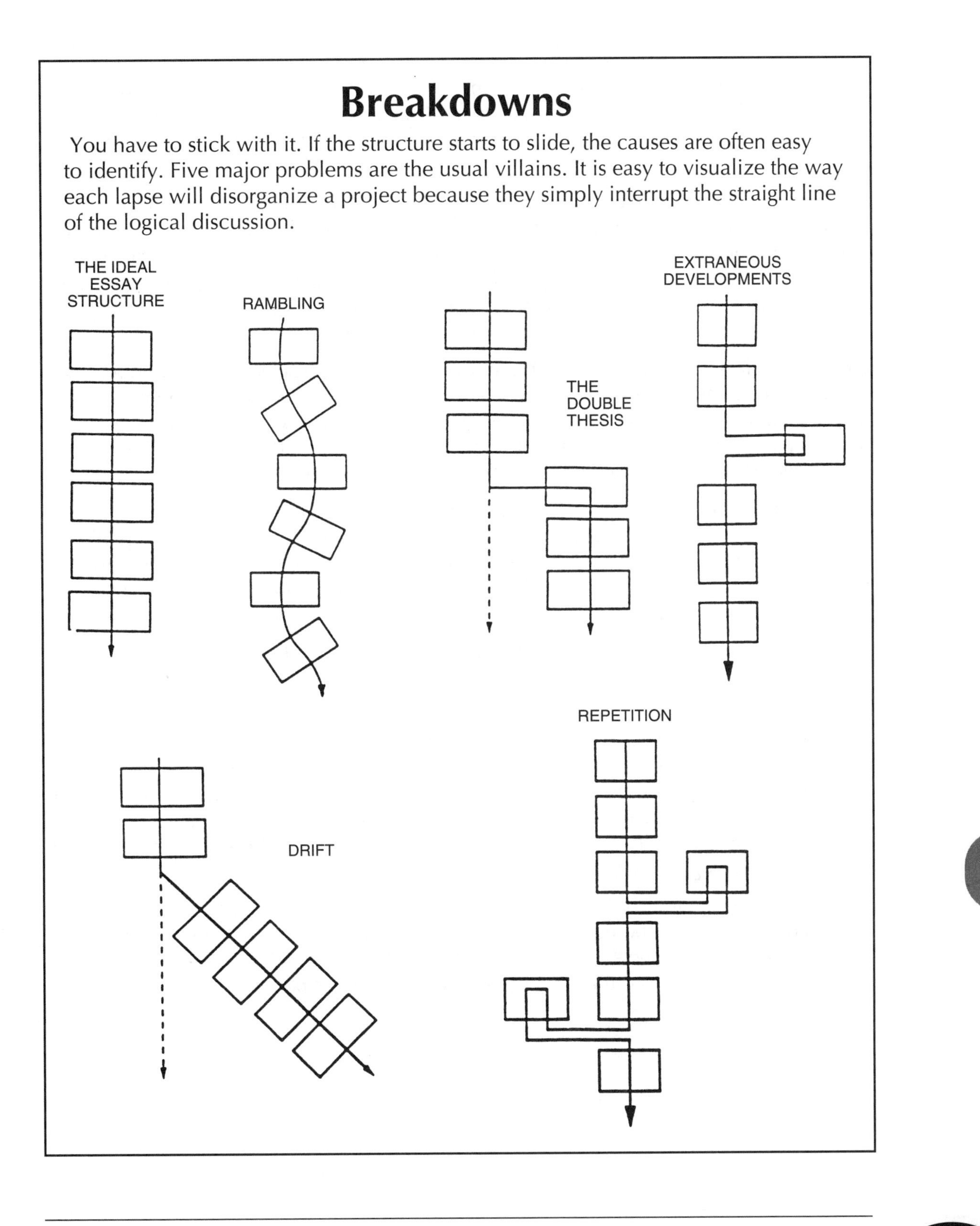

crime in the streets, an author may have two really good halves of two really good—but different—projects: the first half of the substance abuse proposal and the second half of the crime-in-the-streets issue. Of course, there is not a whole document for either project. Focus on the outline to make sure you have one issue on your mind. You can have fifty parts but only *one* issue.

Extraneous means outside of your concerns. Material can be irrelevant but still show up in a document. Why? There are probably two reasons. If a writer gathers a lot of material, it will often show up if only to show the reader the paces of the writer's *efforts*. However, efforts do not always translate into meaningful documentation. In addition, if the extraneous material is at the end—let's say pages upon pages of "extra" material—it is called "data dumping" and is a common device for trying to overstate a case.

Drift is the classic. People "go off on a tangent" when they speak. It is a notorious complaint about college professors, as I recall. In writing, drift is usually caused by a poorly understood proposal. If you do not want to write a project but you do it anyway, watch for drift. You need to *believe* in your proposal to follow it through. The outline is the best tool to help the draft along and to help the document stay on course.

Repetition is a device I use more frequently than you will. Do not hesitate to use repetition if you are writing on a learning curve, but you usually will not have this particular need. I am writing a textbook, so key concepts can be repeated. Repetition, as a tool for me to use, is similar to student drills. Repetition is a learning pattern. In general, however, what is said is said. Do not, under most circumstances, repeat data.

There is only one tool that restores order to the organizational plan for a project: the outline. It is the most practical solution for finding a straight road through the development of a complex project.

In earlier chapters, I suggested that you can avoid all the wrong roads with a clear understanding of the project, plenty of preparation time, plenty of development time, and a good outline. All these tips will keep you on track, but the outline is the best tool to use as a structural control feature. You need to see the project *structurally* in terms of its logic units. The visual architecture of an outline is very effective at controlling the problems illustrated in the diagrams on the preceding page.

If you see the paragraphs of your projects as building blocks, then the diagrams of the last two chapters can easily be utilized. Of course, they are ideal models. Real projects will vary the logic of the patterns I have structured. It is also likely that real efforts will go awry and produce the deviations that are commonly the results of poor control.

Outline Structure

The outline is very helpful because it organizes your writing before you start to write the text. It is not just a preview of a writing strategy that then reverts to front-to-back writing. The outline provides structure to follow and gives you a springboard for your ideas. An outline reflects preparation, whereas the front-to-back method lacks any such foresight. With an outline, you have written ideas to refer to as you work on the rough draft.

Outline Principles

An outline does more, however, than put your major ideas in a logical sequence. There are specific internal functions that are performed by the outline. If I remove the organizing symbols of numbers and letters from the outline and leave only the lines, I have a structure that looks like this:

SIGNS OF INTELLIGENT LIFE FORMS:
COORDINATION AND SUBORDINATION

Now, if you found this pattern scratched on a rock on Mars you would puzzle over it. If the lines were not fracture lines, could they show signs of intelligence? They appear systematic, do they not? There is, of course, logic in the structure itself. The secret of the outline is that it creates *coordination* and *subordination*. The larger lines are coordinate. The smaller lines are coordinate. The lesser lines are also subordinate to the greater lines. An outline is an organizational tool; it allows you to divide subjects and cluster other subjects together. It provides boxes for storage. For a long project, the outline is indispensable as a device that can help you organize material.

The concepts of coordination and subordination are universal. In mathematical systems, the analogous concepts are *sets* and *subsets,* and the structure is usually visualized with brackets as illustrated in the following figure. Obviously, the logic is similar to the logic diagrammed in the preceding figure. In fact, because the mathematical concept underlies the logic concept of coordination and subordination, they are very dependable tools.

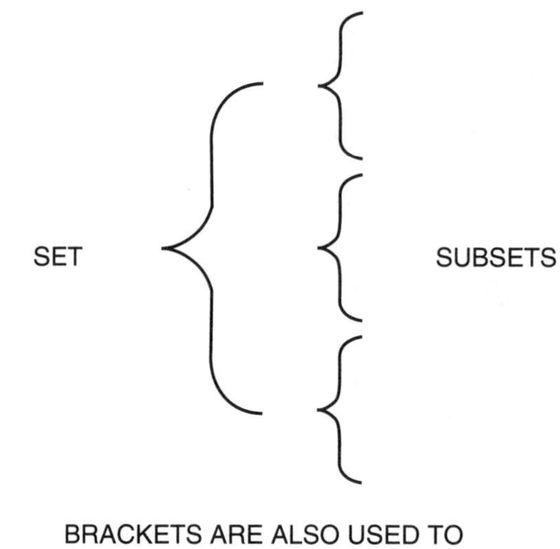

BRACKETS ARE ALSO USED TO COORDINATE AND SUBORDINATE

Do not forget that your world is organized because your *mind* organizes it. The mind organizes everything into systems. Trees that lose their leaves in the autumn all behave the same way, but there are hundreds of varieties that do so. Some are oaks. Some are maples. But you organize them all together and say they are "deciduous." Situation-based writing—writing to meet a unique set of demands at work—calls for another effort to organize. You must organize the subsets or points of an issue or problem, and then you may have to organize potential responses or solutions. Because the task is situation based, it will be unique. Use an outline. It will organize your perceptions and structure the writing you have to do.

Outline Types

You have your choice of standard formats for outlines: the full-sentence outline, the topic outline, or the less-common command outline. You can also mix the formats, since I notice that many working outlines (created by writers for themselves) tend to be constructed of full sentences to state the main points and sentence fragments, or even just a word or two, to establish subheadings. The sentences and fragments can also be used the other way around (see p. 230). The *full-sentence outline* makes you state your case with a great deal of logical clarity, and this format may be very helpful to you. If each point is stated as a sentence, you will clearly see the logic in your plan of action. In fact, the challenge of writing full sentences forces you to be logical. In addition, other people can easily understand the full-sentence outline, and this will prove to be useful.

If you want to be sure of what you are doing before you begin to write, show the outline to your supervisor or a fellow engineer or technician and ask for an opinion about the format and the ideas developed in the outline. This tactic can save you time and energy. It is much easier to revise an outline than it is to rewrite a report, even if the report is on a disk.

If you prepare the outline for yourself, it will often involve "action verbs" telling you what your writing chores will be. I call these *command outlines,* and they are a variation of a sentence outline. You can think of the main points of an outline as the concepts or ideas to be proven or demonstrated by the subsections. The main points, then, are usually commands. The subsections often identify other tasks you must perform, and so the lesser points may be commands to "explain," "identify," "verify," "demonstrate," and so on. These commands will help provide you with direction and order.

I often ask to see sentence outlines. The authors do not always see them as time-savers, but as a supervisor, I know that the outline is very cost effective. It may take an hour or two (at the most) to develop an outline, but if I do not see what I expect from an employee or student project, I can change the approach and save ten or twenty hours of wasted labor. The outline in Sample 9 is typical of the sort that I see. You can tell that the author will be in total control of his project because of the thoroughness that is apparent here—and on the additional pages he developed.

The shortest outline format is a *topic outline* and it is the easiest to develop. It is composed of words, phrases, or sentence fragments and the result is quite informal in contrast to the full-sentence outline. The topic outline has two drawbacks: (1) No one can review the outline for you because the logical argument is not thorough enough or clear enough to vividly indicate what you intend to write. (2) If you are not very skilled at writing, the topic outline might be too sketchy, even for *you.* It may simply organize the subjects you want to discuss, the "topics" of a topic outline. This is a useful but limited achievement. If you have not had much practice in writing, organizing the topics is less of a problem than organizing the logical *argument* of your discussion. To structure the logic, you need *sentences* so that you see what you are trying to demonstrate or prove.

Sample 9

TROUBLESHOOTING SEMICONDUCTORS WITH AN OHMMETER

Using an ohmmeter to troubleshoot diodes, transistors, FETs, and other semiconductors is a skill that every technician should be familiar with and is a simple, reliable method for isolating faulty devices.

I. Almost any multimeter, analog or digital, that has a built-in ohmmeter will work for testing PN junctions, but note a few precautions:

   A. Make sure your meter has an internal voltage greater than about 0.7 V so it can forward bias the junction; measure the voltage, or check your manual.

   B. Use a range setting of about 2000 ohms; if your meter has a special range for testing solid-state devices, use it.

   C. Check the polarity of your meter so you can identify the negative and positive leads; some meters have reverse polarity.

   D. Since your meter may have reverse polarity, we will refer to the probes as positive and negative from now on to avoid confusion.

   E. Out-of-circuit testing is best, but in-circuit testing is also possible; just be aware of possible parallel resistances that can affect your readings.

   F. The data sheet can give you the device's pin configuration, semiconductor type, and information about how internal PN junctions are arranged.

   G. In the following, it is assumed that you have isolated the suspected device and have removed it from the circuit.

II. Tesing a diode is the easiest to do and is the basis for testing more complex devices.

   A. Connect the negative lead to the cathode (banded side) and the positive lead to the anode; this forward biases the junction.

      1. A good junction will have a resistance of around 500 to 800 ohms (depending on your meter) for a silicon-type device and about 200 to 400 ohms for a germanium type.

continued

Using the Outline

Use sentence outlines whenever you want to share your plan of attack. Use the topic outline if the strategy is to write without consulting anyone, but remember that the sentence outline is better for shaping your argument. Do not confuse outlines with the task memo discussed earlier. You use the task memo to find out what is expected of you; then you must think out the issue. The outline follows; it is the third phase of the writing process. For you, as the author, the outline shifts attention from *what you have to do to how you plan to do it.*

Divide the project into whatever logical categories are appropriate, then organize them in whatever order they should appear in the project. *Any* technique that works is a tool to get the job done. If you have learned other methods, such as "mapping," by all means use whatever is *fast.* I do not use mapping, because I prefer a system that is vertical or linear (or rigid or highly structured). I use what works for me. You must do the same for yourself.

Place the outline beside you and begin to write the document. The outline provides you with a guideline, and it also helps you keep from drawing a blank. It *tells* you where to go next. It gives you a springboard that is not available in the front-to-back writing strategy. The outline gives you direction *and* confidence. It keeps you going. You still have all the white paper staring at you, but at least you are not looking at the landscape without a map.

Examine the following samples of each of the three basic types of outlines (Models 9.A, 9.B, and 9.C). Develop whichever one seems to be appropriate to your needs. These are followed by two comparison outlines. The comparison is unique in being an outline involving duplication. I encourage authors to structure the outline as a visual tool. In the case of comparisons, the best way to create a visual representation is to structure *two* outlines, such as you will see in Models 9D and 9E.

This first sample of an outline is particularly thorough, which is indicated by the subset details (A through G), but also because it was three pages long.

Model 9.A

PHOTOGRAPHIC/RADAR/LASER TRAFFIC MONITORING

OBJECTIVE: To discuss the legal ramifications of a new policing technology for highways.

PROPOSAL: The benefits from this proposed law are many, but there are those who challenge the effects on their wallets and their legal rights. **1**

I. ARGUMENTS FOR: There are beneficial reasons for acceptance of a traffic speed monitoring system. **2**
   - A. Speed limits can be enforced if motorists know they will be ticketed even though there are no police in sight. The presence of a camera/laser unit will slow most motorists.
   - B. Collisions will be reduced by lower speeds. Lower speeds will allow drivers more time to react to a situation. Collisions that do occur will be less severe. Less severe collisions can translate into lower repair costs and lower insurance rates.
   - C. Police officers assigned to traffic divisions can be reassigned to more serious crime problems.

II. ARGUMENTS AGAINST: There are bound to be arguments against a system that can take money out of a person's pocket. **3**
   - A. Establishing the system involves start-up costs and a support system, both of which spell tax dollars. **4**
   - B. Since violations are also costly, many of our "faster" voters will certainly vote against the concept.
   - C. Police departments often have sided against the system because it violates the motorist's right to confront the accuser. Some city officials have said it could prove "divisive."
   - D. Legal issues will surely be raised in court. If an owner decides to contest a ticket it means the driver will have to prove he or she was not at the wheel, and the driver will have to incriminate whoever was driving. **5**

III. THE NUMBERS: This is not new technology. It has been tested and proven.
   - A. These machines or similar systems are in use in seventy countries.
   - B. In terms of constitutionality, photographic and speed radar readings are accepted forms of evidence in a court of law.
   - C. It has been proven that accidents and the number of fatalities decline where these systems are in use.

The Sentence Outline

This first outline is in the full-sentence format I request in order to preview each project that is to be developed for me. This particular structure involves *two* perspectives toward automated speeding ticket systems, so the project has two major divisions in the body of the text. The subdivisions of this type of analysis may or may not be closely matched. In this particular case the supportive arguments are quite different from the counterarguments.

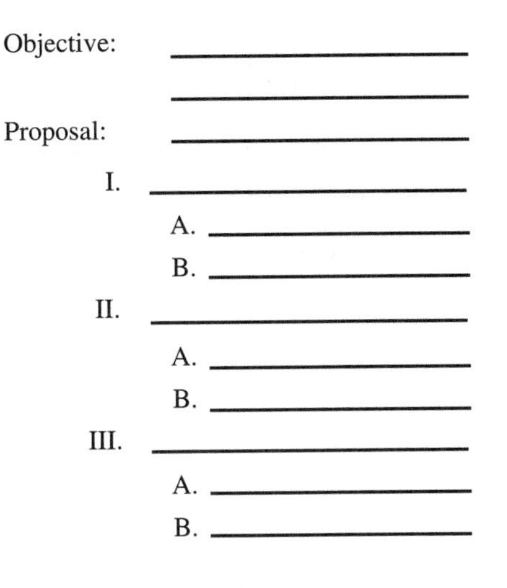
A GENERIC MODEL OF AN OUTLINE

Objective: ____________________
____

Proposal: ____________________

I. ____________________
A. ____________________
B. ____________________

II. ____________________
A. ____________________
B. ____________________

III. ____________________
A. ____________________
B. ____________________

1 The outline starts with a title and a statement that reflects the proposal (thesis) for the project. It is important to state the intention in clear and simple terms. The structure of the outline should be guided by an effort to demonstrate or to prove the proposal. If no opinion is to be rendered, the proposal should explain that fact and identify the project simply as a description or report.

2 The full-sentence outline style is particularly important if an outline is going to be discussed with a supervisor, a committee, or a professor. If the material is stated in topic-outline style, the logic of the argument is not sufficiently clear for a reader.

3 This project is designed as a for-and-against structure, which is a two-sided discussion. Any subject that is even vaguely controversial will lend itself to this organizational tactic.

4 I look to see that the major points *are* major. Of course, a pro-and-con analysis is usually easily identified. The issue, then, concerns the subsets. I look to see if the arguments that are proposed in the subsets are appropriate and if any key arguments are missing.

5 Although Argument A in favor does not have to match argument A against, I look to see whether there is a logic that carries over from one viewpoint to the other. In the case of a discussion of the advantages and disadvantages of a topic, the issues are occasionally unrelated, as we see in this outline.

Model 9.B

HOWTO PLOT A REFRIGERATION CYCLE ON A MOLIER GRAPH

Objective: To provide a set of instructions for entry-level students in Heating, Ventilation, and Air Conditioning Systems Engineering who must have a thorough understanding of the plot process. **1**

I. Introduction **2**

A. Define and explain the refrigeration cycle.

B. Show a graphic of the refrigeration cycle. **3**

C. Label the graphic.

D. Define what a Molier graph is and what it can do.

II. The Molier graph

A. Give a detailed explanation of the functional purpose of the Molier graph.

B. Describe the components of the Molier graph.

C. Show a graphic of the Molier graph and label.

D. Describe how the components relate to the refrigeration cycle. **4**

III. The example

A. Describe a typical problem using refrigerant-12. **5**

B. Show how to plot lines and points on a graph.

C. Describe how to interpret the graph. **6**

D. Build a graphic of the points and lines with labels of the problem.

IV. Conclusion

A. Explain why the Molier graph is useful for engineers and mechanics.

B. Describe an "ideal" cycle.

C. Explain the limitations of the description in this paper.

The Command Outline

This outline is unique in being a set of instructions for writing a set of instructions. In any refrigeration manual, a fundamental skill that must be mastered is the plotting method for a Molier diagram. Notice that this project was intended as a "trainer" to help entry-level readers.

A GENERIC MODEL OF AN OUTLINE

Objective: ______________________
____

Proposal: ______________________

I. ______________________
 A. ______________________
 B. ______________________

II. ______________________
 A. ______________________
 B. ______________________

III. ______________________
 A. ______________________
 B. ______________________

1 The sample outline states an objective without a proposal. Since the title indicates the proposal of the project (that is, the instructions), the proposal line was omitted.

2 Many authors prefer to include the introduction and the conclusion in the outline, where they become the first and last major headings. That approach was used in this particular sample.

3 The use of commands is a particularly helpful tool for a writer because it serves as a set of instructions for the project. The resulting outline is more forceful than a conventional outline.

4 Many outlines are composed of full sentences, partial sentences, or a mix of the two. In this case the main headings are little more than titles.

5 Without exception, every subsection begins with what are called "action verbs." These serve the author by being a group of commands that call for one task after another.

6 Because the Molier graph is a "figure," each major section of the outline calls for the addition of a graphic feature. The graphic feature is, in fact, the focus of the entire text.

Model 9.C

WATER SYSTEM PLAN: SNOHOMISH NORTH DISTRICT

OBJECTIVE: This technical report assembles and presents the key elements of a new water system in accordance with state guidelines.

I. BASIC PLANNING
   A. Location
   B. History of water system development
   C. Future service area map and agreements
   D. Existing population and land use
   E. Future population and land use projections for the next ten years
   F. Existing water consumption
   G. Future water demand for the next ten years

II. SYSTEM ANALYSIS
   A. Inventory of the existing system
   B. Evaluation of the existing system
      1. Hydraulics
      2. Fire flow
      3. Water rights
      4. Water quality

III. PROPOSED FILTRATION IMPROVEMENTS
   A. Direct filtration
   B. The filtration unit
   C. Disinfection and CT time
   D. Schedule for the ten-year planning period

IV. COST OF OPERATIONS

V. OPERATIONS PROGRAM
   A. Routine operation procedure
   B. Preventive maintenance procedures
   C. Sampling procedures including response when sample results exceed state standards

VI. APPENDICES
   A. PREVIOUS ENGINEERING REPORTS
   B. WATER RIGHTS
   C. CT VALUES AND SELECTED EPA GUIDANCE MANUAL REFERENCES

1

2

3

4

The Topic Outline

This outline was developed for a project that was being prepared by a civil engineering firm. I worked with Bob on the draft, and the outline was the first stage. Later, a substantially similar version became a table of contents. I prefer full-sentence outlines, but this outline was going to be long and clumsy in the full-sentence format. We decided on a topic outline, but with reasonably thorough descriptions, as you will notice in Section I, Basic Planning.

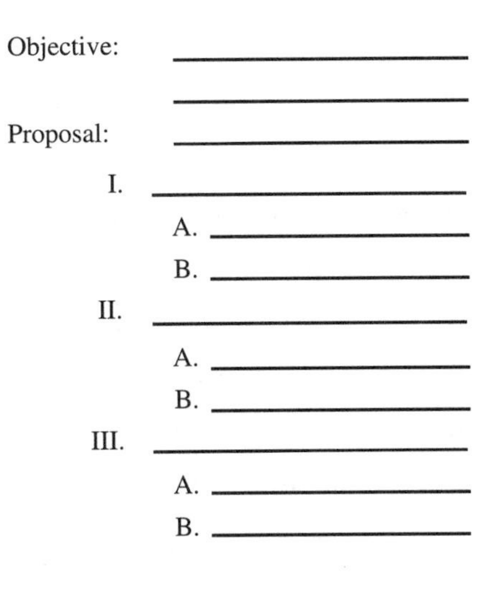

A GENERIC MODEL OF AN OUTLINE
Objective:
Proposal:
I.
A.
B.
II.
A.
B.
III.
A.
B.

1 This outline opens with a background description. For an engineering firm dealing in any sector that concerns impacts on population, an easy and necessary introduction to a proposal will often be concerned with present status and future needs. Consistent with that typical opening, this project examines the existing water system and the expected population growth factors.

2 Sections II and III are expansions of the initial discussion. Here, the existing system and the proposed improvements are the logical follow-up to the introduction.

3 The next two sections involve the costs and the postinstallation procedures for the improvements—because the systems must be maintained at a price. Cost is the bottom-line and this section is the presentation of the bid. Maintenance—the indirect cost—is the other consideration that a development council will look at in detail.

4 A formal engineering proposal can be overly complicated if all the relevant documents are not well organized. The author's strategy, which is common practice, was to construct a group of supporting materials at the end of the proposal. This material can be consulted as needed but is otherwise out of the mainstream of the discussion.

Model 9.D

PRINTER SELECTIONS:

LOOKING FOR THE BEST BUY

Audience: This paper is for someone who would like to know the difference between a laser printer and an inkjet printer. The paper will also help someone choose a printer to purchase. No particular technical understanding of printers is necessary, although I assume that the reader is familiar with computers.

Objective: I will discuss a laser printer and an inkjet printer in separate sections. I intend to compare the two in terms of their printing resolution, output per minute, fonts, memory, and PostScript printing language.

1

Laser Printer	Inkjet Printer
• **An introduction to laser print technology**	• **An introduction to inkjet print technology**
• **Output per minute**	• **Output per minute**
Generally superior	Slower
Depends on product line	Depends on product line
• **Font Support**	• **Font Support**
Yes	Yes, but only by software driver
Hardware upgrades available	No hardware upgrades
• **Printer Memory**	• **Printer Memory**
Yes	Yes, but lower than laser standards
Expansion slots for more	No expansion slots
• **PostScript Support**	• **PostScript Support**
Yes	No
Hardware	Possible only by special software

2

3

The Comparison Outline by Subject

The outline in Model 9.D was designed to discuss printers for word processors. The author decided to use two popular types of printers and did not focus on brand names. The text concerns the general characteristics of the systems. Notice that each subject is handled in one column of the outline.

THE TWO-COLUMN OUTLINE

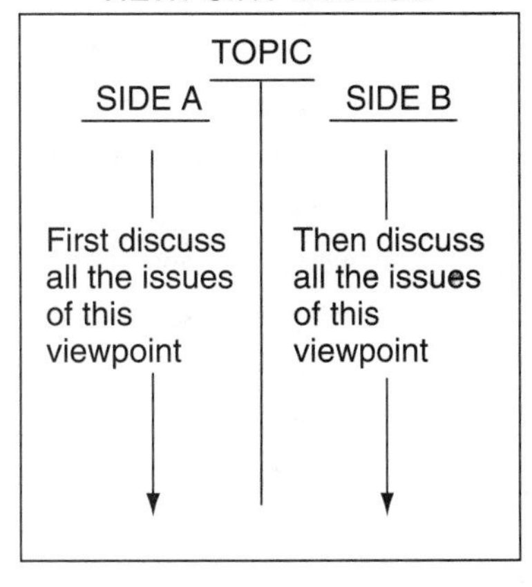

1

If you want to develop a comparison, decide on the method you want to use for the analysis. Do you want to develop the text by subject? Do you want to develop the document by the points of analysis (the criteria)? Here, as the author explained, a discussion of the laser printer will be the first half of the paper. The second half of the paper will concern inkjet printers.

2

Notice that the criteria used in the paper are structured in duplicate. By recreating the first outline, the author keeps the maximum organizational control of her overall document design. Readers anticipate that the pattern of discussion in a comparison paper will be repeated.

This author used bullets to design the outline. If there are only two major categories in the outline—major headings and subheadings—the subheadings can be handled in this fashion.

Model 9.E

ELECTRIC MOTORS IN HVAC SYSTEMS

Objective: Electric motors are the most important load devices in the various types of HVAC units, where they are used in the compressor or the fan blower. There are two types of motors, single phase and three phase, available on the market. This presentation is a comparison of the two motors. It is handled as a discussion of criteria for selecting the appropriate motor.

1

I. Physical description

A. Single-phase motors

1. Two hot wires or one hot wire and ground
2. Low torque, otherwise use capacitor in the start winding to increase the torque
3. Windings not reversible

B. Three-phase motors

1. Three hot wires plus one ground
2. High starting torque and do not need a capacitor
3. Easily reversible (the direction of rotation)

II. Performance

A. Single-phase motors

1. Low-efficiency
2. Derive high currents

B. Three-phase motors

1. High efficiency
2. Derive low currents

III. Application

A. Single-phase motors

1. Used in household appliances and light commercial applications
2. Different types for different applications

B. Three-phase motors

1. Usable in all industries and all commercial applications
2. Two types available: wye type and delta type

IV. Maintenance

A. Single-phase motors

1. Generate more heat and are prone to failures because they use high current
2. For repair and testing see Figure 1.

B. Three-phase motors

1. More durable because they use low current
2. For repair and testing see Figure 2.

V. Power Consumption

A. Single-phase motors

High level of power consumption

B. Three-phase motors

Low level of power consumption

The Comparison Outline by Criteria

The final model outline is a second example of a comparison project that is distinctly different from the first sample. This one outlines a project that will discuss the distinguishing *characteristics* of two types of motors. It is developed by criteria.

THE TWO-COLUMN OUTLINE

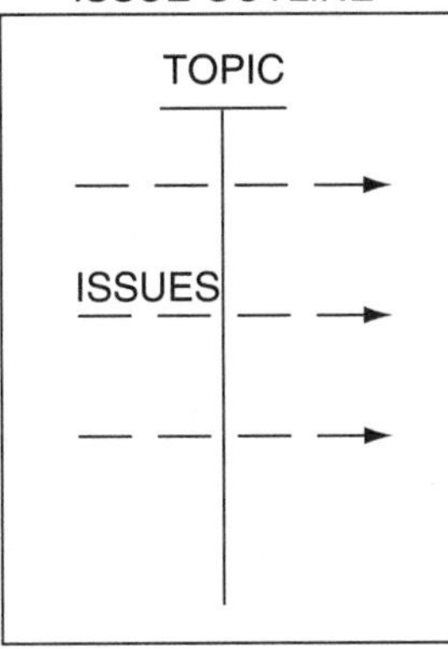

1

Contrast this outline with the preceding one (Model 9.D). This outline is handled in two columns, but the author intends to develop the five main considerations: physical description, performance, application, maintenance, and power consumption. The subject does *not* organize the text.

2

Symmetry is, again, the order of the day. The subsections under each motor type are carefully matched, so that if the author discusses efficiency, for example, he will discuss that aspect of each type of motor in turn.

Summary

- Determine the number of topics to be discussed.
- Decide whether the project should be descriptive or evaluative.
- Decide whether you should use a full-sentence outline, a command outline, or a topic outline.
- State key ideas or criteria and begin to coordinate and subordinate the sections.
- Decide whether to organize by importance (prioritize) or by occurrence (what happens first).
- Determine whether a different organization for the outline may be required by the subject at hand or by your company.
- Use your research to determine points of discussion for the subsets or subtopics to demonstrate or prove the key ideas.
- Be sure to indent subsets; topography is the key to clarity in an outline (see Model 9.C).
- Save your outline as a template for future use (see Appendix B).

Activities Chapter 9

Use the chapter summary as your guideline.

Develop an outline for each essay project you develop. Use it to review the project plan with your instructor. The chapter outline summary will help you design an effective outline. The single-topic outline should consist of at least nine sentences. The comparison outline is likely to double the number of sentences found in a single-topic outline.

- **First outline** *Develop a full-sentence outline for your first project.*
- **Second outline** *Develop a command outline for the second project. Use full sentences.*
- **Third outline** *Review the samples of comparison outlines in this chapter and develop a comparison outline in full sentences. Develop the outline in two columns if that style is helpful.*

Share a Project

Work with two other members of the class to practice and discuss the effectiveness of outlines. Each member will need a recent technical project from another course to use as a project model. An upcoming writing project will also work quite well. Use the project to complete the following:

- *Develop a one-page topic outline of the project. Construct it using words and phrases. Save the outline and develop another one that is written in complete sentences. Give the topic outline to the team members and ask them to explain your intentions. Then give them the* sentence outline *to look at. Briefly discuss the results.*

Work in Progress

Finding a Strategy

When I returned to the hearing conservation material, I didn't know where to start. I asked myself, what would the employees be willing to read? They needed something fast but easy to read, and they needed to retain the information. That meant a focus on paragraphs. My goal was to educate them at a level that would implement training in their everyday work activities. I have inserted the following excerpt from my final document to show the typical paragraph length and content I used to focus the material. It was based on a simple approach: what did they need to know and want to know?

What is noise? The American Speech-Language-Hearing Association considers a decibel level over 80 dB to be potentially destructive. Exposure to sound with a dB level as high as this or higher can cause damage. The higher the decibel levels the shorter the duration necessary before damage can occur. Since noise-induced hearing loss cannot be repaired or cured, Labor and Industries has adopted permissible exposure limits (PEL) of an eight-hour time-weighted average (TWA) of 85 dB (decibels) for noise measured on the A-scale at slow response or equivalently, a noise dose of 50% on a dosimeter. This noise standard requires the employer to establish an effective hearing conservation program for employees exposed to 85dB or above.

Industrial noise is divided into two types: continuous and impulse/impact noise. Continuous noise is described as a noise with a duration of more than one second, such as what is produced in the warehouse, polish room, media stripping area, and the sealant room of PAT. Peaks define impulse/impacts at intervals of less than one second (similar to a typewriter, rifle shot or explosion), which are produced when the air pressure guns are used to clean off stubborn dirt in the assembly area.

What is the NRR? NRR is a rating system set up by the Environmental Protection Agency (EPA) as a guideline that indicates the amount of potential protection a hearing protection device will give in a noisy environment. The higher the rating, the greater the noise reduction capability the product offers the user. The actual noise reduction obtained from any protection device is dependent on proper size, fit, and usage, as well as specific work environment. Because actual noise reduction achieved is usually less than the published NRR, OSHA recommends "derating" the labeled NRR by 50%. For example, the existing noise level in the polish room is 93 dB. We are required to reduce the noise level to 80 dB. Calculate: 93dB – 80dB = 13dB. Derate the number 50% (13dB × 2) = 26dB. For a noise level of 80dB, choose a hearing protector with an NRR of at least 26dB.

I decided it was imperative that management, supervisors, and employees know, at a minimum, the definition of noise, how to determine when hearing protection should be required and enforced, what kinds of hearing protection are available, and how to apply the NRR to an individual employee's needs. After I found the questions I wanted to answer, the document was easy to write as you can see above.

L.C.L.

Sub-structure Logic and Paragraph Design

As we saw in the last three chapters, the logic structure that governs the architecture of most basic forms of documentation is predictable and conforms to a strict and obvious set of patterns. In this chapter, we will examine the logic structures of paragraphs, particularly the paragraphs that conform to the needs of technical documents in engineering technologies. The descriptions or demonstrations or explanations you construct are built into the logic blocks of the paragraphs. The overall structure of the paper *coordinates* the paragraphs, but the job gets done in the body paragraphs.

The structure of basic paragraph logic is the key to your writing strategy. This is not a complex matter, the paragraph is simply a *proof.* Its function as a logic block is to prove a point or at least to describe or demonstrate one. Since the overall document is *also* a proof, you might ask what the connection is. Basically, the essay is somewhat like nested Russian dolls or Chinese boxes. Open a doll and a smaller duplicate is inside. Open the box and yet another is inside. The overall document you develop is a logical argument that tries to describe or prove a point. Inside the structure are smaller logical arguments that work toward proving subordinate points.

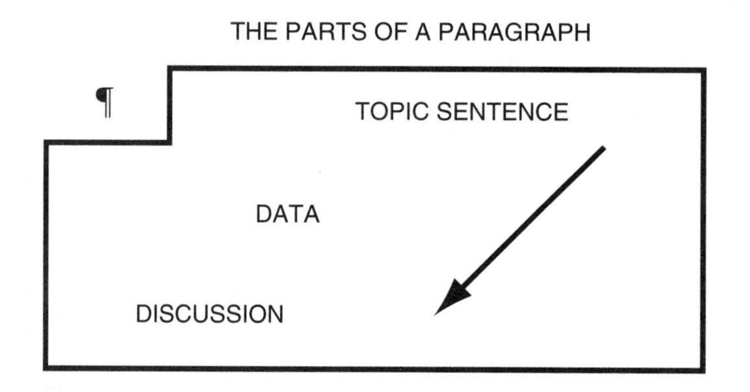

The preceding figure diagrams the phases of logic involved in the reasoning process of a paragraph. At first you may not see the similarities between the paragraph and the overall document, but look at the basic three subsections of the paragraph from several angles. In the following illustration notice that the right-hand column corresponds to the simplified preceding version.

THE PARTS OF A PARAGRAPH ARE
THE SAME BY ANY NAME

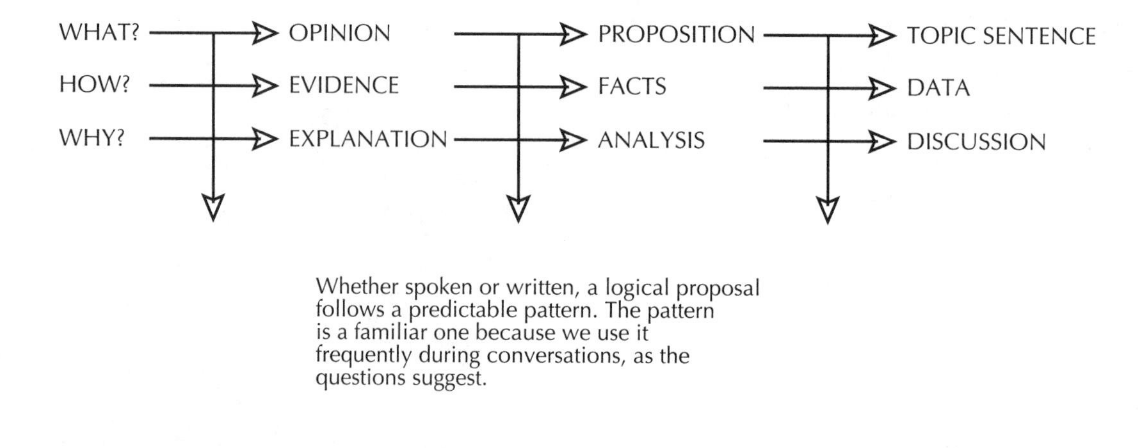

Whether spoken or written, a logical proposal follows a predictable pattern. The pattern is a familiar one because we use it frequently during conversations, as the questions suggest.

There are any number of ways to look at exactly the same pattern of paragraph logic. If you listen to anyone state an argument during a discussion in the lunchroom, you will see the universality of the pattern that is used for developing a logical proposal. If I say the Marlins will take the pennant, I offer an *opinion*. If I then mention the bullpen and the RBI records, I introduce *evidence*. When I explain why the pitchers are strong and why there is a powerhouse at bat, I am developing the *explanation* to prove my logical argument. The pattern that is used to prove something in conversation is no different from the one that is usually used in a written paragraph. In fact, there is even an order in the questions that emerge in conversations, and my baseball friends will surely ask the usual questions to challenge my logic: What? Win the pennant! How? And a little later, Well, *why* do a few good pitchers clinch the title? When you have a good idea for a paragraph, simply start to write: explain why it is how it is what it is. Dialog runs in a predictable pattern.

Let's look at a few student samples of paragraphs so that you see this particular analytical process. Earlier, I suggested that when you develop a document, you should develop a proposal and an outline of the project to use as the basis of your introduction. You can use a similar system in each paragraph. The paragraph usually opens with the key idea you want to discuss or explain (the *topic sentence*). Each topic sentence is a miniature thesis or proposal, and it is also a segment of the overall proposal of the full document. Also, the first sentence or two often outline the intentions of the paragraph. Evidence and discussion follow. Each paragraph is a miniature analysis and a component of the overall analysis of the document.

Introductory Paragraphs

If a paragraph is often a miniature version of the logical process of the document from which it emerged, I should be able to support this observation with evidence. In fact, *any* analytical presentation should verify the idea, but there is an obvious first demonstration that is most appropriate. Since the introduction to an essay or paper is often an outline of the overall presentation, it demonstrates the "miniature analysis" concept of the paragraph in its most obvious form.

As you read the following introduction you will observe that the paragraph does indeed structure an argument, the very argument that is the thrust of the entire document.

> Client workstations often are the least stable part of a network. This is partly because they are not dedicated network devices in the same way as file servers or active network equipment and also because these are the pieces of the network to which users have direct access. Private office desktop PCs are often prone to reconfiguration. The owner may decide to upgrade the operating system or hardware without consulting a network technician, or the owner may try out some completely inappropriate network software. When the computer can no longer access the network, the problem suddenly becomes yours if you are the technical support technician.

> All in all, client faults should be expected to form a significant portion of the problems arising on any network.

Why is the topic sentence in the sample so successful? Because it states the author's opinion, which the paragraph then goes on to discuss and explain. Observe also that the paragraph is quite specific, but it contains no numerical data. The argument is purely logical, and the support consists simply of examples and explanations. Finally, notice that the paragraph has a conclusion: it sums up in the last sentence. Well-designed body paragraphs often duplicate the basic essay logic pattern that moves from introduction to body to conclusion. Concluding sentences in the paragraphs of technical documentation are your guarantee that the reader will get the point.

In the following sample introduction the author prepares to discuss a modern satellite-based global application for marine disaster warnings. Notice the details that develop between the first and last sentences. The sentences in the center of the paragraph identify a large number of topics that the paper discusses at length.

> The Global Maritime Distress and Safety System (GMDSS) is one of the most complex, yet easy to handle, communication systems devised to date for maritime personnel. The system uses satellite as well as advanced terrestrial communications. GMDSS is primarily a ship-to-shore, shore-to-ship, ship-to-ship system. Considering the technology used, the area of coverage, and the automation, the system is really quite simple and easy to use. Once the operator learns some new terminology and the capabilities of GMDSS, everything falls into place and makes very good sense. Under GMDSS, radio watchkeeping is almost entirely by silent, autoactivated receivers. The human element, as with the former Morse code distress system, is the key to the system. The operator can readily ensure that the proper frequencies and equipment for distress alerting are being guarded.

Although less of a logical argument than the first sample introduction, observe that by description—or demonstration—the author indicates his basic point: the communication is complex but easily managed.

In the next example of an introduction observe the presence of the pronouns *I* and *we*. Here an author comes on stage to explain the overview. Although popularly viewed as an unnecessary and "mechanical" approach to opening the curtain on your project, you will see this practice used frequently. Since your supervisors will look at writing in terms of time management, use the technique that gets the job done. It appears that writers who are uncomfortable with writing tasks find the "I will examine . . ." technique convenient. The choice is yours.

> Multimedia technology allows a user to include text, graphics, sound, animation, video, and interactivity in training sessions or presentations. There are a wide variety of multimedia authoring programs. *Authoring* is the term for multimedia presentation

> design programs (for system requirements see Appendix A). This training manual introduces one of the two best authoring tools, *Authorware* for Windows (the other is called *Director*). For the purpose of this manual, we will be limiting ourselves to creating a simple *linear* presentation, similar to a slide show. By *simple* I mean a presentation using only text, graphics, and limited special effects.

Concluding Paragraphs

If, as noted earlier, the conclusion follows the same logic you see in the introduction, conclusions will often fall into the proposition → evidence → discussion pattern also because they, too, can be used to summarize arguments. In the following example you can clearly surmise the content that was analyzed. The overview and the outcome of the discussion are equally clear.

> In conclusion, virtual perception in engineering graphics can be obtained through several new ways of looking at the design process. Creativity must be understood for its importance in enhancing visual perception. Understanding the importance of each process—wire frame, solid modeling, CNC, image processing, animation, and rapid prototyping—unifies every aspect of the design-to-market process. The use of computer-aided-creativity-programs will continue to be a dominant goal for programmers and software developers because these programs truly combine the technical and creative aspects required in visual perception. As these programs continually become more available for a mass market, engineers and designers will be able to design and engineer projects more quickly and efficiently, with greater skill and clearer perceptions of their designs than ever before.

Conclusions and introductions are particularly vivid examples of the analytical paragraph model if they are designed as overviews of the whole document. Since the total document proceeds through the discrete phases of analysis—proposition → evidence → explanation—the miniature versions will follow suit.

Analytical Paragraphs

I do not mean to suggest that body paragraphs are not equally as logical as introductory or concluding paragraphs. Body paragraphs are commonly analytical in the sense that the evidence and discussion lead to a conclusion. Notice that in the following sample the first *two* sentences are the proposal for the paragraph. Also notice that this example is a product comparison—in other words, a part of a double-viewpoint document.

> The majority of consumers believe that CDs offer sound quality that is vastly superior to that of their older vinyl counterparts. In most cases this is true, but there are significant exceptions. The CD is a digital component that reproduces music with greater clarity and less distortion than vinyl records. However, if the piece was orig-

inally recorded using analog technology or was not digitally remastered or both, then the CD is not superior. The CD remains only a digital means for playing music; therefore, the sound quality is only as good as the original recording. It is also dependent on the quality of the remaster work, and remastering is primarily done to remove excess background noise from the original recording.

In the effort to explain CD quality to the reader, the author added another paragraph to provide additional details and to explain to the reader how the code on a CD label can explain CD quality.

> One can determine any CD's sound quality by looking for the symbols AAD, ADD, or DDD on the label. AAD means that the original recording was analog, the master tape was analog, and the compact disc itself is digital. The AAD label is found mainly on older pieces of music, and the sound quality is equivalent to that of a well-maintained vinyl record. Most of the music on CDs today, however, has at least had the original analog recording digitally remastered (ADD) or has been originally recorded digitally (DDD). Sound quality on CDs with ADD or DDD is far superior to music found on vinyl, even though records that are well maintained and played on quality turntables are capable of delivering excellent sound.

Again, you do not see "facts" emerging as numerical data. The evidence here consists of examples or points of information that are then *explained.* The argument is logical and orderly. The pattern is always the same: proposition, evidence, explanation. Notice the organization, also. The first paragraph has a simple balance of sentences that *compare,* so the pattern is simply CD-LP, CD-LP, CD-LP. The second paragraph opens with a handy outline, and then the sentences follow it through: AAD, ADD, DDD.

As observed much earlier, a technical paper always has a mission—to explain or demonstrate or propose or analyze or serve some similar purpose. At the heart of the effort there will be analytical paragraphs of the sort presented here. The overall document will follow some variation of a predictable pattern:

opinion → evidence → explanation (proof model)

The key paragraphs that support the effort will follow a similar pattern:

proposition → facts → analysis (proof model)

These are simply commonplace logical patterns for developing a proof. The system is as rudimentary and fixed as any basic mathematical proof. Paragraphs are easily designed to serve this analytical model once you understand the concept—and have the data.

Paragraph Progression

Four additional paragraph types are commonly used in technical and engineering documentation. Although analytical paragraphs are perhaps the glue that holds technical projects together, they are important well beyond their number. Of the primary paragraph types in technical writing, the analytical paragraph is probably the one writers use the least.

There are four other basic paragraph logic structures that are important in technical work:

- **temporal progression**
- **spatial progression**
- **temporal-spatial progression**
- **causal progression**

These are fundamental tools for developing any project that involves writing about technical or engineering matters.

Temporal Progression

Whenever you write a paragraph of instructions, you are using time as the ordering device for the progression of the paragraph. Time, of course, has a logic of its own. It is clean and simple. Corporations use hundreds of thousands of manuals that are largely composed as instructions. All the material in these manuals is constructed in simple time progressions of activities. Either paragraphs or outlines are used for the subsections of the manuals.

The following short paragraph concerns a few details of Windows. It was developed by a computer lab assistant to help walk people through a process.

> Now select the **My connection** icon with one mouse click (not a double click!) and select **Properties** from the popup menu. When you see the window (as in Figure 8), enter the telephone number specific to your service provider and select the appropriate country from the drop-down list. In this example, we will use the phone number from the college. If all the information is correct, click on the **Connect** button. You can hear the modem dial the phone number that you just typed in. If everything is going right, you should see the following screen (Figure 9). If something goes wrong, try again. If nothing at all happens, you can always disconnect. If worse comes to worst, you can turn off your computer and dial right back in.

If you have a strict set of tasks to perform, the paragraph is not always the tool of choice. A *list* is often a superior alternative. In the following paragraph, there are instructions, but notice that the directives are blended with explanations and outcomes. In this case, a paragraph is superior to a simple list of tasks that are to be performed.

> The following figure has a single voltage supply, three resistors and one load. First, we must remove the load resistor and short the voltage supply. R_1 is now shorted. This procedure leaves two remaining resistors in parallel. How are they in parallel? If you were to use an ohmmeter to find the total circuit resistance, you would want to measure from terminal a to terminal b. Current enters the circuit at terminal a from the meter, travels through R_2 in parallel with R_3, and finishes its path at terminal b. At the top of R_3, the current also travels the other path toward R_2. R_2 and R_3 are combined into one. This single resistor is now known as the *Thevenin resistance.* Next, we restore the voltage supply and calculate the Thevenin voltage using the voltage divider rule (listed in the glossary). E_{TH} is measured with a volt meter. The current through the load resistor can be found by using the formula $I_L = E_{TH} / (R_{TH} + R_L)$. Without Thevenin's theorem, we would have to recalculate the entire circuit for every different value of R_L, which would take an enormous amount of effort and time.

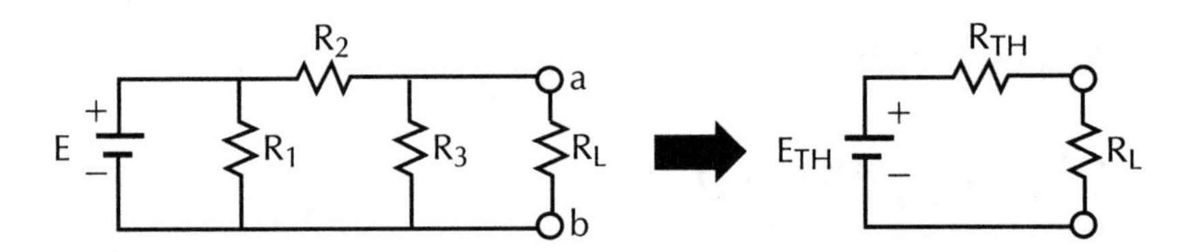

Procedures, a variation on instructions, have a similar function but are not intended as step-by-step instructions of a sort that a reader is expected to perform. Often the intent is to describe a process that involves group activities.

> Once the CSG modeling is complete, the engineer exports this information into a Cam program. Cam experts then create a computer numerically controlled (CNC) code, a software programming language in the Cam program. CNC code orders the steps required to run automated milling equipment or to unfold the solid model in order to create a flat pattern, and orders the steps required for punch presses. Sheet metal manufacturers, through Cam, now receive CNC code directly from the solid-model database and proceed with the manufacturing processes.

Although temporal progression can also run backward, most chronological applications tick forward. A paragraph structured by temporal progression is simply a description of a sequence of events.

Spatial Progression

Spatial progressions are physical descriptions and are equally as common as temporal progressions; however, in engineering and technical areas, pictures are often substituted for words. In other words, given the opportunity, you will illustrate an object rather than describe it. A paragraph that describes a physical object develops an image of space with words. Such descriptions might consist of a sequence of comments about the simple appearance of something or about the component parts of an item.

Notice that the following description of a computer menu lacks a specific focus on its parts. Since the menu is not a traditional physical object, it is analyzed in terms of its electronic functions.

> One of the most important components in the development of GUIs has been the evolution of menu systems as a means of offering command options to users. A **menu** is a list of commands that the user can issue by selection. Menus have evolved over the years from numbered or lettered lists of commands in a text environment to the drop-down type provided by the GUIs we have today. A drop-down menu system has a menu bar across the top of the screen that displays the names for types of commands available beneath each option. When you select one from the **menu bar,** the menu drops down below the command you have selected.

A much more literal spatial progression is obvious in this description of a fluid power system. In this case the parts are tangible.

> The basic fluid power system, whether it is liquid or gas, consists of six basic parts. All these parts are depicted in Figure 4. The first part is a tank or reservoir (part A), used to store the extra oil or air at a certain pressure. The second part is a pump or compressor, used to force the oil or air through the system. The pump is used specifically for hydraulic systems (part B). A power source or electric motor, the third part, is needed to drive the pump or compressor (part C). Two parts of the basic system have to do with the flow of fluid or air through the system. Piping holds the fluid or air for the length of its trip, and valves control the direction of the flow. Piping is shown as part D, and various valves are shown as part E. The last part of the system has to do with motion. Actuators are the parts that react to the pressure and compress, expand, or move in response.

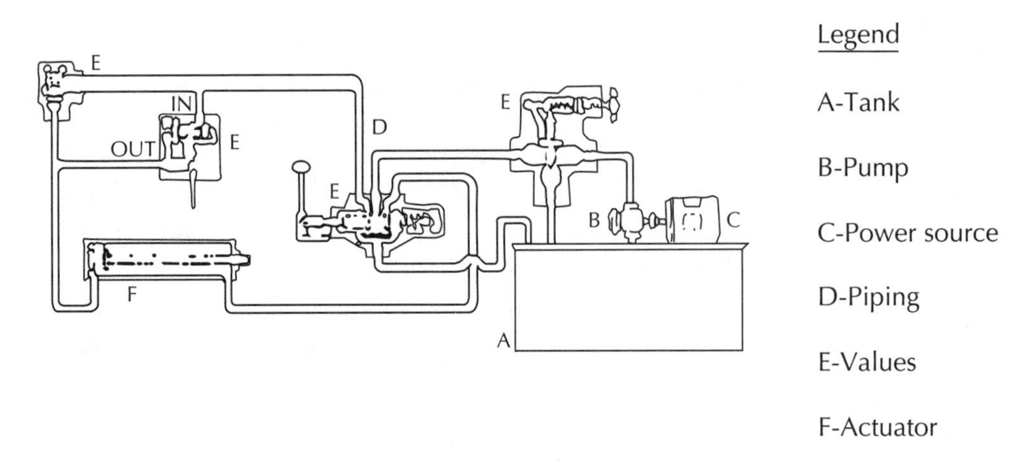

Figure 4 *Basic Fluid Power System*

Very commonly, the spatial progression of a paragraph is directly linked to a visual image of the object, as in the preceding case.

The descriptions for both temporal progressions and spatial progressions are managed in paragraphs and graphics. The writing and the graphics mutually support each other.

Spatial-Temporal Sequencing

You can look at temporal and spatial sequencing as two separate logical progressions, but in practical applications they are often brought together. Maintenance manuals are a common case in point. Maintenance manuals are one of the engines of the literature of industry. Without them everything grinds to a halt. There are surely millions of them. They are composed of instructions, and those instructions are structured in temporal sequences—possibly in paragraphs. However, manuals cannot function very clearly without the instruments under discussion—whether it is a heart pacemaker the size of a dime or a hydroelectric dynamo the size of a house. Graphics, schematics, a host of alternative visuals—and descriptive paragraphs—are used to visualize the subject. These manuals bring together the two paragraph types. The use of one or the other or both is determined by the intent of the writer: Is the issue procedural (time)? Is the issue an object (space)? Is the issue an object in motion (time and space)? Is the issue an object that is being altered (time and space)?

In the process of describing a mechanism, manuals combine the two progressions—temporal and spatial. Any device that performs tasks must be described from a mixed perspective that involves time and space:

> When you turn on a computer, the first task it performs is a self-diagnosis called a **power on self test (POST).** During the POST, the computer identifies its memory, disks,

keyboard, display system, and any other devices attached to it. Then the computer looks for an OS to **boot.** A PC looks for the OS on the primary floppy drive first; if it finds a bootable OS there, it uses the OS; otherwise, it looks on the primary hard disk.

As you know, not much happens in a computer that can be discussed in traditional terms of "actions." The exceptions, of course, are the laser systems that scan for data. Here is a brief description of a CD-ROM laser.

> The CD-ROM optical system that reads the data and converts the light impulses to electrical impulses consists of the laser, focusing lenses, a one-way mirror, and a photodetector (Figure 2). The laser light, after it goes through the nonreflective side of the mirror, is focused through the collimating and objective lenses onto the uneven surface on the underside of the disk. Then the reflected light goes back through the lenses and is reflected off the reflective side of the mirror onto the photodetector. The whole optical system is mounted on an oscillating platform that can be quickly positioned so that the laser is precisely focused on the part of the track that contains the desired data.

Causal Progression

Troubleshooting—a particularly commonplace activity in technical and engineering fields—is an appropriate way to draw this discussion to a close. *Troubleshooting* is a popular term for causal analysis. Determine the effect. Determine the cause. The procedure takes experience and a thorough knowledge of the field in question. Expertise is the key. Consider the role of troubleshooting in the U.S. space flight programs. Many NASA projects seem to go into orbit with failures of one kind or another that need attention. The best engineering never seems good enough—if only because of the sheer impact of the engines that hurl all the delicate instrumentation into space. Engineers monitor every flight. The engineers use computers and simulations—both electronic and real—to try to determine the causes and effects of each glitch that occurs. For every Hubble Space Telescope failure that you read about in the popular press there are probably dozens of lesser problems to solve.

If you look at a paragraph on troubleshooting, you will observe that the material is indeed causal in progression. The paragraph is also structured around a sequence of events affecting some object. Thus, the paragraph is also a spatial and temporal sequence, but the focus of attention is the *cause* of the event. You analyze symptoms to determine remediations. The symptoms are the facts, and the remedy is the proposed action. There is a logical argument here. You "argue" that you see the cause and the effects and that you can recommend the best solution.

The following paragraph concerns C program errors and explains what to look for (the specific problem) and the remedy for the error.

> Mistyping a close-comment sequence can cause errors that are very difficult to find.

If the comment that is not correctly closed is in the middle of a program, the compiler will simply continue to view source lines as comment text until it comes to a close comment (*/) that closes the next comment. When you begin getting error messages that make you think your compiler is not seeing part of your program, recheck your comments carefully. In the worst case, the executable statements that the compiler is viewing as comments may not affect the syntax of your program at all and the program will simply run incorrectly. Mistyping the open comment (/*) will make the compiler attempt to process the comment as a C statement, causing a syntax error. Your strategy for correcting syntax errors should take into account that one error can lead to many error messages. It is often a good idea to concentrate on correcting the errors in the declaration part of a program first, then recompile the program before you attempt to fix other errors. Many of the other error messages will disappear once the declarations are correct.

Notice that there is more to this complex causal environment than merely an analysis of data and a proposed solution. There is also a *sequence* of events. The sequence is apparent in the following sample as well. A problem is an event that precedes its solution. The order is unmistakable. The paragraph progression will be logical, but it will also be temporal.

Shut down the engine using the fuel mixture control and remove the top engine cowling. Throw a drop of water on each cylinder; if a violent sizzle is not produced, you have located the bad (cold) cylinder. Remove all the spark plugs from the engine and put the magneto switch in the position of the malfunctioning magneto—either left or right. Locate the spark plug wire leading from the bad magneto to the bad cylinder—either the top or bottom wire on the cylinder. Using a cloth, hold the end of the wire an eighth of an inch from a good ground on the engine case and have someone spin the propeller in the normal direction of rotation. If a good, healthy spark is produced, the problem is the plug. In such a case, a good cleaning or replacement will take care of it. If no spark appears, the problem is in either the magneto or the wire. Switching either the magneto or the wiring harness with its counterpart on the opposite side of the engine and repeating the test will isolate the problem.

You can clearly see a chronological structure in both of the preceding samples. Chronology is an inevitable feature of an if-then analysis of any kind because causal logic is sequential. Further, a causal analysis will usually discuss some physical object, whether it is a cancer cell or a microfracture on a beam or a faulty microprocessor circuit the size of a carpet mite. This means that the paragraph progression will be logical (causal) *and* temporal *and* spatial. Troubleshooting, or causal analysis in general, demonstrates the way in which our most popular methods of ordering our technical worlds can be combined to serve our needs. The five paragraph progressions can easily intertwine.

Paragraph Proportion and Scale

You can see the logic structure at work in the sample paragraphs. However, since many people have difficulty writing paragraphs, you might look at a paragraph from another perspective. Many writers seize up and cannot get much down on paper. One common cause is inadequate evidence for discussion. The usual cure is thought to be a stack of resources, but there is another way to attack the problem.

A paragraph is popularly defined as a group of sentences that relate to a topic of some sort, but that description tells us what a paragraph *is* and not what it *does*. A paragraph does the work of *describing* or *proving,* which are logic functions. Thus, you cannot write a paragraph if you do not have the specific details or evidence to develop it. However, a paragraph is not just a logical argument; it is usually an amplification or *magnification* of evidence.

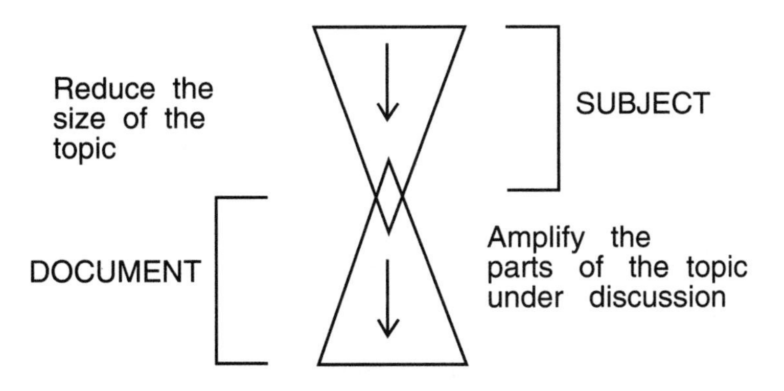
AMPLIFY THE ISSUE IN THE DOCUMENT
Reduce the size of the topic
SUBJECT
DOCUMENT
Amplify the parts of the topic under discussion

Scale the subject to fit the length of the project. The subject is just the right size if you can be thorough about it in the body paragraphs.

Any document you write is limited in its capacity to cover a subject. It is a sampling. Usually it is limited by its specific proposal and by its length. The document is "thorough" only if enough paper fully describes or proves the proposal. Problems in paragraph development will occur when the subject of the overall project is not in proportion to the writing project. Briefly dealing with a broad-based issue will lead to superficiality.

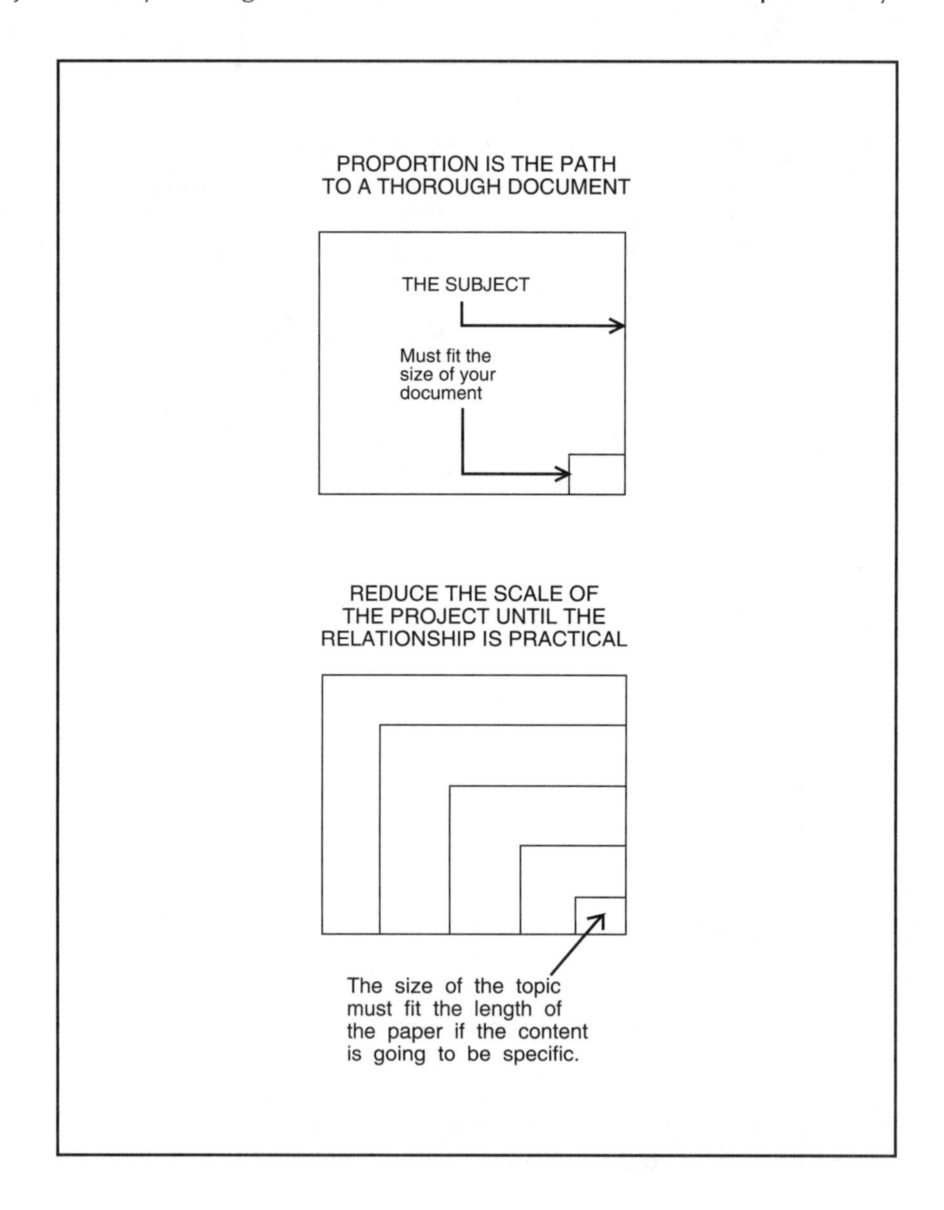
PROPORTION IS THE PATH
TO A THOROUGH DOCUMENT
THE SUBJECT
Must fit the
size of your
document
REDUCE THE SCALE OF
THE PROJECT UNTIL THE
RELATIONSHIP IS PRACTICAL
The size of the topic
must fit the length of
the paper if the content
is going to be specific.

The preceding graphic illustrates the single most devastating situation faced by many a writer. An author cannot build a paragraph properly without scale. The writer must reduce the subject of the overall document to the point at which he or she can be thorough. And at what point is it thorough? At the point at which each logic block can *magnify*—that is at the point at which details can be identified and discussed in the body paragraphs. You cannot write a proposal on nuclear disarmament in two or three pages. The subject and the scale of the paper must be proportional. If a writer is frustrated and thinks all his or her ideas have dried up, the first place I look for the cause of the problem is in the scale of the topic. It is usually too big. *Specific ideas do not come to mind if the topic is too large. The paragraphs cannot magnify.*

Proper scale is easy to achieve if you properly understand the use of scale. Many writers have told me they intentionally select large topics for projects so that they "will have enough to write about." The fact is that the exact opposite situation will be the result. I cannot handle nuclear proliferation in a couple of pages, but the author of the discussion of CDs and LPs did quite nicely on two sheets of paper. The project must be proportional so that the paragraphs amplify the material under scrutiny. An author achieves this end by providing evidence and discussion, but only if the subject is properly scaled so that the paragraph cores develop small, precise details of the discussion and not large generalizations.

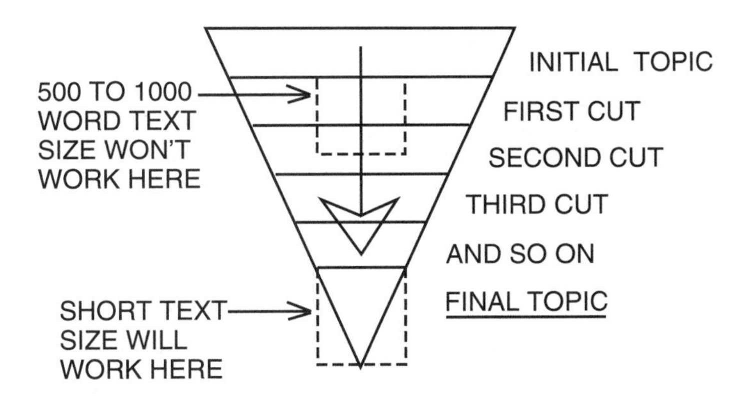

Neither engineering professors nor your coworkers are in the business of teaching people to write, so they are not likely to understand the issue of subject/project proportion the way I am presenting the problem to you. I am explaining the *cause* of a weak product, which will, in turn, result in a weak outcome. You can spot a problem document immediately if it earns a C in college. You might say that the B and A parts of the paper got left out somehow! If your supervisor says the project is "too general" or "superficial" or "generalized," you have the same result. Readers are not in the business of identifying the cure, but they will certainly see the warts. These criticisms simply translate as "something is missing." A large topic will lead to superficiality, and large issues tend to be generalized. Write just the right amount for your topic. Do not write a little about a lot; and do not write a lot about a little.

It is a good practice to overwrite and cut rather than to underwrite and add. In other words, provide as much evidence and discussion as you can. You can always edit the paragraphs to remove material. The opening chapters of this text were two to three times larger than they are now. I removed extensive and detailed discussions that I did not need. If the subject is too big, however, you will underwrite the paragraphs and find yourself scratching for specifics to develop. This is the more common problem.

PARAGRAPHS OFTEN DEVELOP IN INVERSE RELATIONSHIP TO TOPIC SIZE

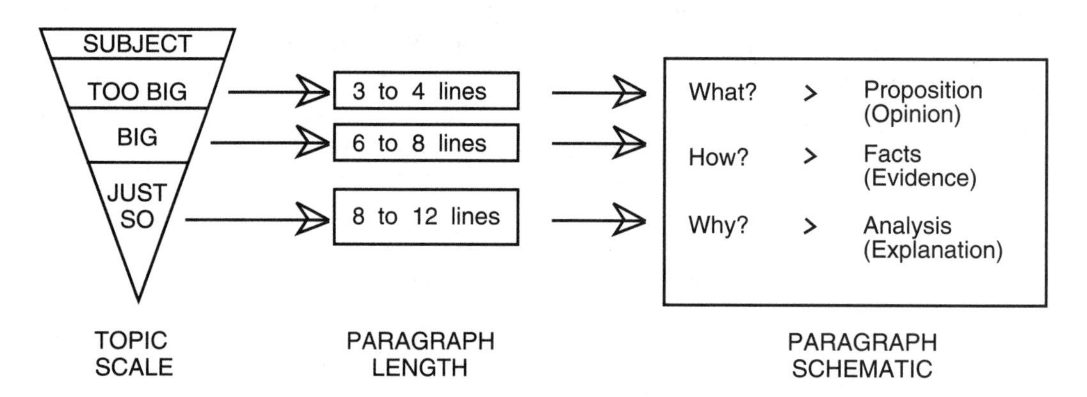

If the document you write is criticized as "vague," "superficial," "nonspecific," or "general," the paragraphs did not develop evidence or explanatory discussion. The problem is likely to be a topic that is too large for the project.

The preceding illustration is my answer to a common question: How long should the paragraph be? I usually respond dutifully: as long as it is supposed to be. This is no answer at all, of course. The concept of paragraph length is subject to interpretation. Some authors probably see the process as bean counting. Say it all. Other authorities tell us to be sympathetic with the readers and limit paragraph length on their behalf. Recently, an article argued in favor of half a typed page in double-spaced text as the ideal length on the grounds that people favor this length statistically—and for no apparent reason. I catch myself falling into this last group, but the chart indicates the real reason well-designed paragraphs frequently run in the 8– to 12–typed lines range. This length seems to be a comfortable volume for introducing the evidence and developing a discussion of the main point of the paragraph. In other words, you can often magnify an issue and get the job done in that range. Short paragraphs, to the contrary, simply do not say much.

Another sample from the classroom illustrates this point. Don wanted to do a project on computers. You do not develop much on that topic unless you have a few thousand pages available. He narrowed the topic down to desktop word processors. That still seemed too big, so he focused on a comparison of a few interesting issues concerning two popular models: the IBM and the Mac. At this level of reduction, the scale started to balance, and he set to work. Each paragraph was an amplification of some key facet of the discussion. Here you see a brief discussion of the popularity of the Mac.

> The Mac "point-at-what-you-want" system is very popular among new computer users. These first-time buyers generally do not purchase a specific computer because they need to be compatible with systems at work or school. They want a "helper," a computer that will help organize, arrange and edit writing, accounts, and visual aids. The Mac offers this help with a minimum amount of effort on the part of the user. The Mac is fast and well supported by software programs and peripheral manufacturers. Many Macs use a 64-bit processor chip that can handle information as fast as the comparable PC's 32-bit chip. The Mac disk drives are state-of-the-art and can hold more than twice the data per disk of a comparable PC disk. Some of the most powerful and sophisticated software in the industry is available to the Mac user. Leaders in software developed these programs, indicating the industry support that Mac can offer the Mac owner.

You can see the concept of magnification at work. The author takes a close look at the specifics and discusses the evidence. The subject is sufficiently small for the paragraph unit to get the job done.

Samples 10.A

1

The first step in decimal-to-binary encoder operation is selection of a numeric value. The ten push-button switches have been assigned labels, in this case decimal-numeric values. When we use a calculator, we are asking the processor to keep track of a specific value and to convert that information into binary form. The ten keys of this schematic are used to switch the supply voltage of 5 volts along a specific path toward electrical ground.

2

Then, the designers placed the parts inside in separate sections to provide easy replacement. The laser is mounted horizontally across the bottom, power supplies are placed on the side, and the digital circuits are positioned flat along the base. The optical mirrors and lenses are mounted securely in the center so sudden jolts will not disrupt any of the settings.

3

Fuel subsequently flows around the annulus to a connecting passage, which delivers it through the metering valve and charging ring to the mechanical advance.

4

Once all settings have been properly assigned, the power switches to both the modem and CRT can be turned to the "on" positions. To activate the process of "dialing" to Site B from Site A, the software documentation running the communications program must be consulted for specific syntax.

Transitions

Good writing is fast reading. If writing is an obstacle, it defeats its purpose. I frequently use the expression "take the reader by the hand." Guide the readers. Give them a map, by developing an outline in the introduction to your document to guide them through the text. Once you are motoring through the body of the document, you should then help your reader by using another convenient device: the transition. Writers are usually taught to look at a paragraph block as a convenient unit that will help the reader by shaping and limiting concepts inside the block. The blocks also define a little breathing space between the paragraphs. The breathing space marks an unstated transition to the next issue. At every one of these intersections, the reader must recall your map or else depend on you for help. Transitions are connections that help the reader bridge the issues.

The simplest method for developing transitions is to open a new paragraph with connectors that link the logic of the new paragraph to the logic of the former paragraph. Notice the paragraphs in Samples 10.A and 10.B. The usual or common transitions include the following:

Then,	**Moreover,**
Second,	**For example,**
A third consideration is	**However,**
Finally,	**Accordingly,**
Next,	**Nonetheless,**
Similarly,	**On the contrary,**
Thus,	**Consequently,**
On the other hand,	**In addition,**
In other words,	

Although I use all these transitions from time to time, I frequently construct a transition sentence to do a more thorough job. A sentence can more fully provide a transition than a single word or phrase. Notice the first half of the sentence I used in this paragraph: "Although I use all these transitions. . . ." It is a transition is it not? Sentence transitions, unlike simple word or phrase transitions, give readers more material to make the logic of the transition clear to them.

These brief paragraphs are unrelated, but notice that transitions will persuasively control the reader's attention. Here the samples could almost be construed as a set, if an odd one.

Samples 10.B

1

The answer to this problem came when the designers realized that the torsion springs could, in another mode of operation, also be made to serve as shock absorbers to dissipate the energy that would have gone into the bowstring. This was done by extending the ends of the arms inserted through the torsion springs (which are called heels), reinforcing them and positioning pads on the inner frame uprights so that the heels struck them before the bowstring straightened.

2

Similarly, when we read the mileage on the binary odometer, we multiply the digit that is showing by the appropriate column's power of 2 and then add these totals together. Showing all 1s, the four-column binary odometer has traveled

(8) + (4) + (2) + (1) miles

or, in decimal terms, 15 miles. Table 2 lists some decimal-to-binary equivalences.

3

After the ROM-BIOS takes care of the shift and toggle keys, it needs to check for some special key combinations such as the [CTRL]-[NUMLOCK], which makes the computer pause. **Finally,** if a key action passes through all this special handling, it means this key is an ordinary one.

You might think that most transitions are attached to the first sentence of a paragraph, and that usually is the case. The topic sentence is usually the first sentence, also. Obviously, the two functions—transition and new topic—are commonly combined to open a paragraph. All you have to do is take a topic sentence, let's say

> NASA budgets should not be cut,

and add a transition to it so that it will help the reader move along:

> However, NASA budgets should not be cut.

Notice that this transition could just as well appear at the *end* of a paragraph rather than at the beginning of a new one.

Take the readers by the hand. Help them follow along. Use transitions as frequently as you think they are necessary. The need is not always easy to determine, however, because you as a writer *always* knows what the plan is. You have what is called the "omniscient viewpoint." You know what is going to happen next and what will happen after that. This awareness can make transition decisions a little difficult. Any reader can tell you where a transition should go if you have the opportunity to ask someone to read the document.

Three types of transitions are not very effective. First, avoid the everyday connectors *and, or, but, nor* to start sentences, much less paragraphs! The second taboo is much more serious. The novice will frequently open a paragraph with a construction such as this one:

> This is why the NASA budget should not be cut.

Avoid opening a paragraph with words such as *this, that, these, those,* and *it.* The problem with such words is the issue of omniscience I just mentioned. You know what you mean when you say "hand me that," but nobody else understands. A moment ago I said, "This awareness can make. . . ." Notice that I put the word *awareness* in the sentence to explain what I meant by the word *this.* Nevertheless, for the first sentence of a paragraph I usually avoid these vague words altogether unless I am certain the phrasing is specific. I do not want to confuse the reader.

The amateur may also open the very *first* paragraph with the word *this.* Readers may get a first sentence that reads, "This is a major cause of pollution in the Midwest" This? Readers do not know what "this" is; it does not work. Here is a real power failure for readers, and it is only the first paragraph, in fact the first sentence. After some head scratching they realize that the writer was referring to the title with the word *this.* Perhaps the title was "Pesticide Abuse," for example. The sentence was supposed to mean that pesticides are a

These additional sample transitions demonstrate popular ways to move the reader along. For example, problems lead to solutions and sequences involve events, some of which occur before others.

Sample 10.C

The first, and least complex supply is the single-diode half-wave rectifier. See Figure 1. It conducts on the positive pulse of the ac and gives a 60 hertz dc pulse. A center-tapped full-wave rectifier, see Figure 2, divides the transformer secondary into two parts. The center tap is grounded, and the two remaining leads are connected to the anode side of the two diodes. As the transformer changes polarity, first the top and then the bottom diode conducts, providing a pulsating 120 hertz dc. It is an advantage to have full-wave rectification because 120 hertz pulses are much easier to filter to a constant dc than 60 hertz. A third type of supply is the full-wave bridge. See Figure 3. It also provides full-wave rectification, but because no center tap is used, the voltage output is doubled over the center-tapped full-wave rectified.

The power supply, as we have discussed it so far, is unusable except for battery charging. Most electronics require a steady dc to operate properly. Filtering helps provide steady dc. A large filter capacitor will smooth out the pulses and hold up the voltage peaks. See Figure 4. The filter capacitor is usually very large and will hold enough current to supply the load until the next pulse comes from the supply. There are many variations of capacitor filtering, such as multiple capacitors in combination with inductors and resistors, but they operate in much the same way.

After the pulsating dc is smoothed by the filter, it is generally necessary to regulate the voltage to a specific level. Two ways discussed here are to use a shunt-zener regulator and a series-pass regulator. The zener regulator (Figure 5) uses the characteristic of the zener diode to operate in its reverse-breakdown region, which maintains a constant voltage drop across the zener.

Regardless of the load (within reasonable tolerances), the rated value of the zener will control the voltage across the load. In the series-pass regulator a transistor is used in series with the current flow. See Figure 6. To control current levels, and thus voltage to the load, a zener is used to regulate voltage on the base of the pass transistor, and since the emitter is 0.7 volt below the voltage of the base, the load is regulated with the zener as the reference.

major cause of pollution in the Midwest. It is too confusing to try to make a reader link a title with a sentence of your text, particularly with words such as *this* or *that,* which are inherently confusing. Use the traditional outline introduction and always avoid transitions that refer to the title or subtitles of the text.

You can see the enormous variety of transition tools available by looking at the preceding brief samples. Sample 10.C shows one author working on the text continuity over the length of several paragraphs. Here, the transitions are very systematic, and they help the reader see the coherence of the component paragraphs.

As you may have noticed, most transitions are brief; they do the job fairly inconspicuously. In other instances transitions are more generous. At times a text calls for a major shift of attention from one large section of discussion to another. At other times there are many points under discussion, and the writer occasionally has to exert some control over the reader so that the reader does not become lost in the text. In these larger situations a writer needs larger transitions, which can take the form of an entire paragraph or a brief outline.

Transitions should be designed in proportion to the text. If you shift from one paragraph to another, a word or two can be a perfectly adequate signal for the transition (or perhaps you need no transition at all). When you need a somewhat larger transition between a page or two of text and new upcoming pages, then you may need a sentence or more to bridge the two sections. In a longer document you may need to develop the transition into an entire paragraph of explanation of where you are headed, or design a brief outline of the subsequent discussion. A large transition of this type is quite common and functions as a handy organization tool for both the reader and the author. In Sample 10.D the author develops a paragraph transition consisting of a brief outline of the upcoming subject areas of the project.

In sum, short transitions mark the intersections of paragraphs. Sentence transitions bridge larger sections of text as long as the discussion is not too complicated. Finally, paragraph or outline transitions join more complex and generally longer discussions.

The boldface highlights indicate the paragraph transitions in the sample on the left. The first sentence of a paragraph is the most convenient location for transitions.

Sample 10.D

7

At some point in the heating cycle, the internal pressure of the container will exceed the strength of the container. **The seriousness of that failure will largely be dependent on**

1. **the material in the container;**
2. **the volume of material;**
3. **the mass of the container;**
4. **the pressure at which container failure occurs.**

Let's consider each of the problems.

1. The material in the container.

If the material is water, then the result will be a steam and hot water explosion, with the possibility that the container will fragment into schrapnel. If the material is a poisonous liquid insecticide or a liquefied petroleum gas (LPG), the problem escalates significantly.

2. The volume of the material.

Are we referring to an aerosol can of Raid garden spray or a 10,000-gallon tanker truck of LPG? Incidents of a critical nature are common at almost any volume level. It is typical for the purity of the product to increase with the volume. Diluted products are most likely for the domestic market.The larger the volume, the greater the potential for toxicity.

Introductions

Regardless of the project you are developing, you will need an introduction for the document. Generally, it will take one or more paragraphs (or one or more entire pages) to explain the project. There are many types of introductions you can use, but for the workplace and for technical material, I would suggest a particularly handy system that develops quickly when you try to compose. The fastest and most useful introduction is the *outline.* Simply sketch the plan of action in a paragraph. This will provide the mapping mentioned earlier, and the reader will then know what to expect. Do not think that you are giving away the story. The technical document is not a mystery but quite the opposite. If you *do not* provide the game plan, the mystery will indeed begin immediately!

Outline Introductions

Although we examined several outline introductions as examples of analytical paragraph models, this additional sample shows the outline at work in the introductory paragraph.

> Even though a network administrator meticulously plans the installation and management of a local or wide area network, problems will inevitably occur. When problems do occur, following a structured approach to problem solving will minimize downtime and costs. The structured approach to network problem solving involves following important steps. As the initial task, identify the problem's priority. Then collect the available data. Identify possible causes. Isolate the cause and test a solution. Evaluate the results. Although the structured approach to problem solving can take time, it usually results in finding an effective solution to a network problem.

The repetition of following in the text the promises that you make in an outline introduction are valuable to both you and your readers; it helps you commit yourself to a set of guidelines, and it helps them follow your material. Because technical work is usually quite complex and often obscure, these traits easily creep into technical writing. The outline helps control the obscurity and complexity. Basically, all you need to do is turn your preliminary outline, or your table of contents, into reasonably smooth paragraphs. Omit the numbers you would use in an outline and phrase the intentions of the project in one or more easily read paragraphs. The result will be a statement of intent as well as a guideline to the text.

You can, if you wish, speak as the author while you present the introduction. Although the technique is considered a little mechanical, you can develop the paragraph by using such comments as "The paper will develop . . ." or "I will then explore. . . ." An example of this technique follows.

Notice the brief outline in the sample on the left. The outline is a transition that was inserted in the middle of a long project. It introduces subsequent discussion.

Although ASCII is the same for most computers, scan codes and control-key combinations are usually different; therefore, this paper will deal with IBM and compatible computers using the BASIC language under the Microsoft Disk Operation System (or PC-DOS). There are three major sections to the discussion. The first one, "The PC Character Set," covers ASCII proper, control codes, and extended ASCII. Its purpose is to cover "what" the characters are and "how" they do what they do. The next section, "Keyboard Operation," discusses how the keyboard works and what it does with the character set. This section also talks about how to generate and use the characters while in DOS. The last section, "BASIC's Keyboard Operation," deals strictly with the BASIC language. It covers character generation at the keyboard and deals heavily with the function keys.

Comparison Introductions

Introductions to comparison papers need care and attention because there is more material to balance. If you clearly identify the items to be compared and highlight the issues, even quite technical subjects will be clear to the reader, as in the following example.

A field effect transistor (FET) is a unipolar voltage-controlled device. Field effect transistors operate with either electron current in an n-channel device or hole current in a p-channel device. Like bipolar junction transistors (BJTs), FETs can be operated in an amplifier circuit or operated in a switching mode. Field effect transistors differ from BJTs in many areas, and FETs are superior in many aspects. Field effect transistors have a much higher input resistance. This allows FET circuits to be much simpler by not requiring bias resistors in the circuit configurations. Also, FETs are more immune to radiation and are less noisy than BJTs. FETs also have greater thermal stability and occupy less physical space than BJTs. One disadvantage of FETs compared with BJTs is that FETs have a lower transconductance. This makes BJTs better for some circuits in which a large amplification is required.

The Statement of Importance

Sometimes it is helpful to precede the outline introduction with a rationale for the overall project. You can open with a paragraph or two (or pages) that explains why the subject is important in the first place. In sum, open the project with two missions: to explain the importance of the topic (the project) and then to sketch the project so the reader can see the outline of how you intend to develop the topic.

In the following sample, notice the statement of importance. Although use of the word *important* obviously is not critical, this type of introduction can clearly show the reader the relevance of the material.

> In the electronics industry, technicians involved in troubleshooting and repair are daily confronted with the possibility of semiconductor failure. Since these devices are frequently the cause of trouble, the technician must be able to find, test, and replace a bad semiconductor. Using an ohmmeter to troubleshoot diodes, transistors, FETs, and other solid-state devices is a skill with which every technician should be familiar and is a simple, reliable method for isolating faulty devices.

Sample 10.E

NAVIGATOR AND INTERNET EXPLORER

There are many software packages available to the surfer for browsing the World Wide Web. Two of the most widely used of these packages are *Navigator* and *Internet Explorer.*

Internet Explorer is produced by Microsoft, which is currently the largest of the computer companies producing such a program. Because *Internet Explorer* is the product of such a well-recognized company, this mean that you do not have to worry about availability or future upgrades to the program. Also, since your operating system is also likely to be a Microsoft product, there should be few or no compatibility problems when setting up the program. This convenience will be discussed in detail below.

Navigator was created by Netscape Communications Corporation. Although Netscape is not as prominent a company as Microsoft, the Netscape package has set the standards for all Web browsers. This is possibly the most widely used of the browsers, mainly because it was one of the first and it is the most versatile. Netscape focuses mainly on communications software, whereas Microsoft creates a variety of hardware as well as software. Netscape is also the most widely recognized company throughout the information superhighway field.

For most people, the deciding factor for selecting a program is not what the software can do but how easy it is to use. Both software packages are very user friendly, but there are some minor differences. Microsoft's *Internet Explorer* is fully customizable and has many features. It supports all HTML and JAVA formats, includes nearly all the available plug-ins and includes its own e-mail reader. *Explorer* is made primarily for those who want everything in one bundle. Almost everything is very simple to configure, including resizing the toolbars.

Other Introductions

You might think about adding additional interest to the text by using one of the more imaginative openings you will often see used, usually in articles that appear in the popular press. These devices are less common in corporate or academic work. You might consider a paragraph that opens in one of the following ways:

Provides historical background

Gives preliminary definitions

States a problem

Starts with statistics

Opens with questions

Relates a striking incident

Gives an opposing viewpoint

Extends a proposal

Narrows the issue

Any of these devices (explained in more detail in *Technical Document Basics*) will create a functional and also effective start for a document. Do not, however, omit the outline introduction in most cases. Simply *add* one of these openers before the outline introduction to build interest.

If you develop a document that is primarily a comparison, you will need to identify the points of comparison early on in the presentation. The introduction then takes on the special role of explaining the importance of the comparison, and it should briefly outline or highlight the items to be compared. In sample 10.E the author uses four brief paragraphs to identify the issue, explain the two software applications under discussion, and briefly mention the criteria for his evaluation. All four paragraphs are part of the introduction.

Although I imply that introductions are one paragraph long, there are many opportunities for developing longer opening commentaries. Comparisons tend to increase the length of a document and may call for longer introductions.

Summary

- Prepare to develop draft copy for your document by focusing on the structural diagrams of Chapters 7 and 8.
- Compose the rough draft and focus on producing a quantity of material.
- Revise subsequent drafts by focusing on the paragraph logic and detailing.
- Check each paragraph to see whether it begins with a topic sentence.
- Generally, use one of the five basic patterns—analytical, temporal, spatial, temporal-spatial, or causal—to structure the internal order of the paragraph logic.
- Check to see that the pattern you used is orderly and clear.
- Use a concluding sentence for important or long paragraphs.

Activities Chapter 10

At your instructor's request, develop a brief analysis of your findings for the following projects. Attach appropriate samples and submit the report as a memo.

Write a paragraph summary (an abstract) of each paper you have written for your instructor to date. Organize the paragraph summaries by following the topic sentences of each paragraph in the paper.

Revise the introductions to the papers you have developed to date. Develop outline introductions for them if the existing introductions do not provide the reader with an overview of the project; then add a statement of importance as a new opening paragraph to one of the projects. Submit the revised first page of each text as a project.

From your first two or three essay projects, see if you can identify the primary logic structures of the paragraphs. Try to find at least one of each of the following types (see p. 247):

- *temporal (chronological)*
- *spatial (physical description)*
- *temporal-spatial (time and space)*
- *causal (cause-and-effect analysis)*

Using new subjects, develop sample paragraphs that use the pattern logic of the four types of paragraphs identified in the preceding exercise.

Using your first two or three papers, examine the projects for scale. Were the topics "too big" or were they appropriate in terms of the length of the essay projects? Were the paragraphs thorough? Did each body paragraph sytematically present a proposition, develop evidence, and provide an explanation? Were any of the paragraphs cut short?

Copy a feature article from a magazine of your choice, preferably a technical magazine. Use a highlighter to highlight key-word transitions, phrase transitions, and sentence transitions.

Share a Project

Work with two other members of the class to practice and discuss the effectiveness of introductions. Each member will need two recent technical projects from other courses to use as project models. An upcoming writing project will also work quite well. Using these projects, complete the following:

- *Develop* outline introductions *for both projects. It will be helpful if you have three extra copies of your introductions so that each member (and the instructor) has a copy of the exercise.*
- *Develop* a statement of importance *to use as an introduction for each of the same projects and place each statement above the matching outline introduction. Show the introductions to the team members and discuss.*
- *Try your hand at the* alternative introductions *if you wish.*

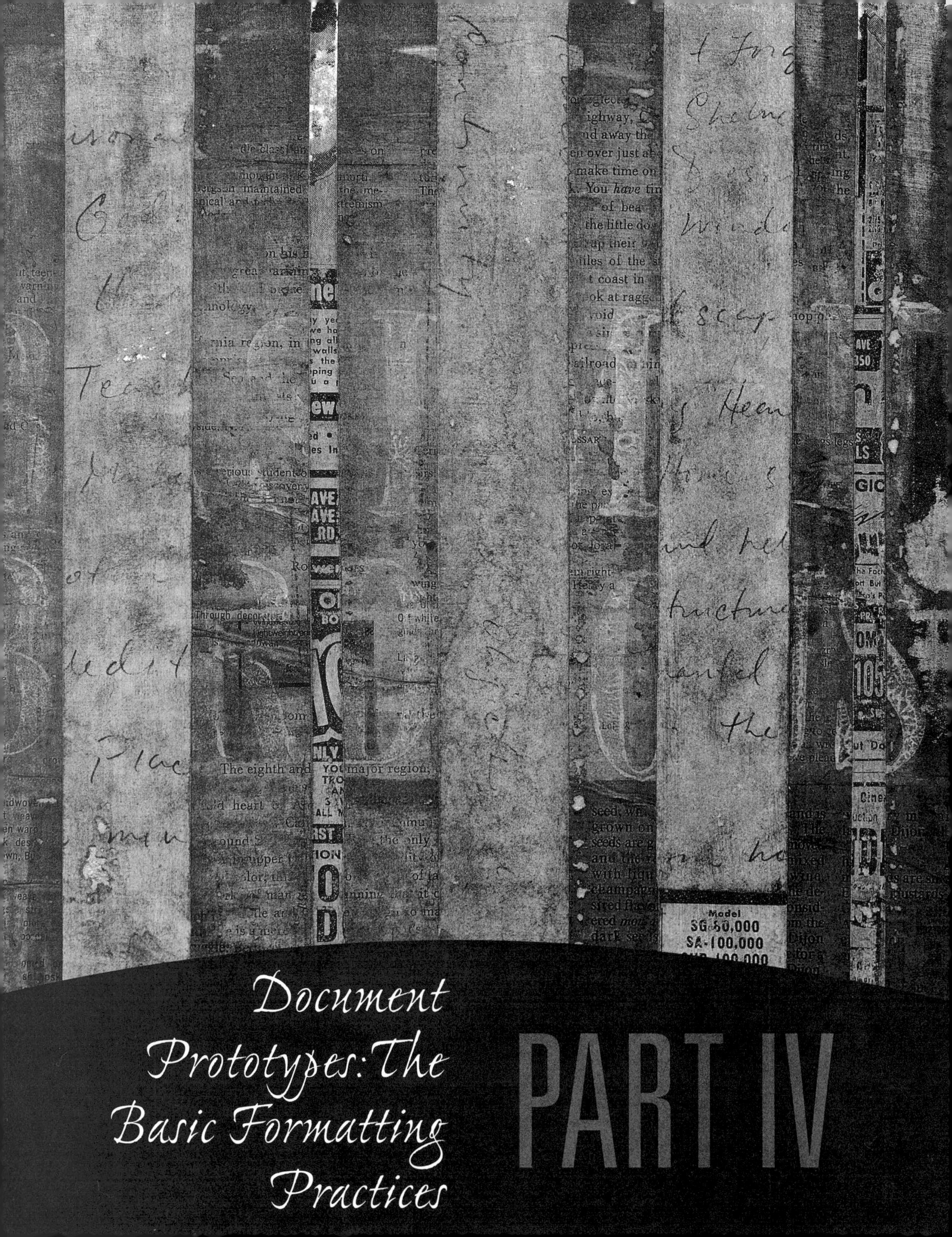
Document Prototypes: The Basic Formatting Practices
PART IV

Work in Progress

Organization is the Secret

When I completed the hearing safety program, I returned to the issue of respiratory safety. This document, with all of its sections, took me the longest. Nonetheless, it was the easiest because I didn't have to scale down the information—I just divided it up into pieces.

Part three consisted of worker activities and respirator use. This was the area that divided individual workers into categories based on the type of work they did and for what period of time. The following is an excerpt from part three of the text.

Respirator User categories—the degree of training, method of issue, and maintenance of equipment may be directly affected by how often respirators need to be used. There are two respirator user categories: occasional and routine.

- **Occasional** use defines employees who may be required to use respirators infrequently or at unexpected intervals. Due to infrequent use, planning ahead is the key.
- **Routine** use is considered a situation in which respirators are worn as standard personal protective equipment as in operations such as regularly performed spray painting and sanding. Regular employee exposure monitoring and work area surveillance are important in routine use areas to ensure the continued effectiveness of selected respirators.

The fourth part described the Approved Respiratory devices allowed by OSHA, and their maintenance and care.

Respirators will be from among those approved as being acceptable for protection by the National Institute for Occupational Safety and Health (NIOSH) under the provisions of 30 CFR Part 11 or 42 CFR Part 84.
Respirator care and maintenance will be conducted to ensure the availability of a clean and properly functioning respirator. Respirators may be cleaned, repaired, and stored by the individual wearer or by a designated individual. All reusable respirators require routine inspection and maintenance before each use. Typically, maintenance will include: 1) washing, sanitizing, rinsing, and drying; 2) inspection for defects; and 3) proper storage.

Recognizing that a refresher course may be needed, Karen, president of the company, considered sending the supervisors to an OSHA training class geared specifically to respiratory protection so they would have up-to-date information to guide their subordinates. The courses were eventually taken by Matthew and Woodrow. They reported their impressions in memos that Karen requested. The memos were addressed to me. Since Matthew was to be the program administrator, he felt quite motivated after the OSHA class and he finally believed that it was especially important to protect his employees from the potential respiratory hazards that are created at PAT. I was glad that others were finally seeing the value of the OSHA guidelines. I no longer felt that the project was mine alone.

L.C.L.

Standard Document Types

Most of the documents you are likely to develop will be presented in one of five basic formats:

Formal project submissions

Memoranda

Business correspondence

Laboratory test reports

Proposals

You have looked at formal project submissions, in the examples of papers and essays. The memorandum—or memo—is the first choice for internal communication between employees in a company setting. E-mail is conveniently formatted to reflect the memo tradition because of the convenience of sending memos electronically.

All five document types are presented in standard formats. Each corporation, in turn, establishes a set of particular practices for its own documentation. These corporate standards may be minor matters concerning the typing conventions practiced by a small secretarial staff, or entire manuals of documentation procedures, such as the one used by the John Fluke Corporation, a manufacturer of electronic equipment.

The general standards for documentation formatting reflect traditional office practices in business and industry. It is usually efficient and convenient to use these standard layouts. As a result, you should not encounter a very dramatic difference between the samples discussed here and the documents you prepare in a company setting.

Your task is to duplicate the formats of the models. The rules of the road are quite precise but quite simple. You have seen all these layouts many times, but sometimes confusion sets in when it is your turn to use them. The task is not difficult.

You will see several samples of the memo, the business letter, the formal laboratory report, and bids and proposals. The samples reflect a variety of the daily applications of these documents. Additional models from engineering-related fields will serve to show both the variety and the continuity in the document types.

Consistent with the overall policy of the *Wordworks* series, most of the models were composed by men and women engaged in college-level engineering and engineering technical programs. The mixture of documents will include some that were composed for entrepreneurial or small business practices, a number that were written for corporate settings, and still others that were projects for college programs. I see a great many documents that are developed for all three of these sectors because working students are actively pursuing their degrees while spending much of their time working in technical environments. They often have the opportunity to use the classroom to develop projects they can submit at work.

The models are usually accompanied by a running discussion structured in commentary boxes. Diagram prototypes that approximate the architecture of documents are also provided to suggest the structure that is under consideration.

The Internal Network

There is more than one information highway. For anyone working in business and industry, the most important terminal points of communication are "in-baskets" and e-mail stops. The highways that deliver information to and fro in company environs are simple and straightforward: the memorandum and the business letter. The memo, probably produced in the billions annually, is the *internal* communication highway of most businesses. The outbound business letters transact the larger communications between companies. Communication specialists could easily argue that the memo is what holds corporations together. Communication *is* the infrastructure of a company, and the memo is the medium that supports the infrastructure.

Although the memo has a highly standardized and recognizable format, it is not the traditional to-from-subject-date organization that marks the document. What is unique about the memo is its style. There is a popular expression—"quick and dirty"—that partly captures the sense of the memo. The memo gets to the point without fanfare. It is usually very direct—and often blunt. Consider this sample:

> Per Dr. Williams, Dr. Mallory PRNs will be converted to PMDs (per MD). THIS MEANS, PER MD., SCHEDULE ONLY WITH DR. MALLORY'S OK. If he asks you to schedule an appointment into a PMD slot, you MUST note (on the note line) "per MD" AND the day he authorized this—OR—if a patient schedules a return appt at the front desk and Dr. Mallory offers a PMD by way of the green or blue worksheet, the note should state "per worksheet" plus your initials.

This message is perhaps too cryptic for someone outside of this office, but it serves the purpose of logging practices and procedures, a typical function for memos. It certainly gets right to the point.

Employees write memos because a substantial volume of their transactions must have a history. Although the memo serves their communication needs, much of the information is primarily for the record. For example, businesses track their total input and output for accounting purposes. This is an obvious practice that calls for historical documentation. By the same logic, businesses use the memo to establish employee practices and procedures or to improve them or to respond to new external forces such as laws and regulations.

At times, posted announcements are handled in memo format by specific organizations such as unions. You will notice that new federal, state, and local regulations are always posted in writing. The format is often in the memo style. In many industries these memos are simply tacked on bulletin boards. In other words, any changes that are mandated by regulatory groups are usually widely distributed for employees—by memo. Such matters as OSHA guidelines are subject to review, and the best possible compliance depends on the best possible awareness of personnel. For this purpose, memos are clearly a safety net. They provide information or procedures and, in themselves, are seen as part of compliance with regulatory agencies.

Other internal information might be forwarded to each employee by personal memo. A typical use would be to announce new medical policy regulations, new tax deductions, or similar changes that are of concern to large numbers of employees. If the announcement merits individual attention, the memo may be made available to employees by enclosing the document with their paychecks. This is a common practice for announcing economic situations that affect employees—particularly if the employees do not have personal mailstops. The memo is always the medium of choice for distribution of information within a company.

The ways employees use memos are quite similar to the practices you see around you at work. You might write memos of a less impressive or forceful character, but the memo format is just as indispensable for you as it is for important company departments and for executives. One-to-one memo correspondence is a vital link of no less importance than a memo that is intended for mass distribution. *In fact, employees use the tool for exactly the same purposes as their supervisors: to inform, to identify procedures, and to track activities.* The simplicity of the format and its lack of formalities make the memo a particularly useful highway of communication.

E-mail is now also a regular part of daily activities at many larger companies. The internal e-mail networks are designed to preformat memos according to company specifications. The only functional differences between conventional and electronic memos are the inclusion of the time of day (to the second) and the complicated e-mail "addresses" of those identified as recipients and senders of the memo. It is important to give e-mail the same care and attention that you would give a conventional memo. An author would not want a printout to reflect careless writing. The style should also be as formal as the style of the conventional memo.

To Send Internet E-mail
Consult the help desk at your company for protocols and directory information. There are a number of widely used systems (Rumba, Lotus, Pine, Outlook, and others).

The speed and convenience of e-mail deserves special attention. E-mail is as convenient as the telephone; however, unlike a phone call, e-mail is a written medium. It should *not* be thought of as a casual channel of communication. Any document that can be filed—

especially if you plan to put a transaction in writing with your name on it—must be taken quite seriously. One practice that is common is to write a memo one day and send it the next—after a revision. This procedure might be doubly valuable for e-mail, since the editing can slow down or correct any hasty behavior or hasty comments. Caution is a safety feature available to you but not to your computer. Additionally, avoid broadcasting your memos. Keep the recipient group as narrow as possible and avoid any temptation to be a publisher. Broadcasting increases readership, and there are risks that are involved as the number of readers increases. More eyes then test the documents—and with more sentiments. You cannot always predict the result.

Finally, be sure to file a hard copy (printed on paper) of conventional and e-mail memos that are important to you, such as work-related memos and documents that may explain your medical benefits, changes in pay, personal achievements, and so on. You might also save your e-mail in a file on your office computer, and you might, in addition, want to keep a floppy version at home and update it from time to time.

Samples 11

A.

MEMORANDUM

TO: All computer operators
FROM: Doris Ritter DR
DATE: November 30, 200X
SUBJECT: Norton Utilities

On 11-24-99 I received an updated Version program for the Norton Utilities program that is installed on all our computers. Please update all your files and discard the outdated Version 3.0 disks. Although we did not experience any of the problems that caused them to update this program, I have installed the new version on my computer and found it to operate properly. If you have any problems installing this update, please let me know.

B.

DEPARTMENT OF TRANSPORTATION
UNITED STATES COAST GUARD

Commanding Officer
USCGC POLAR BEAR
(WAGB 11)
FPO Seattle 9879
1520
24 October 200X

From: CWO2 (ELC) R. G. Smith 034 39 4114. USCG
To: Commandant (G-PTF-1/TP42)
Via: (1) Commanding Officer, USCGC POLAR BEAR (WAGB 10)
(2) Commander, Thirteenth Coast Guard District (P)
Subj. Specialized Training in Advanced Electronics Technology

Ref: (a) COMDT (G-FTF-1/TP42) letr 1520.30.1 of 790CT10
(b) Article 3-F-2 CG PERMEN

(1) Reference (a) directed me to make application to a list of approved schools for the subject training. In lieu of those schools listed, in accordance with reference (b), I request permission to attend Seattle Community College in their program leading to an Associate of Applied Science Degree in Digital Computer Electronics Technology.

(2) Enclosures (1) and (2) list the courses required by the program, with enclosure (2) listing those for which they will give me credit based on my military experience. The remaining courses could be completed in six consecutive quarters including a summer session. I have contacted Oregon Institute of Technology. Registration indicated they accept the Seattle program toward their Bachelor of Science Degree. Upon registration I will meet the residency requirements for the state of Washington, which will greatly reduce the tuition costs. If this program is approved by the Coast Guard, it will mean that a TONO will not have to be issued.

R. G. Smith

The Memo

The memo is often the shortest of the documents you will use daily. It is strictly an in-house tool that is designed to be a time saver, because employees assume that there is less reason for protocol among their coworkers. The result is the typical memo that you see at the top of sample 11. *Memorandum* is the long way to say *memo*. Usually one or the other word appears on the document followed by four predictable points of information:

The intended recipient

Your name

The date

The subject

Then the document starts immediately, without the usual "Dear Jay." It ends abruptly also. I do not put a signature on a memo. Instead, I initial it at the top by my typed name.

There are three common uses for the memo that are reflected in the three types that you will see almost daily.

1) **Memos that inform** The technical updates of corporations are often posted or routed, and they usually are logged in updated filing systems that incorporate the memos.

2) **Memos that request, regulate, or direct activities** New regulations often appear in memos. Task descriptions are common, also.

3) **Memos that document events or commands** Like the minutes of a meeting, the memo is often simply a record of conversation and of decisions that were made.

As the uses indicate, the memo is the tool of choice for leaving paper trails and for recording necessary information for convenience, efficiency, and evidence.

The memo at the bottom of Sample 11 illustrates how national corporations and agencies may have to increase the information on a memo head in order to organize internal correspondence. Such a memo, like many you will see, is very formal, and can run for pages.

A brief analysis of memo samples follows (see Models 11.A, B, C).

The two sample memos on the left represent typical internal uses. Brevity is a common element in a memo but notice that memoranda can, by necessity, be quite complex.

Model 11.A

E. I. Division COMPUTING SERVICES

MEMORANDUM

TO: Jack Backstrom

FROM: Gil Lundquist, Computing Manager *GL*

SUBJECT: Network and backbone installation, per request

DATE: September 15, 200X

Three of five buildings at DCI are connected to a fiber optic backbone that connects to central via one 10 megabits/sec digital line. Administrative Computing has 280 computers, of which 75% are connected to the backbone through either the outdated UNIX network or a file server running Novell 4.xx. Please note that the UNIX network will be replaced by three Windows N.T 4.0 servers as part of the 200X capital budget.

Facilities Computing has 750 computers, of which 135 are connected to four file servers running four different versions of UNIX. Thirty of the 750 computers are connected to the corporate backbone.

I hope this is the information you are looking for. I might add that DCI went through a five-year period without a single computer upgrade. Last year they spent most of their available resources upgrading old machines to this network performance level.

1

2

2

3

4

5

An Information Memo

The memo format of Model 11.A is typical of a simplified style that is used by many businesses. The logo or company identification is usually printed at the top. The word *memorandum* usually appears at the top left or in the center. Smaller businesses simply type the word on their regular stationery. Larger companies often print a specific memo banner for in-house use.

MEMORANDUM

TO: ____________

FROM: ____________

SUBJECT: ____________

DATE: ____________

1 This particular style of address is the most common, although the organization of this material will vary somewhat from company to company. There is, however, always a minimum of four points of information: receiver, sender, subject, and date. Notice that the document is initialed (not signed) at the top. If you ask a favor you might sign below the memo, closing with a "thank you" and your signature.

2 Memos can be widely distributed. This one was not. This particular sample was addressed, in-house, from one director to another in response to a request for highly specific information. Notice that the subject notation line is blunt and direct. It also acknowledges that this memo is a response to a request.

3 The information memo is the most common of all memo forms. It has probably been the number one choice for dissemination of information for the better part of this past century. Very often there are meetings to announce the exact same material that is distributed in memos, but the memos will always go out for posting nonetheless.

4 The memo opens without ceremony and starts to provide information. This is acceptable because the document is in-house. Even a memo will usually provide a modest introductory remark, often by opening with the traditional comment you will see frequently: "In response to your request . . . ." This author omits the device because it appears in the subject notation.

5 There is no secretarial notation on the bottom. This means that the author created the document.

Model 11.B

AFG Systems Group

Thomas Quillian, President

October 9, 200X
5-3010-09-7033

To:	W. Young	D-3017
cc:	G. Lankamp	D-3013
	D. Quist	D-3064
	G. Hayes	G-7092
	K. Fuller	G-2098
	L. Doran	D-4062
	L. Larson	D-2120
	S. Tontoni	D-9931

1

Subject: RECOMMENDATION FOR STARTUP OF MULTICONDUCTOR INK-JET MARKING MACHINE

Reference: Memo 8-3010-03-4298, dated September 14, 200X, J. Porter to K. Douglas, subject: "Recommendation for use of No. 9 Ink-Jet Machine for Marking Multiconductor Wire"

2

An investigation into ink-jet marks that lacked durability in final assembly resulted in an investigation of JDT 120-41 wire and ink-jet machine marking consistency. The investigation resulted in the following recommendations for ink-jet marking of the wire:

3

1. The No. 9 ink-jet marking machine should be used for marking JDT 120-41. (Reference)

2. The viscosity and the temperature of the ink and the running time must be recorded once each day. The ink shall have a viscosity of 1.8–3.0 centipoise at a temperature of 20.5 ±1.5°C when measured with an Eaton LVG viscometer. (Responsibility—Quality Control, Doran)

4

3. The ink drop spacing must be observed by the operator—approximately every 2 hours with a microscope—and shall be machine adjusted by Maintenance as described in the T&I Technical Support Manual for the Video Jet Printer series 7200, if necessary.

5

4. All ink must be viscosity checked per D. Quist when shipped from Salem for use in the Bellevue ink-jet systems. Also, a sonic test per 9J-403211 should be performed on makeup ink shipped to Bellevue. (Responsibility—Quality Control, L. Doran)

A Memo of Directives

The sample from AFG Systems (Model 11.B) is somewhat more elaborate than the first memo. Larger companies often develop more complex memoranda. Partly, the complexity has to do with filing systems, as indicated in the file number reference under the date. Also, the documents often have a historical frame of reference indicated by the notation "Reference."

MEMORANDUM

TO: ______________

FROM: ______________

SUBJECT: ______________

DATE: ______________

1

The method of address always contains the basic four points you observed in the first memo: receiver, sender, subject, and date. Multiple distributions are common for memos. This one is handled by distributing a memo addressed to Young to the other concerned parties. Sometimes this format simply addresses all the parties under the receiver notation (To:) and omits the copy notation seen here. Notice that this memo identifies no sender notation because it is from the office of the company president.

2

The reference notation signals an ongoing dialog. This memo is a response to another memo. That document, in turn, can be a response to yet another document, and so on. The identification of one or more of the background documents helps streamline the memo style. Otherwise there would be a lot of explaining to do.

3

The first paragraph of the memo is short and to the point. Memo paragraphs are usually rather brief, but they must be thorough. Extensive details or analysis is often rendered in frequent paragraph divisions to add clarity to complex material.

4

The introductory paragraph identifies "recommendations" that are then outlined below. The first of the recommendations uses the verb *should*. The second is more assertive and shifts to the verb *shall*. Directives are often stated in both styles.

5

The remaining directives continue the pattern of *shall* and *should*. This sort of document often finds its way into a practices and procedures manual where it can be referenced as standard practice. The use of *shall* is very common in this environment and is also typically used in bids and proposals also.

Model 11.C

INFORMATION SYSTEMS INC.

NETWORK SERVICES

MEMORANDUM

TO: Bill Flores DATE: August 3, 200X

Roy Lord

FROM: Bailey Glendale *BG* THRU:Thomas Clark, Supervisor

Computing Manager

SUBJECT: Assistance and supplies available from Building Three Computing

Attached are three forms you approved at our July 13 meeting. Employees can use these forms to receive assistance with hardware and software and to order computer supplies. To order more forms call #2069. These are reproduced as revised by the E. I. committee in the minutes.

Boldfaced changes below reflect the approved revisions added to each form; the third form is totally revised.

Hardware Service Order Form: Used to request installation, troubleshooting, or repair of monitors, printers, CPUs, mice, keyboards, file servers, and related computer equipment. **For nonemergency, fill out form, fold in half, and drop in company mail.**

Software Service Order Form: Used to request installation of software or to reconfigure software for new peripherals **(printers, scanners, etc. . .),** and to troubleshoot software. For nonemergency, fill out form, fold in half, and drop in company mail.

Request for Computer Supplies: **Used to request computer supplies. Where it asks for a budget number please write "computer resources" and mail to Diane Stockwell at #2323.**

If these changes do not conform to your understanding of the committee's plan, please notify me.

BG/pb

cc: Christina Diaz

June Hecker

1

2

3

4

5

5

5

6

A Tracking Memo

Memos are frequently intended as a follow-up to a meeting. The meeting can be a supervisor's discussion with an employee or a complex situation involving, for example, teleconferencing for a committee of many people. Very typically, a committee will use memos to either institute group decisions or to announce the decision. Either instance is a method of implementing or tracking outcomes.

MEMORANDUM

TO: ____________

FROM: ____________

SUBJECT: ____________

DATE: ____________

1 The address method of this memo is commonplace but differs from the previous two samples. Notice that the document can be sent directly to more than one person. All primary recipients should be identified at the top; all other recipients can be identified at the bottom. It would be a discourtesy to distribute copies of a memo that someone thought had been sent to him or her in confidence.

2 The protocol at ISI is to identify supervisors who are aware of the substance of a project that is addressed in a memo. The "Thru" notation (by way of) is common and is particularly prevalent in the military sectors because of the rank system and its effect on chain-of-command decision making.

3 The first sentence of the memo is well designed. It provides an immediate fix for the intent of the memo. It is the outcome of a meeting. Wherever possible, use this springboard device. Mention the event and the date that has prompted your correspondence.

4 The committee was evidently updating three simple descriptions for internal services. Each is vividly identified in headings on the left, and boldface was used to indicate new phrasing.

5 Hardware services are evidently a little slow, since the form routes by way of company mail. This order form was redefined as a "nonemergency" request as a result of slow delivery speeds. Software services had not formerly explained the word *peripherals.* The committee inserted a few words of clarification. Supply requests had to be updated because the procedure changed; in addition, Diane was new and so was her phone number.

6 Bailey had his secretary complete this document, and he routed three copies. Notice that memos use all the notation devices found in business letters, such as the carbon copy line and the secretarial identification line.

Summary

- Use the memo to inform, to regulate, or to document events.
- Address only the appropriate recipients.
- Whether you use hard copy or e-mail, write the memo one day and mail it the next, if time allows.
- Use the heading format that conforms to company standards.
- Open by referring to any appropriate events or correspondence that led to your memo.
- Get to the point of the memo very promptly.
- Prioritize the contents in paragraphs that move from most important (first) to least important (last).
- Keep the paragraphs reasonable in length if possible (four to eight lines).

Activities Chapter 11

Use the chapter summary as your guideline.

Create a package of three original memos that are in some way related to your field of study or your employment. They can be authentic or fictional. Write one memo to inform. Write the second to request or direct activities. Write the third to document a recent conversation or meeting.

Look for samples of memos at work. Are there certain conventional memos used each week or each month? Do they inform? Regulate? Describe events? Gather samples in a package (of three or more) and briefly explain their use in another memo to your instructor.

When the projects are returned to you:

At your instructor's request, resubmit *the first preceding project with the corrections to the errors that are indicated on the original.* Boldface *all corrections so that the instructor can see the revisions. Use the* Writer's Handbook *as your reference.*

Share a Project: First Option

Work with three other members of the class to discuss and edit the set of memos from the first activity.

- *Consult Appendix A. Before holding the meeting, review the fourth editing checklist for memoranda.*
- *On the day the projects are due, have each member of the editing group explain his or her project in terms of objectives and project development.*
- *Hand the projects around for a critical reading and editing.*
- *Have each member edit the texts for both writing errors and technical errors.*
- *Have each member write a one-paragraph critique at the end of each project.*

Share a Project: Second Option

---

1) *As a team alternative to one of the preceding exercises, consider the following:*
   - *Work with two to four other members of the class. Look for samples of memos at work that reflect the three basic types identified in the earlier exercise:*
     - ✔ *memos that inform*
     - ✔ *memos that provide directions*
     - ✔ *memos that document conversations*

   Collect other types of memos also. Bring the samples to class for discussion of both layout styles and content.

2) *Try your hand at developing memos for your team. Complete the initial discussion and then develop a sample of each type of memo in the second half of the meeting. After you have composed the three memos, share them with the team for discussion.*

Share a Project: Third Option

Try to follow the following complicated memo transactions. If there is some confusion, that will add a touch of realism to the office practices. Work with two other members of the class to construct three memos. It will be helpful if all members of the group are from the same engineering technology program.

- *Use your campus labs as a setting and decide on a familiar technical situation that will allow you to develop all three memos with three distinct intents:*
  - ✔ *a memo to inform*
  - ✔ *a memo to provide directives*
  - ✔ *a memo to document a recent conversation or meeting.*
- *Have each member complete a set of three memos under the following conditions:*
  - ✔ *Agree to the objectives of the documents and conclude the first meeting.*
  - ✔ *Write an information memo to a team member before the second meeting.*
  - ✔ *Assign a meeting date and bring the information memos to the meeting.*

- *At the second meeting, read the memos of each author.*
- *Write a memo of directives in class (related or unrelated to the first memos) and route it to the team members. Read and discuss each others' memos.*
- *Adjourn to write the third memo, which summarizes the second meeting.*
- *Provide a set of the projects for the instructor.*

Work in Progress

Dealing with Outside Agencies

Here are more observations about my OSHA proposal. Remember, I'm a technician and not a writer, and this was a major project and learning experience for me.

Since my pattern of dividing my individual topics into several separate units was working, I was pleased. Still utilizing my goal to let Pacific Aero Tech's employees "read it, get it and go," I plunged into yet another one of my safety programs.

In the fall of 1987, OSHA extended enforcement of the Hazard Communication Standard, or "Right to Know" regulations, to businesses and operations of all sizes in the United States. If an employer has hazardous materials in any quantity, and any number of employees, these rules apply to the company. PAT certainly falls under these stipulations. I had to write a number of letters to Washington D.C. to resolve vagaries about hazard standards. I then had to split the training document into two parts.

- *Rights and responsibilities of management (which in this case included supervisors), employees and the program administrator*
- *Recognizing, and labeling hazardous materials and material safety data sheets (MSDS).*

The "Right to Know" responsibilities are also presented in poster form in a prominent break area for all employees to see. This and all safety material must be periodically updated. Some of these responsibilities include the following, which is an excerpt from the HazCom text:

Managers and supervisor shall:

- Maintain a list of hazardous chemicals used at the location as well as corresponding MSDS for each hazardous chemical used.
- Determine how MSDS sheets will be maintained for all locations.
- Ensure all identifiable exposures to chemicals are evaluated and communicated to employees.

Employees shall:

- Use materials only for their prescribed purposes and only in the manner for which they are intended.
- Not use materials until they receive proper training and understand the hazards involved.

The Program Administrator shall:

- Maintain records of employee training, employee exposure, monitoring results and illnesses related to chemical exposure.
- Upon request, coordinate training programs specific to employee needs.
- Provide hazard assessments and monitoring services on request.

OSHA dictates the material to be taught in this section of Hazard Communication, so the work was self explanatory, but sorting through is was time consuming. HazCom affects all of the employees working at PAT and became an important consideration. Although I found the web to be a convenient source of OSHA information, telephone requests definitely were ineffective. I had to address formal letters to both state and federal offices to resolve specific concerns at PAT.

L.C.L.

The Correspondence Loop

One would think that the electronic highway would have replaced the enormous volume of paper correspondence through which companies have usually done business with one another. The invention and popularity of the telephone roughly coincided with the invention and popularity of the typewriter, yet the former never outpaced the latter. As recently as the 1980s there was an enormous nationwide effort to upgrade office practices with electronic communications, but nothing much changed. From "office-of-the-future" ballyhoo to chic teleconferencing on the fiftieth floor to Web hype, the basic need for hard copy, especially in the form of a business letter, has never been challenged. Why?

The answer is as simple as the fax machine. Companies *do* upgrade and adapt to the conveniences of their electronic offices, but the written documents have not changed. Companies send them by fax or optic fiber or satellite or bicycle hot-run as long as there are documents. This single fact separates the day-to-day role of the business letter from the much less tangible world of cellular telephones. There must be a historical record, and paper is considered the most appropriate record of business transactions.

The business letter serves much the same purposes as the memo except that it assumes greater importance because it is the out-of-house link to other parts of the economy that are vital to a company's success. The business letter is equally at home in serving as a legal document that seals a small binding contract or an order for half-a-million microprocessors. Employees must record and file vast numbers of these links for obvious reasons. What is important for employees to realize is the role they have in contributing to the company's success with each modest letter they draft. Employees may not be signing million-dollar orders, but any letters they create are going to be on letterhead, and they must maintain high standards for effective and appropriate correspondence. E-mail is a likely link in well-established corporate connections, but the same care and attention is critical in e-mail correspondence.

A great deal of correspondence will address some aspect of company output. Whatever products or services are sold will be the focus of a large volume of business letters. If you follow the loop of correspondence for a product or service, the loop begins with "sales and promotions" correspondence. These marketing features include colorful literature, but there is usually a business cover letter included. Direct inquiries are then addressed through personal correspondence to respond to a potential client's interests and questions.

Letters are used to confirm the shipments or services, and then additional consideration enter the picture: delivery problems, production queries, claims against the product or service, information requests. Each of these personalized issues calls for an attentive response—one that is always addressed in formal business letters. Even if dozens of phone calls enter the loop, the business letter is the last word because it is the most acceptable documentation of business activities. At the least, shipments large and small often contain a letter of transmittal to accompany the order.

Perhaps less apparent than outputs are the vendors and services that provide inputs of raw materials for an industry. Companies obviously cannot manufacture a product if they have no raw materials. A company depends on the resource pool of agents that provision a manufacturing or engineering activity. This dimension of a company is perhaps the less colorful part of the plant, but the arrival gates determine the success of production for any business. The paperwork is no less dramatic in "receiving" than in "shipping." If you look at the business letter links at either the receiving gate or the shipping gate, you will find the same concern for courtesy and precision and protocol at both ends of the production line.

Where do the engineers fit into the overall scheme? I would position them squarely in the middle of the process. Engineers and engineering technicians may be involved in a wide

variety of activities that call for business letters because engineering work and research work are at the center of many industries. However, engineers and engineering techs are less concerned with the input services and output services than with manufacturing procedures or design engineering.

Engineering activities call for a full spectrum of writing supports. If a product is developed in-house (a company patent, for example), the history of the development is handled in memos. If the product is for an allied industry (the automobile or aviation industries contract out thousands of parts and assemblies), formal correspondence is the link, and business letters are routine. Since engineering firms are often looking at development more than at production, the engineers must use business letters to communicate with clients. Similarly, the engineers involved in the development of a site plan must continually correspond—and document their relations—with contractors, municipal groups, regulatory agencies, and the like.

You can use e-mail links in the same manner as routine correspondence, but always with the same regard for courtesy, precision, and protocol. Remember that e-mail will not project your paper image if you have carefully crafted letterhead stationery, numbered ordering forms, and other company documents. In this sense fax is an excellent tool, since it represents both electronic speed and convenience *and* hard copy in facsimile. E-mail will basically dump you into whatever indistinguishable format is allocated to the e-mail procedures of a system such as the popular Pine e-mail. Essentially, when you use e-mail, you are conducted along the computer highway in a very generic looking style.

Until a contract connection is well established, it might be a superior strategy to use your stationery and other forms designed to meet your company's specifications for working on contract relations. Fax, which transmits your exact image is perhaps a logical second step, since it offers e-mail speed for transmittal of your logo, business card, specific company forms, or signed contracts. E-mail is probably best utilized for well-established business relations where "image" is no longer critical.

The craft of business letters is not complicated. Write business letters with a polite tone at all times. Tone is half the letter. Indicate the reason for a letter promptly, usually by the end of the first paragraph, and certainly no later than by the end of the second paragraph. Triple check for accounting errors, parts numbers errors, and grammatical errors. These issues spell profits and losses.

The following sections will explain the basic characteristics of the business letter layout. Study the samples that accompany the text, and the four models. The samples and models illustrate the most common types of business letters:

- **Letter of request** **(Sample 12.B)**
- **Letter of reponse** **(Sample 12.C)**
- **Letters of adjustment** **(Models 12.A and 12.B)**
- **Letter of agreement** **(Model 12.C)**

Sample 12.A

Metropolitan Bank
Computer Services Division
401 West Chester Blvd.
Kansas City MO 40000-1234

April 12, 200X

Mr. Gordon Hastings
Installation Division
ProCount Corporation
3019 Walnut St. SW
Chicago IL 60000-1234

Subject: Conversion Compatibility/
Metropolitan Bank Inc.

Dear Mr. Hastings:

During the first phase of the new ProCount network accounting system installation, our technicians and programmers have indicated that the conversion will definitely prove to be a cost-effective tool. Of course, as you will recall from our telephone conversation of last Thursday, April 8, they have run into specific problems within the network.

Most of the glitches are in payroll where we cannot afford error. Also, some departments are having difficulty because they cannot access the complete network.

I suggested to the board that we postpone the complete conversion until the incompatibilities are resolved. I assured them that ProCount is making every effort to complete an outstanding system for Metropolitan, and the board members want to assure you of their cooperation.

Enclosed are the hard-copy bugs in the files we discussed. Please call me as soon as your analysis of the problem is complete.

Yours truly,

Lloyd McCarthy
Services Analyst Director

LM/cf

cc: John Sloan, Board Director

Enclosures: Accounting file spreads
Program access bugs (6)

HEADING

(Double-space)
Date
(Double-space or more)
INSIDE ADDRESS

• *subject insert*

(Double-space)
SALUTATION
(Double-space)
BODY OF LETTER

(Double-space between paragraphs)

COMPLIMENTARY CLOSING
(Adequate space)
SIGNATURE

(Double-space or more)
• *Identification insert*
(Double-space)
• *Carbon copy insert*
(Double-space)
• *Enclosure insert*

Business Letter Practices

The conventions of business letter practices are not likely to be new to you. Regardless of, your company setting, the standards are similar nationwide. The letter in Sample 12.A illustrates a typical use for a business letter. On the right edge of the letter I have indicated three concerns you need to review:

> **FIRST, THE PARTS ARE IDENTIFIED IN CAPITAL LETTERS.**
>
> (Second, the usual spacing between the parts is indicated in parentheses.)
>
> *Third, four convenient inserts are indicated in italics.*

The spacing of a business letter is going to vary depending on the length of the letter. The margins will typically be one inch or more on all sides. A short letter is usually "stretched" to fill the page by adding another half inch or more to all the margins. The individual sections are single-spaced, and double spacing is used between all the sections. A few areas at the top and bottom can use wider spacing as needed to fill the page. One of the marvels of skilled secretaries is the speed and skill with which they can make any size of letter—short or long—fill the sheet of paper. The letter is not supposed to sit high on the page with a blank bottom half.

The left-hand or right-hand placement of the parts of the letter will be a convention established by your company. I prefer to place the heading and signature on the right (see Sample 12.C). A more "legal" look is attained by placing all the parts of the letter against the left margin. The business letter consists of six major sections, which we will look at in turn.

Heading

If you are using company letterhead stationery, there is often no heading because a heading consists of the return address (yours), followed by the date. Since the address is part of

In the sample on the left, note that the heading (the address) does not include the name of the author. The date is often aligned with the heading.

Sample 12.B

City of Auburn
4221 Renton Blvd.
Auburn WA 90000

May 20, 200X

Mr. Loren Jump
Sales Representative
L. N. Curtis and Sons
326 S. Industrial Way
Seattle WA 90000-1234

Dear Mr. Jump:

Our department is currently investigating the purchase of up to ten S.L 20 Mobile Chargers to install on our two main engines.

I would appreciate all the information you can provide concerning the items listed below:

1. Amperage drawn while charging
2. Types of connections
3. Cost for each charger
4. Types of mounting brackets
5. Availability of chargers
6. If ordered, approximate delivery date

If you can provide me with this information within the next few days, and if it is not cost prohibitive, I will be able to submit this to the budget committee by the end of June.

Thank you for your consideration and assistance.

Yours truly,

Dwight Sawrey
Auburn Engineering Department

DS:kel

cc: James Fisk, Supervisor
James Gintz, Assistant Supervisor

the letterhead, your company may use only the date and perhaps a routing stop number or department name. (It is a strange national inefficiency that businesses use their internal mail-stop numbers on memos and *not* on out-of-office correspondence.) Be sure the date is correct, by which I do not mean today's date, but the date that fits the logic of the document. As recently as last year, I had a trademark application held up in Washington for two months because a dated letter did not "match" the date of an application I sent the Department of Commerce.

Inside Address

This is the full address of the person you are writing to in the letter. Be sure to use *Mr.* or *Mrs.* or *Ms.* to be polite. Ms. is now quite popular and solves the problem of having to determine whether a woman is single or married. If you know the job title and the department, these items of information help route the mail (since you will not usually have mail stops, as I mentioned). Use the full address, including the U.S. Post Office two-letter state abbreviation and the nine-digit zip code, if you know it (see Sample 12.B).

Salutations

Always use the word *Dear*, even if you think it seems wimpy. It will seem worse without it. It will seem rude. After the name put a colon (use a comma in personal letters to friends). If you do not have the name of a specific person to address and you judge the letter to be highly important, make a phone call to establish direct contacts and get the name you need. If you are satisfied with a generic reference—to a sales division, for example—then there are new rules. Do *not* say "To Whom It May Concern," "Dear Sir," or "Gentlemen." The sales team may be 55% female, which is more or less the current makeup of today's labor force. If you use a generic salutation, simply write, "Dear Members of the Sales Division." If the result seems too clumsy, omit the salutation style and use the Attention insert:

Attention: Sales Division

The attention insert is very convenient at times for replacing salutations, but observe that it is cold and blunt. For purposes of illustration, I created an attention insert as a substitute salutation in Sample 12.D, but this style is uncommon.

The inside address is important for your record keeping. It provides the name and address of the person you are writing to. This important information will be available to you whenever you need to review the correspondence.

Sample 12.C

ComputerFix Inc

Dodge Centre, Suite 1230
420 North 5th Street
St. Paul MN 50000-1234
(612)371-0000
FAX (612) 371-0000

July 10, 200X
RMA# 90-910603-1D

SHIP TO:

Ralph Mitchell
Instructional Computing
Educational Software Incorporated
9600 Meridian Way North
Seattle WA 90000-1234

Dear Ralph:

We received your fax and Susan transcribed your phone message. The problems you were having with the Mac Rescue were due to substandard 74F263 multiplexer chips on the Mac Rescue board. After replacing these chips, the unit worked perfectly.

I also replaced the MRC9 chip on the Rescue with the most recent revision of the PAL chip. This brings your Mac Rescue up to our current standard. The board will now work with RamDisk+++. I have left the board installed for your convenience.

The Abaton QuickStep printer was worked on twice by an Avery-Dennison technician who was not able to repair the problem. Upon his recommendation, the unit was sent to Everex (the manufacturer) on their RMA number AO 3261-092, May 27, 200X.

A new drum assembly was installed upon the Avery-Dennison technician's recommendation. Note that this did not fix the problem. I'll let you know as soon as we hear from Everex.

As always, if you have any problems or questions, do not hesitate to call. Thanks again.

Sincerely,

John Sand
Design Engineer

Enclosures: 1 pc.—XXXKE Mac Motherboard
1 pc.—Mac Rescue board
4 pcs.—64 Meg SIMMs

COMPUTER PRODUCTS AND PERIPHERALS

The Body

The body is the content of the letter, which may be as short as one paragraph or run to several pages. For clarity, the paragraphs often are on the short side, perhaps as few as four to six lines. The longer paragraphs that you would use in engineering reports are more likely to conform to the what-how-why format of an analytical project. Correspondence is always informational, but might not be analytical. The paragraphs are shorter when there is no need for detailed discussion. Letter paragraphs are often abbreviated to explain only *what* and *how,* and many letters explain *only* what (directives or information). The letter in Sample 12.C is a typical example of brevity.

The opening of a letter is often handled with a historical introduction of some sort to provide a setting for the document to the reader. This is the reason so many letters open with the comment "Regarding your letter of May 10 . . . ." This opening technique provides the historical background in very few words. If the letter is going to discuss a problem, the historical introduction provides the background for the problem (see Sample 12.C).

People who seldom write will often write *only* if they are angry about some matter. If you are going to discuss a problem, do not allow a letter to reflect a bad temper. Purge any anger from the document. In fact, one national authority on writing suggests that authors should always use a strategy he calls the "good news–bad news" approach. Open with a kind word about a business, product, agency, or whatever. Show some support and some appreciation. Then say, "however," and explain the bad news. This device is well known as a way to massage services.

Complimentary Closing

As a courtesy, always close with "Yours truly," "Cordially," or a similar phrase. Capitalize only the first word and use a comma after the phrase. Sign your name below the closing.

Signature

Sign the original document and forward it. If you sign a laser copy, that is fine if it is printed on the letterhead paper stock. It was traditionally a taboo to send off the carbon and keep the original. Some companies are very strict about "originals," but in the age of laser copies it hardly matters. *Always* type your full name below the signature so people do not have to read your writing. This practice may seem odd to you, but many people do not use a legible signature, and you would be glad to see the name typed.

I end this section with a personal tip. Needless to say, all the significant documents of your company are typed. Any legal or otherwise important matters in your personal life that call for correspondence should *always* be typed as well. If you fight your automobile insurance claim with lined paper and a blue ballpoint, you will not present a very strong image. Writing is leverage. A properly presented document helps show that you are in command.

The body of the typical business letter is brief and is often composed of brief paragraphs. Note that this author used several sentences for each paragraph to keep the complex information orderly and readable.

Sample 12.D

City of Clairton

Fire Department
1010 Miller Ave.
Clairton WA 90000-1234

May 24, 200X

PERSONAL

Insurance Division
DMF Corporation
15110 N.E. 110 St.
Renton WA 90000-1234

Subject: Code Conformance

Attention: Hazardous Materials Committee

Enclosed are copies of the Uniform Fire Code pertaining to storage of flammable liquids. Also included are copies of past inspections by our department and a copy of your hazardous materials permit as well.

Please refer to your copy of the first inspection made on May 10, 200X, concerning code 7914. Your company is required to meet the requirements of the Uniform Fire Code, Section 79-403 as stated on page 233.

Your prompt attention to this matter is appreciated.

Yours for safety through fire prevention.

Cordially,

Harold Philips
Lieutenant

HP:DL

cc: Herb Martin, Captain

Enclosures: 5

HEADING

(Double-space)
Date
(Double-space or more)

• Personal insert

INSIDE ADDRESS

(Double-space or more)
• Subject insert
(Double-space)
• Attention insert
(Double-space)
SALUTATION (here omitted)
(Double-space)

BODY OF LETTER

(Double-space between paragraphs)

COMPLIMENTARY CLOSING

(Adequate space)

SIGNATURE

(Double-space or more)
• Identification insert
(Double-space)
• Carbon copy insert
(Double-space)
• Enclosure insert

Insert Conventions

There are several very handy devices that you will want to use with regularity. I refer to them as "inserts." They include the following six convenient additions that you frequently see in correspondence:

Personal

Subject:

Attention:

Identification

Carbon copy:

Enclosure:

All these inserts are commonplace, and all are matters of convenience. They are used in the sample letters so that you can see the positioning of the inserts. Sample 12.D utilizes all six of the insert devices.

Personal Insert

There is not much privacy in today's world. If you address a letter to your supervisor that is going to discuss private matters concerning your medical leave, be sure to type "Personal" *above* the inside address *and* on the lower left side of the envelope. If the envelope does not say "Personal," nothing inside will be personal for long. The second problem is that once the envelope is gone, the privacy of the letter is not likely to be honored unless it has a reminder—"Personal"—at the top. This is, by the way, the only insert that is not regularly used on a daily basis.

All the conventional business letter inserts (the internal notations) appear in the sample on the left. The conventional locations for them are also indicated.

Subject Insert

You realize from seeing memos at work that the insert for referencing subject matter is a convention borrowed from memos. A letter is simply easier to read or expedite if it states the subject before it starts. This insert is quite commonly used by businesses in correspondence.

Attention Insert

Again, this insert is borrowed from memo practices. As explained earlier, this reference device is a handy way to direct correspondence to generic locations: agencies, bureaus, departments, and the like. Notice that you omit the salutation when you want to avoid saying "Dear Sir" or "Gentlemen" (Sample 12.D). You will see the attention line occasionally used to identify the recipient by name, but this is not a very courteous practice for business letters.

Identification Insert

Notice that the first three of the six inserts appear at the top of a letter. The remaining three appear at the bottom. The identification line indicates that there was a typist other than the author who created the finished document.

Letters are quite precise by intent, even if they are not usually "binding" in any legal fashion. As a result, any slipup in, say, the number of zeroes in a debt you have, is a serious matter. Typists create typos. The initials of the identification insert signal that the author was *not* the typist. The first initials are those of the author; the second initials are those of the typist. If you type your own work, do *not* add ID initials—unless you have reason to pretend to have staff on board. (Incorporation laws have led to a lot of comical "family" corporate boards on which all the family members sit as executives.)

Carbon Copy(cc) Insert

Of course, carbon copy means photocopy nowadays. Several important matters are of concern here. First, be sure *you* always keep a copy. Your copy is *not* indicated on a letter; it is assumed. File your correspondence copies for a year, or two, or ten or however long you think you may need them. I recently had to refresh a supervisor's memory concerning a job description, with fifteen-year-old documents. You will find that nothing gives you less credibility than being inconsistent about what you thought you said in a letter. *Always* keep a copy.

Another rule of thumb is to use the carbon copy insert to indicate all those who are going to receive the letter. At times this list will have a dozen or more names on it. As a matter of courtesy, the person you are writing to needs to realize who else is seeing the document.

Finally, it is important to note that the two letters *cc* can be more powerful than anything else in the letter and must be used with care. If you write a letter to your boss in which you are somehow critical of her, there are probably certain risks you are taking. However, if you *also* send a copy of the letter to her boss, there will be sirens and blue lights everywhere. As another example, if you are at loggerheads with a company or agency that will not respond to your fair and entitled requests—on a medical insurance claim, for example—it will do little good to be hostile in your writing even if they are abusive. Keep the document historical and accurate and neutral. Send copies of the letters to the Better Business Bureau, and perhaps the Insurance Commissioner, and perhaps the state Attorney General's office. This tactic will usually end certain kinds of stalemates. A word of warning though: do not abuse this tool. You must follow through and address a cover letter and a copy of the letter of adjustment to any such agency you threaten to contact.

Enclosure Insert

If you send a check, it is usually identified as an enclosure—by amount—so that the letter documents that a check of that amount was in the envelope. Putting that fact in writing is more important than stapling the check to the letter. Of course, the common use of this line is for the identification of whatever paper documents will be found in the envelope.

We conclude with a look at additional samples (Models 12.A, 12.B, and 12.C), but first I should add a few useful remarks. Do not staple the pages of a letter. Instead, include information on each page for identification. Use the name of the receiver, the date, and the page number. Do not say "page 3 of 9" unless it is a practice of your company. There are two standard formats:

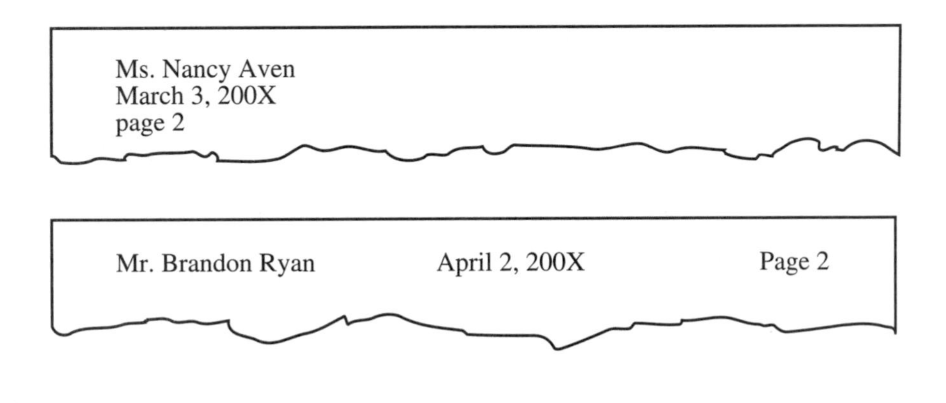
Ms. Nancy Aven
March 3, 200X
page 2

Mr. Brandon Ryan April 2, 200X Page 2

Sample 12.E

The Envelope Layout

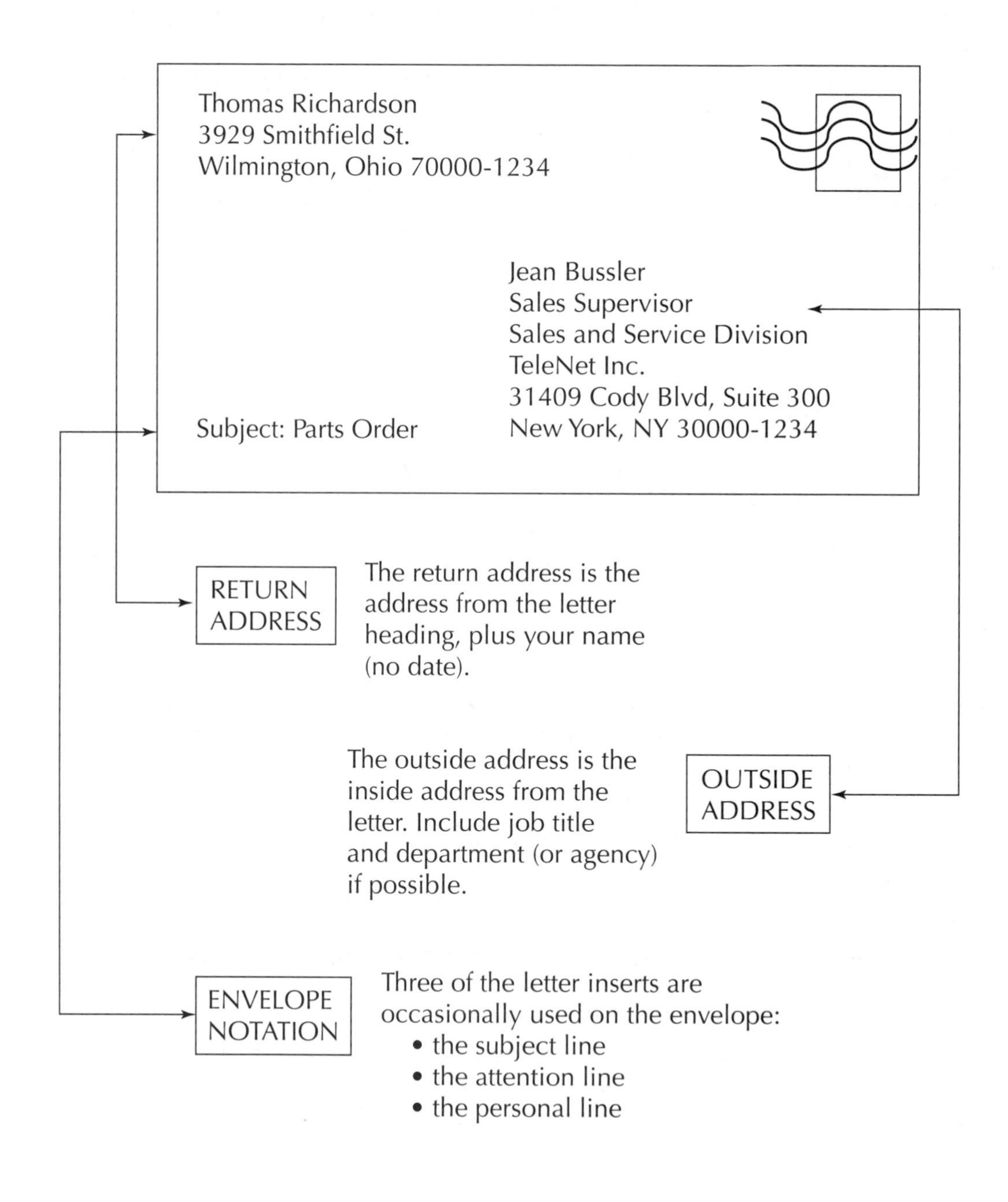

Thomas Richardson
3929 Smithfield St.
Wilmington, Ohio 70000-1234
Jean Bussler
Sales Supervisor
Sales and Service Division
TeleNet Inc.
31409 Cody Blvd, Suite 300
New York, NY 30000-1234
Subject: Parts Order
RETURN ADDRESS
The return address is the address from the letter heading, plus your name (no date).
The outside address is the inside address from the letter. Include job title and department (or agency) if possible.
OUTSIDE ADDRESS
ENVELOPE NOTATION
Three of the letter inserts are occasionally used on the envelope:
• the subject line
• the attention line
• the personal line

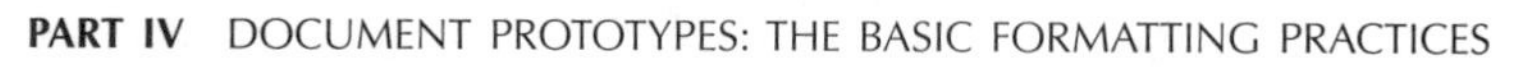

There is a final detail: fold business letters correctly. Use a bottom-up-first, top-down-second method (see the illustration) and place it in the envelope so that the center section is upright. If you have a small envelope, you are not supposed to simply crease a flap on the side of the fold I just mentioned. There is a simple trick: fold the sheet of paper in half, and then fold it in thirds.

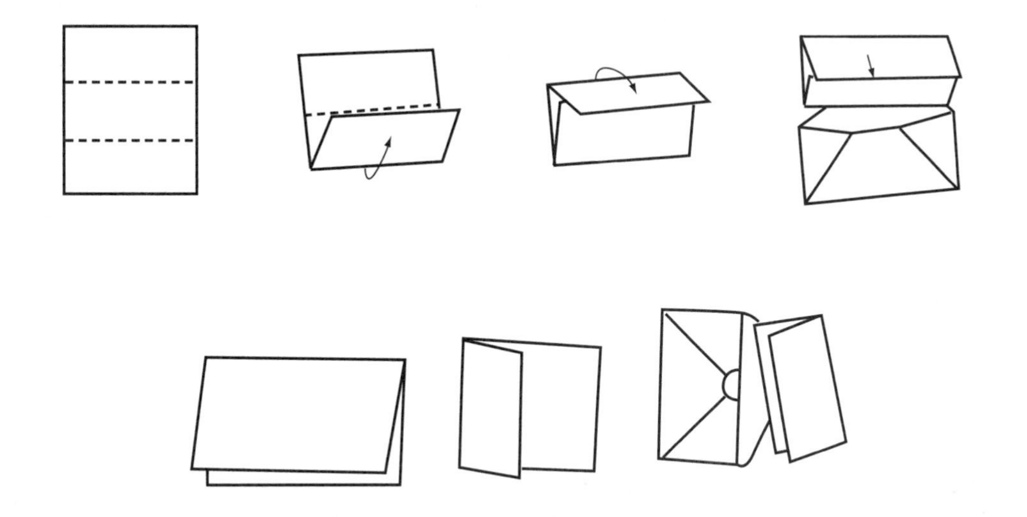

The completed letter is then placed in the appropriately addressed envelope (see Sample 12.E).

Model 12.A

Network Solutions Inc.
2031 34th N. W. Suite 6
Seattle WA 98000 (206) 786-0000

December 4, 200X

Walter Buchanan, Manager
Sales Department
PC Delivery, Inc.
954 W. Washington St.
Chicago IL 60000

1

Dear Mr. Buchanan:

On November 28, 200X, I placed an order with you for a PDCom 500 MHz. Pentium computer complete with 64 MB RAM, a 556 KB Type II cache memory board with Weitek math coprocessor socket, and a Mitsubishi 1.2 MB 12 ms ESDI hard drive.

2

I received the order December 3, 200X with one exception: the hard drive formats out to only 540 MB and the average access time (with settling) was approximately 60% slower than stated. Since no part number for the hard drive was mentioned either by yourself or in your advertisement in *Computer Shopper* magazine, I am unsure whether you have sent me a defective unit, or if the hard drive is simply the wrong one. In either case, the hard drive I received from your company is unacceptable, as you realize from the speed problem.

I am certain that this problem will be solved as quickly as possible. Please send me the correct hard drive via Overnight Express along with the authorization number for the defective drive.

3

Yours very truly,
NETWORK SOLUTIONS INC.

4

Denise Harnitt
President

DTH:pb
cc: R. J. Adamson, Purchasing Division, Network Solutions Inc.

Business Letter 1: Claim Adjustment

The business letter is the standard instrument for all purposes that involve communication external to a company's environment. Model 12.A includes a letterhead, which always adds image to your writing if you own your own business, as you see here in Ms. Harnitt's case.

The inside address is a standard element of business letters, although it is seldom a component of personal letters. For business correspondence, it is important because this is the only location on the document that provides critical information regarding the recipient once the envelope is gone.

Business Letter Practices

HEADING
(Double-space or more)
- *Personal insert*

(Double-space or more)
INSIDE ADDRESS
(Double-space or more)
- Subject insert

(Double-space)
- Attention insert

(Double-space)
SALUTATION
(Double-space)
BODY OF LETTER
(Double-space between paragraphs)
COMPLIMENTARY CLOSING
(Adequate space)
SIGNATURE
(Double-space or more)
- *Identification insert*

(Double-space)
- *Carbon copy insert*

(Double-space)
- *Enclosure insert*

1 Notice the practice of writing to a person by title and by department. When possible, add one or both points of information.

2 The body of the letter begins with the customary "geography." It positions the letter in a historical context and explains the focus of the document. The middle paragraph explains that there has been a problem: the hard drive is malfunctioning. Notice that although the author is firm, she constructs the argument that there may also be a mistake. Mistake or mechanical failure, the last sentence seeks a solution.

3 The final paragraph is curt and calls for a somewhat pricey response, but notice the neutral tone throughout. This is a "claims" letter, but the feathers never ruffle. That is important. However, as an initial response to the problem it is probably overly assertive.

4 The closure is always cordial, regardless of the contents. The company name can go below the complimentary closure or below the author's title. It can also be omitted, since it is part of the letterhead.

Model 12.B

DESIGN SYSTEMS

12304 23rd Ave. S.E.
Bothell WA 90000-1234
November 26, 200X

Award Roofing and Gutters
7761 Aurora Avenue
Seattle WA 90000-1234

SUBJECT: Roof installed for Show Homes Northwest
at 346 NW 89th, Seattle

Dear Mr. Lynn:

Your company has installed a roof for my company at 346 N.W. 89th in Seattle. The roof was in very bad shape as you know, but through the efforts of your company and Show Homes Northwest, the structural deficiencies have been corrected, the new roof has been installed, and it looks wonderful. However, during the recent storm I was at the house and noticed that a small leak has developed around the chimney area.

1

2

I called Show Homes Northwest and they evidently contacted you regarding this leak. Show Homes Northwest called me back the next day and informed me that one of your people had inspected the roof, found that the chimney was leaking, and indicated that I need to call a chimney repair company. I went up on the roof and checked it myself. To my surprise, the flashing of the chimney is missing.

3

I have enclosed a copy of the contract for the work that you contracted to do. I have highlighted the portion of the contract that states that you would replace all flashing when the roof was installed. I am sure you will agree that the responsibility for the repair is in the hands of Award Roofing.

The new roof looks great and I appreciate the work you did. I am certain that the minor problem with the roof can be properly taken care of before we market the property in January.

4

Sincerely,

Donald H. Foll

DHF/PBB

cc: Show Homes Northwest, Karen Henry, Owner

encl: Highlighted copy of contract

Business Letter 2: Claim Adjustment

The letter from Design Systems involves a banner, but the address has been positioned at the upper right. Stationery can be expensive, and this author may have left the address off the letterhead during a period of address changes. This allows him to type in any address he wishes.

The subject line is popular in business correspondence. It is, of course, borrowed from the memo convention. It facilitates routing, filing, and quick referencing. At times the subject line is elaborated, and at times it may be little more than a part number.

Business Letter Practices

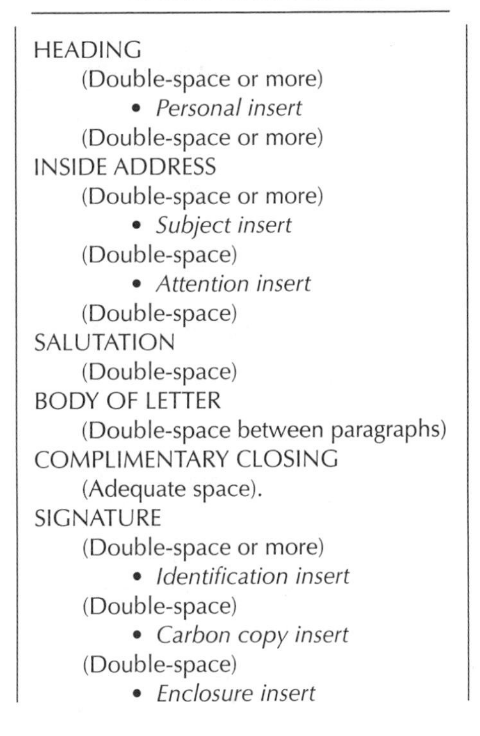

HEADING
(Double-space or more)
- *Personal insert*

(Double-space or more)
INSIDE ADDRESS
(Double-space or more)
- *Subject insert*

(Double-space)
- *Attention insert*

(Double-space)
SALUTATION
(Double-space)
BODY OF LETTER
(Double-space between paragraphs)
COMPLIMENTARY CLOSING
(Adequate space).
SIGNATURE
(Double-space or more)
- *Identification insert*

(Double-space)
- *Carbon copy insert*

(Double-space)
- *Enclosure insert*

1 This letter is another claims letter but it is handled in a different style. The first paragraph employs the popular "good news–bad news" strategy. It is a very common tactic to set the stage for an adjustment with a few compliments. The theory is that flattery will get you somewhere. At the least it shows good form.

2 The twist in the good news–bad news style usually is little more than a single word. In the fourth line of the paragraph there is a reversal. Our author says the job "looks wonderful" and then adds the classic wet towel: "however. . . ."

3 Considering that this is a second-stage complaint that is a follow-up to phone calls, notice how polite the author is. Essentially, this letter is more polite after a rebuff than the previous letter was from the outset. This shows better form in business relations.

4 As the author closes he again extends his appreciation for a job well done. Although this is not a necessary element of the good news–bad news tactic, it is a good practice to end on an energetic or upbeat note. Notice that this author did not repeat his company name or provide his title. These are common but optional practices that may be included or omitted depending on company policy.

Model 12.C

E. T. CONSTRUCTION

Tony Sokolich, Owner **9700 Second Avenue, N. E.**
Seattle WA 90000
(206) 784-5731

December 3, 200X

Ms. Judy Dacola
5311 23rd Avenue N.E.
Seattle WA 90000-1234

RE: New Roof Installation/residence phone 329-0000

Dear Judy:

Thank you for selecting E.T. Construction to complete the installation of a new roof on your home. I have listed in detail below all the construction specifications that were listed on the initial proposal submitted to you on September 18, 200X. The contract will be completed as follows: **1**

(1) Tear off existing roof and dispose of all materials.

(2) Lay half-inch CDX plywood as a base.

(3) Lay saturated felt #15 tar paper.

(4) Install 30-year Tallwood composition shingle.

(5) Install four AF vents to code.

(6) Install brown dip-edge flashing to match house.

Please note that item 2 above has been added. Upon inspection of your roof, dry rot was discovered in several places. It will be necessary to replace all plywood on the roof because most of it has been damaged by periodic leaking over the past few years.

Taylor Gutter Company has been contacted regarding the replacement of your gutters. A separate proposal will be submitted for this phase of construction. The first phase of construction is scheduled to begin December 10 at 8:00 A.M. Enclosed is a modified price list and invoice reflecting the changes both in labor and materials resulting from the addition of item 2 to the contract. **2**

If you have any questions, please feel free to contact our office at any time. Thank you again, and we are looking forward to doing business with you. **3**

Sincerely,

David Burbank
E.T.C. **4**

DB/pb
Encl.: price list
invoice

Business Letter 3: Contract Agreement

After a proposal or bid or estimate has been accepted, it is appropriate to communicate a plan of action. The proposal establishes what will be done and at what cost. A letter of acceptance is equally important. Once finances are agreed on, the plan of action must be agreed on as well. Essentially this letter must establish the timelines for the contract details in much the same fashion as the initial letter and estimate (not seen here) established the costs for the contract details. The sample is a proposal from a roofing contractor to a homeowner. He addresses her by her first name and opens with a "thank you." Notice that he uses a subject notation.

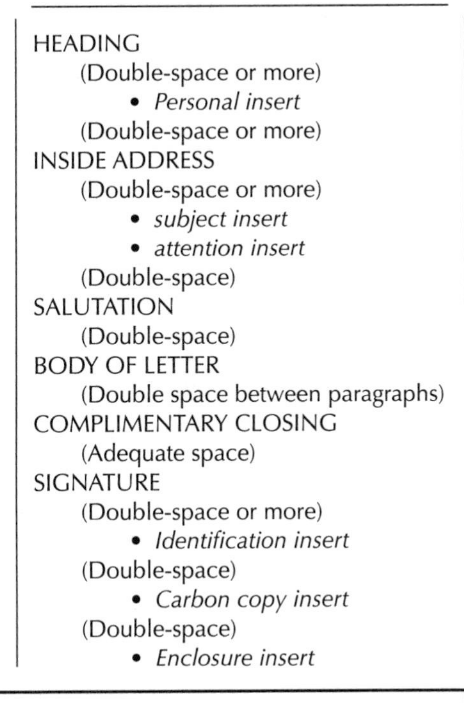

Business Letter Practices

HEADING
(Double-space or more)
- *Personal insert*

(Double-space or more)
INSIDE ADDRESS
(Double-space or more)
- *subject insert*
- *attention insert*

(Double-space)
SALUTATION
(Double-space)
BODY OF LETTER
(Double space between paragraphs)
COMPLIMENTARY CLOSING
(Adequate space)
SIGNATURE
(Double-space or more)
- *Identification insert*

(Double-space)
- *Carbon copy insert*

(Double-space)
- *Enclosure insert*

1 Consistent with each sample we have looked at, there is a historical opening to give the reader a geographical fix on what is at hand. The letter repeats the highlights of ETC's bid, particularly since there has been an addition to the original proposal.

2 The author provides a dateline for the subcontractor and important details concerning price adjustments. Since prices rarely adjust downwardly, this is the time to explore and settle such problems.

3 Notice again the polite closing paragraph. Thanking clients for their business is the usual procedure, and the comment is commonly joined with some reference to an avenue of communication. These comments are polite, but in fact they are critical to success on any contract.

4 In this sample, notice that the company name is repeated at the bottom under the author's typed name. It is not redundant, since the letterhead contains a different form of the company name Obviously the "E.T.C." abbreviation is catchy, and the owner wants to use it in correspondence.

Cover Letters for Resumes and Proposals

Cover letters are letters of introduction. There are many documents which a reader may not have an immediate sense of function or place. If a secretary places your resume on the desk of a potential employer, the employer may not understand why the document is there. If a laboratory director finds an elaborate ten-page lab report on her desk, she might call the lab tech for a reminder about who requested the project and where it is to go. A proposal that arrives at the in-basket of a facilities supervisor may belong in one of ten folders for ten projects that are up for bids and proposals. The supervisor will have to figure out where the proposal belongs. In cases such as these, a cover letter serves to explain the larger document.

The Letter of Application

The letter that you compose to apply for a job is probably one of the most important letters you will ever write. It must look sharp; it must explain your credentials; it must impress the potential employer. These are not easy tasks to accomplish when your letter will, in fact, look surprisingly like ten or twenty other letters that are also forwarded to the employer. They will all use the same traditional layout. They will all open with a request for consideration, and they will all close with a hope: the request for an interview. The challenge is to make the material between the first and the final paragraphs distinctive—and this center section is usually a modest two to four paragraphs.

The easiest way to handle the letter of application is to design it as a cover letter for a resume. If you have a resume, then you almost have the cover letter written. If you do not have a resume, type out a neat and orderly list of your technical skills, your employment history, and your technical education. List the skills in order of importance, and list the employment and education in *reverse* chronological order. Neatly package the highlights of the resume or the background lists in the letter. The difference between the documents is that the letter is written in sentences and paragraphs and has a more cordial style than the matter-of-fact approach of the resume. The cover letter introduces the resume, but it also introduces *you* to the potential employer. Beyond this basic introduction, however, you want to fine-tune the letter to be as winning as possible. To draft the *best* of the ten or twenty letters on the supervisor's desk is the challenge.

Strategy

"Cold calling" is the name given to the process of dutifully marching to each potential employer even though no jobs have been announced. This approach to looking for employment is not likely to produce results that are worth the time invested. First, larger companies announce their openings with regularity, and these are usually posted through appropriate channels: the company personnel office, newspapers, and professional journals. In other words, the jobs are not concealed in any way; on the contrary, companies publicize openings. Second, the same companies will have a protocol for interviewing and hiring procedures, and a cold call is likely to be outside of the process. Perhaps there was a time when dropping in on an employer might produce a job, but larger companies deal professionally with professional people.

Similarly, using any kind of broadcasting is also a type of cold calling. If you fax, or e-mail, or send fifty stamped letters to companies in your local area, the results are likely to be disappointing. The responses will usually be form letters—in response to your form letter. In other words, written inquiries for employment are probably even less effective than a drop-in approach because the correspondence will be generalized and artificial, even though the correct name of the company is at the top of the letter. If you forward a letter of application without knowing whether there are openings, the letter functions as an inquiry and lacks the direct point of contact that is involved in an application letter that addresses an available opening. *Your strongest letter of application will be one that can target a specific job announcement.*

Watch for job announcements in newspapers and professional journals. They may also be announced at local engineering society meetings or posted in engineering society local bulletins. Also watch the bulletin boards of the campus employment office and attend all campus job fairs (nearby university job fairs are also a possibility). Local, county, state, and federal agencies also maintain bulletins of openings. Do not think that newspapers are *the* place to look for employment. In fact, because of their massive circulation, large urban newspapers open the floodgates of competition. If you locate postings at these various other resources, you may find employment prospects that have had less exposure and fewer applicants than the newspaper ads.

Once you are on the trail of a prospect, request a job description of the position (see Sample 12.F). Not every employer will provide job descriptions, but most state and federal agencies use them (and are often *required* to develop them). It is in the interest of most companies to have complete job descriptions. They use them to designate specific skill needs when hiring, and they depend on them as a weeding tool. Potential applicants are far less inclined to apply once they see the specific details that were not part of a twenty-word listing in the newspapers. The job description saves the potential applicant time, and money as well.

With the job description in hand, place it next to your resume and see if there are matching points of interest (see Chapter 11 in the *Writer's Handbook*). If you do not yet have a resume, use lists of your skills, your employment history, and the key elements of your technical education as described earlier. The company job description is probably prioritized. The most important skills are identified first, desirable skills will be farther down the list, and ancillary or helpful skills will appear last. See what you have to offer. If you have none of the important skills that are a priority for an employer, do not apply. If you do have some of the technical skills that are required, prospects are a little brighter, but remember the competition. You need to complement the primary skills with whatever secondary and supplemental skills were identified on the job description.

After the analysis is complete you are ready to write your letter. In the letter identify the skills you have to offer. Prioritize them to *exactly* match the order on the job description. In fact, you can take key words from the job description and use them in your letter so that the readers see what they are looking for. This is the process known as "targeting." Many varieties of writing—including all advertising copy—are designed to fit a specific group of readers. Your letter is an advertisement that is designed to sell you. The job description is your market research; it tells you what the reader is looking for. It is this match between potential applicant and potential employer that makes targeting a superior method for drafting letters while seeking employment. The letter of inquiry is a poor runner-up, since it does not allow you to be job specific in your selling strategy.

Procedure

For the layout of the letter use all the conventions of a formal business letter. If you have someone in particular to whom you can address the letter, identify the person in both the inside address and the salutation. This tactic gives you a point of contact that may be useful, and it immediately places your letter ahead of the pack if the others did not address the recipient by name. In the case of larger companies there may not be an individual employee in charge of the correspondence, and you can use another salutation style:

Attention: Personnel Department

Attention: Director of Personnel

Sample 12.F

POSITION AVAILABLE

COMPUTER MAINTENANCE TECHNICIAN III/COMPUTER SERVICES
NORTH CAMPUS Job #98-160
Opening Date: August 27 Closing Date: September 15, 200X

Seattle Community College District is recruiting to establish an eligible list to fill a full-time position. This is a North campus promotional only position.

DESCRIPTION: Assist the network administrator with design, installation, maintenance, and troubleshooting network services throughout the campus and branch campuses using netware, NT, Unix, and Linux.

RESPONSIBILITIES AND DUTIES: Accounts administration on student interactive login server, student web server, and student mail server (HP 9000); create and support intranet service applications (bootptabman, dhcptabman); train CS staff in intranet service applications use; provide primary support for instructional file servers, instructional application servers, Linux-based infrastructure servers, and Linux based intranet servers (Netware, NT, Linux, HP 9000, bootpd, dhcpd, nsccweb, etc.); provide technical support and advanced cgi programming for Campus Web Presence; backup to Tech II on classroom Macintosh workstations and associated file servers.

MINIMUM QUALIFICATIONS: One year of experience as a Computer Maintenance Technician II, OR five years of experience as a computer maintenance technician OR equivalent education/experience.

PREFERRED QUALIFICATIONS: Proven ability to organize and prioritize tasks in order to meet deadlines. Two years experience accessing and manipulating-data on the administrative HP 3000. Understanding of the state K-20 WAN. Solid background in Netware, NT, HP-UX 10.x and Linux.

SALARY: $2992–$3833 per month.

HOURS: 8:00 a.m.–4:30 p.m.

AFFIRMATIVE ACTION: Supplemental Certification from the eligible list may be invoked to meet the requirements of the District's affirmative action and corrective employment programs.

THIS IS IN A UNION SHOP BARGAINING UNIT. AS A CONDITION OF EMPLOYMENT. YOU MUST WITHIN 30 DAYS AFTER APPOINTMENT (1) BECOME A MEMBER OF THE WASHINGTON STATE FEDERATION OF STATE EMPLOYEES (WFSE), OR (2) PAY A REPRESENTATIVE FEE. OR (3) PAY A NON-ASSOCIATION FEE. NON-PAYMENT OF SUCH FEE IS GROUNDS FOR DISMISSAL. ANY DISPUTE BETWEEN YOU AND THE EMPLOYEE ORGANIZATION AS TO THE AMOUNT OF THE REPRESENTATION FEE CAN BE RESOLVED ONLY UNDER PROCEDURES PROVIDED BY THE EMPLOYEE ORGANIZATION, NOT THE EMPLOYER.

APPLICATION PROCEDURE:
YOU MUST SUBMIT ITEMS #1–#2 OF YOUR APPLICATION PACKET IN ORDER TO BE CONSIDERED FOR THIS POSITION. APPLICANTS WHO DO NOT COMPLETE AND SUBMIT ALL REQUIRED MATERIALS BY THE PUBLISHED CLOSING TIME WILL NOT BE CONSIDERED FOR THIS POSITION. APPLICATION PACKET INCLUDES:

1. A completed official Seattle Community College District application form.
2. A completed supplemental application for the Computer Maintenance Technician III position.
3. Resumes are encouraged.

ONLY COMPLETE APPLICATION PACKETS RECEIVED IN THE HUMAN RESOURCES DEPARTMENT BY 4:30 P.M., SEPTEMBER 15, WILL BE FORWARDED TO THE SCREENING COMMITTEE FOR REVIEW.

Persons with disabilities needing assistance in the application/examination process may call (206) 587-0000 or, for the hearing impaired, call the Relay Service at 800-833-0000, or for hearing and visually impaired, call Braille Relay at 800-833-0000.

If you use the "attention" device, you may also want to use a subject line to immediately indicate the point of the correspondence.

> Subject: Computer Technician Opening, Computer Services Division

First Paragraph The body of the text is very important. Each paragraph is designed to convince the reader to meet with you for an employment interview. The opening paragraph must state the obvious fact that you want to apply for a job opening. There is no one particular way to compose this statement. I would suggest that the paragraph consist of two or three sentences that get down to business. The first sentence or two should state the background of the letter by identifying the position, the employer, and how you heard about the position.

> I would like to apply for the recently announced opening in the Post Testing Department at Comtron Electronics. The Bench Technician Three position was advertised at Oakland Valley College last Thursday.

You identify the position and the company to immediately state your purpose in terms that are meaningful to them. These points of reference also keep your letter from looking like a form letter. It is customary to identify where you saw the job announcement but this is not necessary. It does, however, keep the first paragraph from being too abrupt, since the next and final sentence or two state your request to be considered for employment.

> I would like to be considered for the bench technician opening. I have read the job description and I think I am qualified for the position.

Here are three variations on the theme:

> I am very interested in the bench technician position, and I think my qualifications are appropriate for the opening.
>
> The bench technician position is similar to one I held with the TM Corporation for five years, and I am now seeking work in the local area.
>
> I hope to locate a position with Comtron, and I would request a review of my qualifications.

Job descriptions can be very thorough. They will provide you with an immediate gauge against which you can measure your skills. Notice the preference for resumes (item 3) identified in the right column.

Sample 12.G

HEADING
(Do not use current work address.)
Phone
E-mail
Fax
Date

INSIDE ADDRESS
(Address the personnel officer by name if you know it.)

SALUTATION:
(Use a person's name if possible.)

FIRST PARAGRAPH: Identify the specific position, and the company. Explain that you are qualified for the job and want to be an applicant.

SECOND PARAGRAPH: Develop your strongest point: education, employment history, or skills.

THIRD PARAGRAPH: Develop a secondary strength by discussing one of the remaining areas of interest (identified in the paragraph above).*

FOURTH PARAGRAPH: Explain that you are available for an interview and that you have enclosed a resume.

COMPLIMENTARY CLOSING

Signature

* Additional paragraphs may be needed.

The first paragraph also sets the tone of your letter, so keep it precise and modest. The rest of the document will shift the tone to reflect a competent and confident image in the body paragraphs, and it can conclude with a sense of anticipation and a hint that you are excited.

Some authorities suggest that an applicant use a more creative approach to the opening paragraph, perhaps one that opens with questions or a description of the author and his or her skills. Feel free to try these openings, but they may tend to appear contrived and over-confident. One alternative that is known to be effective is "leverage." If you are referred to a personnel officer by someone the officer knows and trusts, then you are in command of a very powerful opener. Instead of noting where you read about the position, identify the person who told you about it.

Second Paragraph Depending on which is more important, use the initial body paragraph to highlight either your employment or your education or specific skills that reflect your capabilities. For the young graduate the third prospect is the least likely option until hands-on experience supports the technical abilities. Make the paragraph from four to six sentences long. Do not feel compelled to be thorough. The idea is to *highlight* to gain emphasis. If the letter is too long it will conceal the main points.

Third Paragraph Select the second most important of the three considerations—employment, education, skills—and touch upon those elements that address the job description. The younger graduate is likely to have used the second paragraph to highlight technical education. In this paragraph the focus would then shift to work experience. If the work experience is unrelated, include the paragraph anyway. Employment always serves to prove that you have "workforce training." Besides, a great many skills from the workplace have value. Leadership, management, teamwork, and other experience can be crucial secondary skills for an engineering tech. In other words, if you were a supervisor at the local Burger Bar franchise at the age of nineteen, that is good news—even if it is low tech.

Do not identify any shortcomings that you have. If a job description identifies a critical skill that you do not have, let the employer find the weakness. Do not offer it or otherwise state anything that will create a negative image of you.

Fourth Paragraph You may add additional paragraphs. At the minimum, a letter of application is likely to contain four paragraphs, in which case the fourth paragraph draws to a close the request for consideration. Like the first paragraph, the last paragraph follows conventions, and it is not likely that it will be unique. In fact, you must not be too clever or express too much confidence. The closing *must* explain that you are enclosing a resume and that you would like an interview.

> I have attached a resume and a list of references from the local area. Please feel free to contact me at (201)301-0000 if my qualifications are appropriate for the position.

Because the letter is formal, there is little opportunity to color the text in order to project a sense of personality into the message. You can review the body paragraphs and see if an

Sample 12.H

Kay Smith
2245 28th Ave. NW
Seattle, WA 90000
(206)810-4523
e-mail: ksmith@sttl.uswest.net

November 1, 200X

Jennifer Rhodes
Orca Networks
2765 Third Avenue
Suite 402
Seattle, WA 90000-1234

Dear Ms. Rhodes:

I would like to be considered for the Computer Operator position that is currently posted at the Campus Placement Center for the Seattle Community Colleges. The graveyard shifts fit my schedule, perfectly! As a graduate of North Seattle Community College's Technical Support Program, I would be able to use my technical skills in this position while learning more about computer installations.

My experience and training covers all of your requirements for this position. I am experienced with MS DOS 6.22, UNIX, Windows 95, 98 and NT, and all the applications in Office 95 and 97. Designing web pages for friends has given me the opportunity to learn HTML. One area of particular interest of mine has been learning more about hardware upgrades and wiring. Since completing my course of study at North Seattle Community College I've enrolled in both A+ and Cisco CCNA certification preparation courses.

This summer I participated in the Technology Improvement Project at View Ridge School. This project's purpose was to allow more Internet access in the school's computer lab and to configure the classrooms for supervision by the Seattle School District's technology staff. My contribution in this project included upgrading both PCs and Macs. I also assisted with wiring classrooms.

I am available for an interview at your convenience. My resume is enclosed for your consideration.

Sincerely yours,

Kay Smith

occasional word or two will create tone and emphasis. The tone you want to convey is a sense of enthusiasm or energy, or professional dedication. For example, rather than say you "held" a position and explain *what* you did, say that you "enjoyed" the position and explain *why*. You might use an adjective here and there for mental associations. If you say the college program was "excellent," or that your instructors are "highly respected," the comments indirectly speak about you.

For more direct statements about yourself, the opening of the final paragraph may be the ideal location. Enthusiasm may be seen as contrived if it is given much vigor in the opening paragraph. If, however, the personnel officer is still reading by the time the last paragraph opens, there is good reason to suspect that you show potential. Now you can add a sentence or two to define yourself. Explain, for example, the origin of your interest, or your involvement, or your successes and rewards.

> I am always excited about computers and I think that matters. I am part of the first generation of computer technicians to be brought up with the technology as a familiar part of day-to-day living at home and at school and at work. This familiarity has been a very helpful background for my computer network program, which I will complete next month. I have enclosed my resume for review. I would like the opportunity to meet with you or your representative.

Here is another sample:

> Although I feel that my skills and education demonstrate a background that is appropriate for the position, I also want to point out that I am a dedicated employee and I have received three in-service employee awards during my five years with Media Services. I would enjoy the opportunity to meet with you. My resume is enclosed. Please feel free to contact me at the number listed below.

Before You Mail the Letter

You need to be in full control of the letter of application. That control includes using letterhead, attractive paper, laser printing, and the layout of your choice. Be aware that e-mail may cost you all these features. Fax will at least preserve the layout. The most appropriate presentation is the business letter sent by mail. Also, be sure you save both a floppy disk copy and a hard copy of the letter. This is a difficult document to write, but once the job is done, variations of the same letter can be used again and again—but let's hope that it is not necessary if the job is well done!

This letter of application is designed to open with little fanfare even though it is given a little energetic touch. It is important to identify the job announcement and the appropriate qualifications in the introduction. The letter then presents the applicant's qualifications in more detail.

Sample 12.I

HENDERSON CONTRACTING SERVICES

Commercial Interiors and Exteriors

4011 East Evanston
Seattle WA 90000

July 10, 200X

Dr. James Kent
Northwest Veterinary Hospital
9500 35th Ave. N.W.
Seattle WA 90000

Dear Dr. Kent:

I hope that the enclosed bid is prompt. I realize from our discussion that you would like to complete the interior work while Dr. Savary is on vacation.

My calendar is clear from July 25 to August 10 if that is still a suitable arrangement.

Please review the bid. I think it covers every detail we discussed. If you have any questions, please call. I will contact you next week to see if you had further questions.

Thank you for the opportunity to make an offer for the work. It was kind of Jim Nolan to refer me to you.

Cordially,

Jeff Henderson

Enclosure: bid

The Cover Letter

The conventional cover letter is a letter of explanation that accompanies another, usually larger, document. The larger project could be a two-page bid or estimate for grading and paving a driveway ramp for the service entry of a building, or it could be a fifty-page proposal from a civil engineering firm that is submitting an offer to build a sewage treatment plant. Note the brief cover letter in Sample 12.I. It was attached to a two-page bid (shown on pp. 370 and 372).

The cover letter is conventional in format except that it can be either attached or included in the proposal.

- The cover letter can be *unattached,* in which case it should include an "attachment" insert (the enclosure notation) below and to the left of the signature on the letter. The attachment is then identified by name.
- The cover letter can be *attached* to the cover of the bid or proposal.
- The cover letter can be *bound* in the bid or proposal, in which case it is placed after the title page.

There are several practical reasons for attaching or enclosing a cover letter with a bid or proposal. An underlying motive is persuasion. The letter establishes rapport with the reader and contains a cordiality that is seldom part of any proposed contract agreement. The proposal may be intended to *initiate* a business agreement, or the proposal may have been the outcome of an agreement that is paying for consultation services that *resulted* in a proposal. In the first case, a company specializing in commercial architectural services offers a bid to install suspended ceilings in a supermarket. In the second case a consulting firm is paid to complete site analyses and to propose recommendations for relocating the supermarket. The firm is *not* offering a contract proposal in the second case; the firm has been paid for the proposal itself. In both situations the proposals are professional and specific, and the documents may be analytical. As a result, the only cordial link that connects the authors of either proposal to the reader is the cover letter. Since proposals either offer services or conclude services, large sums of money are often involved, and the cover letters assume importance in closing agreements.

There are additional, more practical reasons for sending cover letters:

- Cover letters are often called "letters of transmittal"; in other words, a note often *has* to be attached to the proposal for forwarding. Otherwise, the proposal may appear unexplained.
- Cover letters can be used to explain a proposal to readers who may not be familiar with the original reason for the proposal.

This sample is brief and to the point. For a contractor's purposes, a cover letter adds a personal touch to a bid submission, but the letter need not explore details at length.

Sample 12.J

MARKET RESEARCH ASSOCIATES
812 Pike Street
Seattle WA 90000

November 30, 200X

Donald Jamison, President
Western Electronics
956 4th Ave. NW
Portland OR 90000

Dear Mr. Jamison:

On October 18, 200X, the Western Electronic Corporation contracted Market Research Associates to develop an analysis of the manufacturing trends of Western Electronics with the intent of suggesting potential recommendations of meeting demands through the year 2015. Western Electronics' growth is expected to more than double by 2006, and existing facilities will not readily absorb such a high factor of growth.

We are pleased to present our study of the situation that confronts your corporation. Although time constraints were severe, our available data were thorough and up to date since our specialization in market research concerns electronics and electronics-related industries.

The study is based on the problem of excess demand for the high-tech speciality products manufactured by Western Electronics. The "problem" is not, of course, demand. Demand, as we suggest in our report, is the precise measure of the prestige of Western Electronics. Demand is also the key to your development potential. The difficulty is that Western is not capable of expanding present production beyond a 15% increase in output. In an anticipated climate of 200%+ growth, the problem is evident: growth can be achieved only by exercising major expansion options to deal with growth trends.

The possible alternatives available to your corporation are limited to the development of a new plant facility, which could be located near Portland, or in the Southwest. The existing facility cannot be expanded in a way that would allow ongoing production during construction. A remodel is neither practical nor cost effective in any case because the existing plant cannot readily absorb all anticipated growth, particularly that of microprocessor manufacturing. Since microprocessor production demands a highly specific plant environment, production could be developed at a different site in order that the existing facility could continue production with no capital investment in remodeling the existing location.

It is our recommendation that a microprocessor plant be developed as a separate facility. Further, due to the support technologies required for microprocessor production and the geographical market for sales of the finished product, we suggest that the plant be located in the Southwest. For reasons of cost, the recommended location for the plant is Phoenix, Arizona.

We would like to thank your executive staff for assistance in providing the data for Western Electronics' planned development. Do feel free to call for clarifications.

Respectfully submitted,

Market Research Associates

Mary Westlin
Chief Analyst

- In the case of lengthy proposals that are developed over periods of weeks or months, cover letters can explain minor changes or last-minute considerations.
- There may be a request for action in the cover letter, in which case the reader can be addressed outside of the context of the proposal.

Procedures

The content of the letter usually includes a brief historical background for the proposal and a summary of the important features. Open the first paragraph with the point of origin of the project. Identify the dates and the agents involved and, if necessary, include attachments of earlier correspondence. It is often useful to establish an enthusiastic tone, and a few traditional sales elements can encourage the gratefulness of a client.

> You will find two additional site analyses in the discussion. Howard felt that you should consider these sites, although they were not part of the original selected sites. We think there are exciting prospects among the recommendations.

The body of the letter can deal with other matters that are useful. Highlight the content of the proposal in the second paragraph to initiate further discussion. In the third paragraph, discuss changes, timelines, actions, explanations. Additional paragraphs are often necessary. Sample 12.J was constructed in six paragraphs. It was attached to a twenty-five-page proposal (see Model 14.C).

Conclude with a final paragraph that encourages prompt contacts. Thank the reader for the opportunity to submit the proposal. If you are offering a bid for a contract agreement, also explain that you will call in a week to see if there are any questions. The cover letter is the link to the contract agreement, and you want a reason to call the potential client so that you can further pursue discussion of a possible contract.

As I pointed out, the letter of application is probably the most important letter you will ever write. The cover letter for bids and proposals is certainly a close second. It is a critical tool in financial agreements, and it is used by businesses of every size. As a contractor I used to receive cover letters for bids on many contract proposals, even though the contractors were usually self-employed workers addressing a single client. On a larger scale the college where I work recently received proposals—with cover letters—from large contracting firms that are offering to build a laboratory facility worth millions. The entire facilities management team of the college analyzed the proposals. At all levels of financial investment, the cover letter is an effective part of the business world. Chapter 14 explores the larger concept of bids, estimates, and proposals.

Unlike the first cover letter, this one reflects a great deal of discussion. It was designed to accompany an elaborate proposal, and the preliminary comments were considered important.

Summary

- Use the business letter for transactions between companies.
- Use correspondence to maintain written records of important matters that involve your company's interests.
- Always use a neutral tone and a polite style.
- Set the stage in the first paragraph, usually with a historical reference.
- Rank your interests and develop the most important points first.
- Make the paragraphs short to medium in length (four to ten lines).
- Be sure the document conforms to standard layout practices.
- Letters of application should be based on job descriptions whenever possible.
- Rank the demands of the position and rank your skills and training. Then design the letter.
- Cover letters for bids and proposals should highlight the primary interests of the client.
- Use the cover letter to develop a cordial touch because the proposal does not reflect the actual client–service provider relationship, except in terms of the formal contract agreement.

Activities Chapter 12

Use the chapter summary as your guideline.

- *Create a package of letters that are in someway related to your field of study. They can be authentic or fictional. Use letterhead if you have access to any that is appropriate or design letterhead of your own. Develop one of each of the following:*

  Letter of request

  Letter of response

  Letter of adjustment

  Letter of agreement

- *Look for samples of business letters at work. Try to locate samples of each of the four types discussed. Include other types that you discover. Gather the samples in a package and attach a cover letter to your instructor. Briefly explain the letters in terms of the company where you work. (See alternative team exercises.)*
- *If you later develop a bid or proposal as an assignment for Chapter 14, develop a cover letter for the proposal. Use the letter as a cordial complement to the proposal, and highlight key features concerning project activities and costs.*
- *Draft a letter of application. Identify a specific position at a specific company and address the letter to meet the requirements for the specific position. (See alternative team exercises.)*
- *Draft a hostile letter of complaint concerning a topic of interest: computer equipment, airline service, cafeteria food, television advertisers, or the like. Develop a formal and polite letter that achieves the same mission. Summarize the differences in a memo to your instructor and submit the letters as attachments.*

When the projects are returned to you

- *At your instructor's request, resubmit your project with the corrections to errors that are indicated on the original.* Boldface *all corrections so that the instructor can see the revisions. Use the* Writer's Handbook *as your reference.*

Share a Project: First Option

Work with three other members of the class to discuss and edit the fifth project, a set of business letters. You may use the same groups that you used previously or new ones.

- *Consult Appendix A. Before holding the meeting, review the fifth editing checklist for business letters.*
- *On the day the projects are due, have each member of the editing group explain his or her project in terms of objectives and project development.*
- *Hand the groups of letters around for a critical reading and editing.*
- *Have each member edit the texts for both writing errors and technical errors.*
- *Have each member write a one-paragraph critique at the end of each project.*

Share a Project: Second Option

As team alternatives to the preceding exercises, consider the following:

- *Work with two to four other members of the class. Look for samples of business letters at work that reflect the four types identified in the first set of exercises:*
  - ✔ *Letters of request*
  - ✔ *Letters of response*
  - ✔ *Letters of adjustment*
  - ✔ *Letters of agreement*

  Collect other types of letters also. Bring the samples to class for discussion of both layout styles and contents.

Share a Project: Third Option

Work with several other members of the class to develop letters of application for employment. For openers discuss your respective ambitions and decide what would be the appropriate content of each team member's letter. Take notes. Develop your letter for a second meeting and make copies for each member. Discuss.

Share a Project: Fourth Option

After completing the letter of application, continue with this exercise if desired. With one other class member, develop letters of recommendation for each other. The other class member should be unacquainted with you if possible.

- *Exchange letters of application.*
- *Exchange resumes if you have them.*
- *Discuss your career interests, skills, and background in employment and education.*
- *Write a letter of recommendation for the other student. The letter should be three or four paragraphs long. Address the document to the instructor, but assume that the instructor is an engineer who might hire your fellow student.*

Work in Progress

Page by Page

I was really getting into the flow of writing these safety documents. I just had to remember my basic goals and my motto, "read it, get it, and go." As long as I followed my standardized organization I would be well on my way to completing the written safety program.

The second portion of the hazards section was titled Recognizing and Labeling Hazardous Materials. *There was a lot of ground to cover here and some company decisions to make prior to establishing the program. I asked for a meeting between Matthew and Karen to discuss the company's opinion about labeling. Since there is not a standardized format for labeling, I wanted to know how they wanted to proceed.*

After receiving some teasing from Matthew regarding something as trivial as labeling, I doggedly explained that it was a requirement mandated by WISHA and OSHA to have a uniform labeling procedure for all hazardous chemicals used. He was still skeptical, but Karen asked me to go ahead with a generic-labeling program. She wanted me to write what I planned to recommend and then present it to her at a later date. Her orders were good enough for me. Below are excerpts I composed for the HazCom part of the document.

Labeling—While a commercially produced label is not required, as long as the program covers the required information, many companies find it easier to use them. Two common labeling systems use colored bars or diamonds to identify the hazard with a numbering system that indicates the severity of a hazard. Another system uses symbols that identify the hazard and the protection required when using the substance. Some labels combine these approaches. **There is no nationally required system.**

All containers containing hazardous chemicals received or used will be properly labeled with the following information: (1) identity of the hazardous chemicals; (2) appropriate hazard warning and (3) name and address of the chemical manufacturer, importer, or distributor. (4) No container will be released for use by the supervisor unless the container is properly labeled with the above data. (5) Employees will not remove or deface existing labels on incoming containers of hazardous chemicals unless the container is immediately marked with the required information. "Secondary" containers, into which chemicals are transferred, must be labeled with the name of the chemical or product and the appropriate hazard warnings and the manufacturer's name and address. These labels must be legible, in English, and easily readable by the user.

Material Safety Data Sheet—accurate, up-to-date MSDS forms will be obtained, reviewed, and updated for each hazardous chemical used in PAT facilities by PAT employees. The MSDS forms must be readily available to all employees during each work shift. Each work location must maintain copies of MSDS forms for products they use.

I felt that Pacific Aero Tech had a lot of work ahead to fulfill this particular set of regulations. Of all the safety programs, this one would be difficult to initiate because old habits and routines are established and new, safer, and "by the book" methods of working are much more of a challenge to learn and maintain.

L.C.L.

Standard Lab Reports

The laboratory report is a frequently required method of documentation used by many industries. The device is equally commonplace at colleges and universities. From infectious disease laboratories to appliance repair shops to college engineering labs, the lab report is popular as a practical and convenient way to record analysis. At times the reports are outlined and long and formal (always to company specs). Sample 13.A illustrates the most common variety, a brief form that a technician fills out. At the service and repair level, a bench tech's "report" may be little more than a clipboard with a few notes that will accompany the billing for a repair.

Sample 13.A

ENVIRONMENTAL CONTROL TECHNOLOGY
TEST REQUEST AND RESULT SUMMARY

START DATE 10/5/200X
COMPLETION DATE 10/9/200X

PROJECT NO. I-4
TEST NO. 18

TEST REQUEST

SUBJECT OF TEST HICKOK #8

PURPOSE OF TEST To chart performance of unit on a Molier chart. To test unit and make sure it is operational. To then plot PE chart for analysis.

PROCEDURE Air-cooled evaporator & condenser, 16 oz R-12, capillary tube, 115 volts semi-hermetic compressor, glass bulb thermometers, stainless steel thermometer. Unit was run for 30 min. before readings were taken. Unit ran for a total of 1 1/2 hrs. (1/2 hour reading time.)

RESULT SUMMARY

SUMMARY OF RESULTS EVAPORATOR TEMP: 37°F (entering)
LEAVING CONDENSER: 89°F

1. Pressure levels: 144.7 psia (high side), 46.7 psia (low side)
2. compression ratio: 3.1 to 1 3. NRE: 55 Btus/lb
4. R-12 Flow rate: 0.925 ppm 5. Horsepower: 1/5 (actual) 1/4 (rated)
6. COP: 6.11 ratio 7. Discharge temp.: 148°F
8. Compressor displacement: 0.856 lb/min. 9. Heat of rejection: 67 Btus/lb
10. Rejection factor: 0.821 Superheat: 28°F Subcooling: 20°F

CONCLUSION

Followed the proper procedure and recorded the above results. Found the unit to be running satisfactorily, as shown by results from the P.E. chart.

ACTION TO BE TAKEN None needed. All checks satisfactory.

TESTED BY Cynthia F. & Mary G. REQUESTED BY Ivan S. 10/1/200X
Date

REVIEWED BY ______ ______ Mary G. 10/10/200X
LABORATORY SUPERVISOR DATE PROJECT MANAGER DATE

11/98:DS-MH

The sections of the lab report are fairly standardized:

- Purpose
- Procedure
- Results
- Conclusion
- Action

A report of this type is typically filled in with data and phrases. You would not necessarily notice that the lab report—sketchy and blocky as it often is—is quite similar to the formal narratives found in the engineering report, the project proposal, or other more complex documentation of the sort explored in Chapters 7 and 8. In fact, the *logic structure* is quite similar, but the *manner* of presentation can be quite different.

Lab report (functional style)	**Formal report (narrative style)**
I. Summary (outcomes)	**Introduction**
II. Investigation	**Evidence**
III. Conclusion	**Discussion**
IV. Recommendations	**Conclusions**

One unique feature of both lab reports and formal reports is the way the documents are formatted for engineering and scientific industries: the outcomes are often presented at the *outset,* particularly if the document is a typed, formal presentation. In other words, the report opens with findings. This section is sometimes referred to as an "executive summary." Whereas many writing projects begin at the beginning, technical reports and lab reports frequently begin at the end. The technique would spoil a mystery story, but technicians and engineers put the end first for economy. Supervisors may want to see only the results and may not examine the research, but the project will often be filed in the event that there is a problem. When it is on file, the archive is available for retrieval as need arises. In this age of litigation, all manner of work histories are kept just in case there are legal problems.

Researchers may not follow a regularized pattern of testing procedures until they are ready to test large numbers of a simulation or model or event or some such item of investigation. To report on experimental test procedures, or any instance of a one-of-a-kind investigation, researchers have to design and compose a presentation. It is practical to

The conventional organization of the lab report is used in the form on the left. Laboratories will usually use a fill-in-the-blank form if a procedure is used repeatedly.

Sample 13.B

Mary G.

1. PRESSURE LEVELS - 144.7 psia HIGH SIDE
   46.7 psia LOW SIDE

2. COMPRESSION RATIO = $\frac{144.7}{46.7}$ = 3.1 TO 1

3. NET REFRIGERATION EFFECT - (83 BTU'S / #) - (28 BTU'S / #) ← 15° SUPER HEAT

   NRE = 55 BTU'S / #

4. REFRIGERANT FLOW RATE -

   Q = .925 ft³ / #   D = 1.08 # / ft³

   Compressor Displacement = 1.07 #'s / min (RATED) -80% =

   (1.08 #'s / ft³) · (1.07 #s / min)

   = (1.16 #s/min) (.8) = .925 ppm

5. HORSE POWER

   HEAT of compression = 9 BTU'S / lb

   (9 BTU'S/#) ÷ (42.42 BTU'S/#) = .21 ($\frac{1}{5}$ Horse power)

6. COEFFICIENT of PERFORMANCE

   55 BTU'S/# (NRE) ÷ 9 BTU's/# = 6.11 RATIO

7. DISCHARGE TEMP - 148° F

8. COMPRESSOR DISPLACEMENT -

   (.925) (.925) = .856 pounds/min
   (Q) (ppm)

9. HEAT OF REJECTION - (95 BTU'S/#) - (28 BTU's/#) =

   67 BTU's/#

10. REJECTION FACTOR - $\frac{55 \text{ BTU's/\#}}{67 \text{ BTU's/\#}}$ = .821

design the document in fairly conventional terms, of course, and most lab reports are variations on standard lab practices. Here are two additional common formats:

Option A	**Option B**
summary	**outcomes**
procedure	**investigation history**
results	**procedure**
conclusions	**results**

The usual practice is to use these standard headings; however, in many technical papers or narratives, the sections are given original titles that fit the specific circumstances, such as those illustrated in Sample 13.C.

If you design your own laboratory report, simply follow the logic of the basic divisions—summary, investigation (or parts thereof), conclusion, and perhaps recommendations—but order the document in whatever specific form is used by your company. Another option is to outline the entire document rather than use paragraphs. In the cassette player lab report (Sample 13.C) the technician structured the document to conform to general specifications but did not use a "form." This method gives lab reports the formal look when such a method is desired. She also reported her outcomes in the conclusion rather than in the opening summary.

As a record keeping practice, a company may request the original lab data as shown to the left.

Sample 13.C

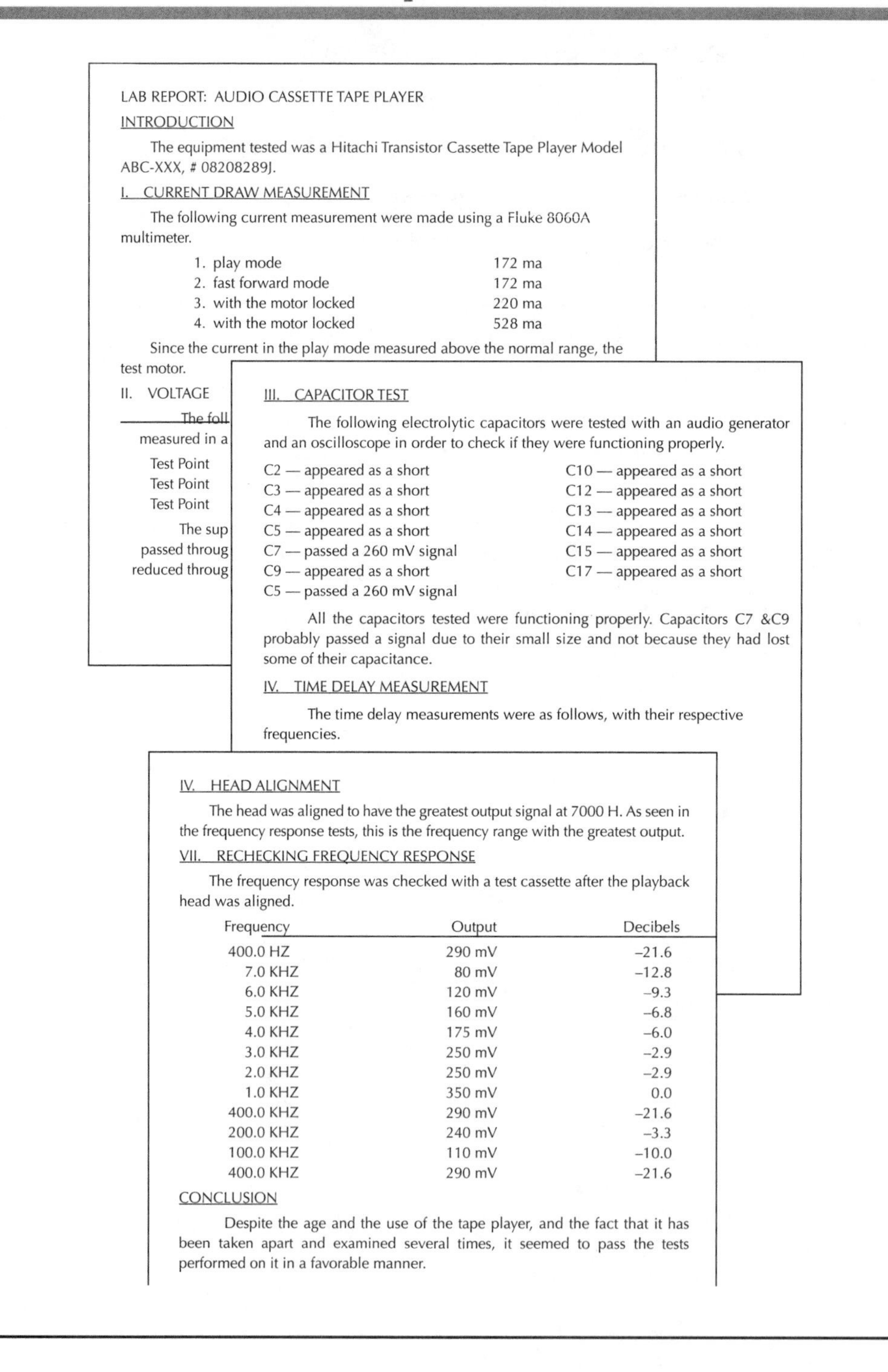

LAB REPORT: AUDIO CASSETTE TAPE PLAYER

INTRODUCTION

The equipment tested was a Hitachi Transistor Cassette Tape Player Model ABC-XXX, # 08208289J.

I. CURRENT DRAW MEASUREMENT

The following current measurement were made using a Fluke 8060A multimeter.

1. play mode 172 ma
2. fast forward mode 172 ma
3. with the motor locked 220 ma
4. with the motor locked 528 ma

Since the current in the play mode measured above the normal range, the test motor.

II. VOLTAGE

The foll
measured in a

Test Point
Test Point
Test Point

The sup
passed throug
reduced throug

III. CAPACITOR TEST

The following electrolytic capacitors were tested with an audio generator and an oscilloscope in order to check if they were functioning properly.

C2 — appeared as a short
C3 — appeared as a short
C4 — appeared as a short
C5 — appeared as a short
C7 — passed a 260 mV signal
C9 — appeared as a short
C5 — passed a 260 mV signal
C10 — appeared as a short
C12 — appeared as a short
C13 — appeared as a short
C14 — appeared as a short
C15 — appeared as a short
C17 — appeared as a short

All the capacitors tested were functioning properly. Capacitors C7 &C9 probably passed a signal due to their small size and not because they had lost some of their capacitance.

IV. TIME DELAY MEASUREMENT

The time delay measurements were as follows, with their respective frequencies.

IV. HEAD ALIGNMENT

The head was aligned to have the greatest output signal at 7000 H. As seen in the frequency response tests, this is the frequency range with the greatest output.

VII. RECHECKING FREQUENCY RESPONSE

The frequency response was checked with a test cassette after the playback head was aligned.

Frequency	Output	Decibels
400.0 HZ	290 mV	–21.6
7.0 KHZ	80 mV	–12.8
6.0 KHZ	120 mV	–9.3
5.0 KHZ	160 mV	–6.8
4.0 KHZ	175 mV	–6.0
3.0 KHZ	250 mV	–2.9
2.0 KHZ	250 mV	–2.9
1.0 KHZ	350 mV	0.0
400.0 KHZ	290 mV	–21.6
200.0 KHZ	240 mV	–3.3
100.0 KHZ	110 mV	–10.0
400.0 KHZ	290 mV	–21.6

CONCLUSION

Despite the age and the use of the tape player, and the fact that it has been taken apart and examined several times, it seemed to pass the tests performed on it in a favorable manner.

The Laboratory Report Narrative

The formally presented lab report documents lab procedures and outcomes using the techniques of report writing. The change involves a shift to paragraphs and a change to a narrative style in which the author must explain and describe the procedure. The logic is organized by practical realities: first, the problem is introduced; second, the problem is investigated; third, the outcomes are identified. Note that outcomes are not necessarily recommendations. The report may or may not be advisory.

Consider a typical lab report format:

- **Purpose**

  The formal report of a laboratory proceeding begins with some explanation of the purpose of the document. The discussion may be largely historical, and it usually reflects the importance of the report. The discussion is likely to be handled in conventional paragraphs throughout much of the report. Depending on the corporation, outcomes may or may not be stated in the initial section of the project.

- **Procedure**

  Unlike a routine lab procedure that is formatted as a set of bench tests (see p. 340), a unique procedure usually reflects a newly developed activity. Thus, the investigator might explain the undertaking from several different perspectives. For example, a historical discussion of alternatives or a narrative concerning the conditions for the investigation may be necessary.

  The procedure is likely to be divided into a number of subsections that may include equipment needed, special conditions, setups, step-by-step activities, calculations of various kinds, mathematical or chemical considerations, and so on. Outlined material is conveniently mixed with discussion paragraphs.

The challenge is to be able to design a laboratory procedure in the absence of a prescribed document. The laboratory procedure on the left is crafted as a one-time single-use report. However, because of the headings, there is little mystery to either doing a procedure or documenting it.

It is fairly obvious from Sample 13.A and the following Eldec checklist that the technician has little writing to do if there are company forms that can be used. However, if the engineering report has to be constructed from scratch, then the need for writing skills and organization skills is critical and must support the technical knowledge.

- **Results**

  The intent of the investigation is usually threefold. A method of investigation must be determined, a demonstration must be developed, and the results must prove the effectiveness of the investigative strategy. In other words, did the laboratory procedure get the job done correctly? The results are one factor in the process.

- **Conclusions**

  Results are not necessarily conclusions. Results can be measured against any number of "controls" or criteria to draw conclusions *about* the results. For example, the results can be examined for

  - ✔ accuracy relative to other procedures;
  - ✔ accuracy relative to desired levels of precision;
  - ✔ accuracy relative to exterior controls or standards of an outside agency; or
  - ✔ accuracy relative to cost.

- **Action**

  There may or may not be a request for action by the investigator. Also, there may or may not be a request for action by the sponsor or supervisor of the report. The findings of a lab report are frequently forwarded without recommendations of any kind.

The following sample is a report sheet for a typical test procedure performed at the Eldec Corporation; it is of the tick-list variety that is popular for bench work in many corporations. The paperwork for the bench tests for this particular set of calibration settings goes on for pages. Compare this "standard form" approach with the subsequent laboratory report samples, and you will see that the pattern of the contents is similar. The method of organization of a lab report is fairly uniform whether the report is a singular event or repeated a thousand times.

Two abbreviated laboratory reports follow in models with commentary (13.A and 13.B). The selected reports were cut by half but still maintain their logical continuity. It is difficult to construct brief lab report samples, but it is necessary to examine the structure of these "papers," since they are frequently the most common document many engineers and engineering techs will produce, both in college and in industry.

TEST DATA

SHEET

TEST: CALIBRATION/FUNCTIONAL — **TEST ITEM P/N 4-013-______**

CONDUCTED PER: FTP 4-013 — **TEST ITEM S/N**

DATE STARTED: — **ENDED:**

5.1	All Inspection Points Approved	Compliant ( )
5.2	Leakage Current (Limit <5.0 ma.)	____________mA
5.3.3	Burn-In Time Started	____________
5.3.5	Burn-In Time Ended	____________
5.4.6	E_0 @ 100% Rated Load (Tolerance: 5.00 ±0.10 Vrms)	____________Vrms
5.4.8	Smooth decreasing brilliance control?	Compliant ( )
5.4.9	E_0 @ Pot. Short (Limit: ≤ 0.75 Vrms)	____________Vrms
5.4.10	E_0 @ Thunderstorm (Tolerance: 5.00 ±0.10 Vrms)	____________Vrms
5.4.11	Smooth imcreasing brilliance control?	Compliant ( )
5.4.12	E_0 @ Pot. Open (Limit: 4.0 Vrms ≤ 5.9 Vrms)	____________Vrms
5.4.15	Output Current, Short Circuit (Limit: 1▷)	____________Amps
5.4.16	E_0 @ 20% Rated Load (Limit: ≤5.610 Vrms)	____________Vrms
5.4.17	E_0 @ External Feedback Disconnect (Limit: >0 .75Vrms & <3.0 Vrms)	Compliant ( )

1▷ 2.0 to 13.0 Amps for 4-013-01
4.0 to 26.0 Amps for 4-013-02
6.0 to 39.0 Amps for 4-013-03
12.0 to 78.0 Amps for 4-013-04

CONDUCTED BY:	APPROVED BY:	CERTIFIED BY:

ELDEC ELDEC CORPORATION FSCM IDENT. NO. 08748 ® LYNNWOOD, WASHINGTON	DOCUMENT NO. FTP 4-013	PAGE 12	REV. E

A Calibration Test Form

(Illustration courtesy of the Eldec Corp.)

Model 13.A (1)

RMA and Water-Soluble Solder Paste

CALTRON

I.N.C.

Introduction:

In order to comply with global and local initiatives, Caltron Inc. is planning to phase out the use of CFCs (chlorofluorocarbons, popularly known as Freon). Currently CFCs are used in volume for cleaning printed circuit board assemblies. To reduce this consumption, water-soluble solder paste technology must be implemented. Tests with water-soluble solder pastes began in 200X, with the Alphametals BetaFacts DT 30JT solder pastes showing impressive results over the rosin-based paste Renton D-301. The objective of the study is to prove that the performance of the water-soluble paste is comparable to or better than that of existing RMA (Rosin Mildly Activated) solder paste.

Procedure:

A total of six test vehicles were assembled for this test: three using BetaFacts DT 30JT (water soluble), and three using Renton D-301 (RMA flux). Test boards 1 and 2 were screen printed using RMA solder paste, and test boards 3, 4, 5, and 6 were screen printed using water-soluble paste. These boards followed specific process parameters and observations, as described in Figure 1.

1

2

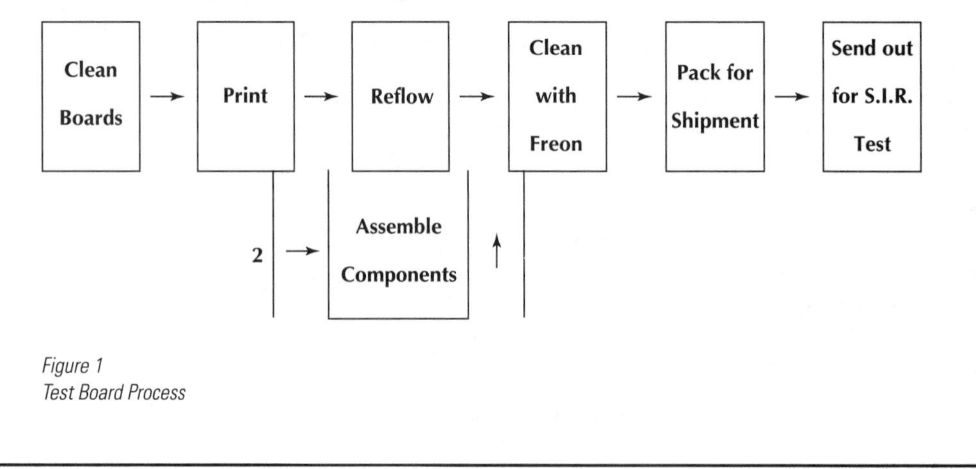

3

Figure 1
Test Board Process

Lab Report 1: Electrical Engineering

As noted earlier, lab reports usually have an introduction. This author's task as an employee of Caltron was to compare the performances of two types of solder paste as part of her company's efforts to phase out the use of CFCs. This will be a report that is kept on file in case it is requested by a regulatory agency.

The subject is clearly defined at the outset: the purpose of the project is to test alternatives to the established practices of using chlorofluocarbons for cleaning. Notice that the author's report is similar to the generic model of a laboratory report illustrated here.

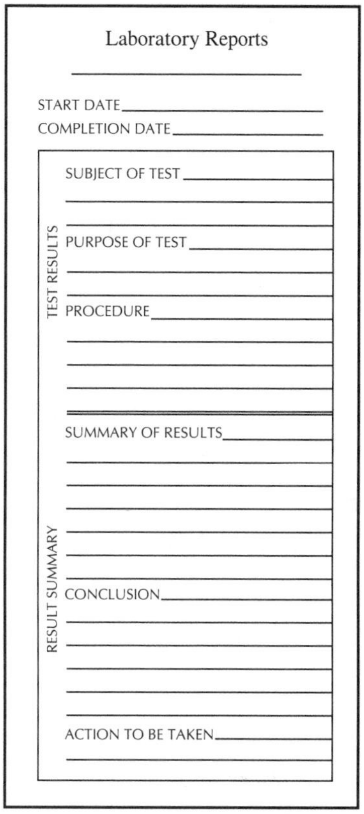

Laboratory Reports

START DATE________
COMPLETION DATE________

TEST RESULTS

SUBJECT OF TEST________
PURPOSE OF TEST________
PROCEDURE________

RESULT SUMMARY

SUMMARY OF RESULTS________
CONCLUSION________
ACTION TO BE TAKEN________

1 The procedure follows the introduction. The procedure was a new one designed by the author to fit the production considerations of her company. She explains the procedure at length. She used six test boards to compare one conventional solder paste and one of the ecologically friendly varieties. (These tests were repeated a number of times in units of six.)

2 Notice that considerable detail is devoted to a precise explanation of the laboratory procedure. She carefully constructed controls that would simulate the precise practices of the company. She begins the explanation with a flowchart and continues the description of the process on the next page.

3 Graphic elements frequently appear in lab reports for a host of reasons. The visual components are important tools for engineers whether they are tables, drawings, graphs, or any one of many other visual assists.

Model 13.A (2)

-3-

Screen Printing Process

Solder paste is a non-Newtonian fluid with thixotropic property, so the viscosity of solder paste decreases with increased shear stress, but upon standing regains its static as well as dynamic viscosity. For my experiment, I opted to use a viscometer that could provide static as well as dynamic viscosity. Refer to Figure 2 for the values of viscosity for both BetaFacts and Renton solder pastes. The screen printing machine that was chosen for this experiment was the automatic MPM AP-24X.

1

Figure 2. Comparison of Controls

	Type	*Alloy*	*Metal %*	*Viscosity**	*Mesh*	*Lot#*
30JT	OA	63/37	90.5	3090	-200+325	31029958
D-301	RMA	63/37	90.0	800+-5	-200+325	31424

*5 rpm on Malcolm viscometer. Measured in poise.

2

Reflow Process

During reflow, solder melts and forms a main single fillet, but occasionally, the paste does not coalesce into a single fillet and some satellite particles get carried away with the flux, which can then cause problems if the flux is not properly cleaned. These satellite particles can cause shorts in the circuit. In extreme cases too many solder balls can cause uneven clearance of components, which can result in poor and inconsistent solder joint formations.

3

For preparation, I hand-soldered two 304′ QFP to a dummy profiling board and attached two thermocouples using Ablebond 77-IS (refer to Figure 3). Then, I measured the mass of the board with the two 304′ QFP and calculated the dimensions of the board. To profile the right temperature settings for the board, I used Datapaq ReflowTracker 2000.

4

Cleaning

Test boards 1 and 2, which were printed using RMA paste, were cleaned in OSL cleaner with FreonTMS. Test boards 3, 4, 5, and 6, which were printed using a water-soluble paste, were cleaned in Fowler cleaner using deionized water. The parameter used in each cleaning machine is described in Figure 5.

1 The pronoun *I* is used here because the procedure is historical. The author constructed and performed the tests, and so she enters the report to relate the story.

2 Tables are used often in lab reports. The table format is a visual convenience that allows an engineer or technician to display a large quantity of data at one time. Here (and in the following lab) the table is a valuable tool for comparison.

3 Here the author describes the reflow process. A technician or an engineer must explain the testing procedure and the calculations. This author explains the procedure in detail because the consequences are important to the company and because the employees must be able to replicate the procedure.

4 In a case such as this, a formal lab report may prove to be very important if local, state, or federal regulations are instituted that will require company accountability in a matter such as chlorofluorocarbon control. The report can be used to determine new, environmentally sound options, but the report can also be filed as part of a company demonstration of compliance with new regulations.

Model 13.A (3)

-5-

Cleaning Assessment

The following tests were used to assess the cleanliness of the test vehicles

- Omegameter test
- Surface insulation resistance test (SIR)
- Electromigration test
- Visual inspection @ 30X magnification

Omegameter Test (Test Boards 5 and 6 only)

After test boards 5 and 6 were cleaned in a Stoelting cleaning machine, the Omegameter test (a contamination test) was conducted using the Alphametals SMD model 600. When the test was completed, the 304′ QFPs were removed from the test board without the use of any flux. The test results appear in Figure 3.

Figure 3. Test (with 304′ OFP)

Item	*Volume*	*Resistivity*	*Time*	*Alcohol %*	*Contami-nation*	*Pass/ Fail*
TB #5	17000	54.92	15	76	0.9	Pass
TB #6	17000	53.81	15	76	1.2	Pass

Surface Insulation Resistance Test

Twenty-four test sites were used for test boards 1 and 2, which were printed with Renton D-301 229D (RMA) solder paste, and twenty-four test sites were used for test boards 3 and 4, which were printed with BetaFacts DT 30JT (water-soluble) solder paste. The test conditions were as follows:

- Relative humidity: 80%
- Temperature: 85∞C
- Applied bias: 100 volts
- Test voltage: 100 volts

1

2

3

4

1 The sections concerning cleaning and cleaning assessment shift the reader's attention to the analysis of the effectiveness of the products. The author used four testing parameters (several have been omitted from the sample). Each of the four tests is examined in turn.

2 Each of the four tests is discussed at length and is summarized in a table, thus generating four tables of test results. An abbreviated form of the Omegameter procedure has been included along with the tabulated findings for this particular test.

3 A table is a convenient way for an investigator to format data, as noted earlier. It is equally convenient for any reader to quickly survey the outcomes that are presented.

4 Surface resistance and electromigration tests and an inspection under magnification were also performed.

Model 13.A (4)

-9-

The SIR measurements were taken between the leads of 304′ QFPs, which measure to 0.020 inch, and from under the components. The SIR patterns under the components were 0.010-inch lines and spaces. Prior to testing, the samples were left in the chamber at the specified settings for 8 hours to allow proper conditioning of the chamber. None of the samples had any dendritic growth or signs of corrosion.

Visual Inspection

In order to assess the cleanliness of actual live boards, 20 active assemblies (Silicon Graphics LG1 030-8202-003 Rev. A) were assembled using both DT-30JT and D-301. The inspection was performed at 30X magnification. The BetaFacts DT30 JT solder paste is a relatively low residue paste, and there were no signs of significant residue. The joint appeared cleaner than on the Freon-cleaned boards.

Conclusion:

The aqueous cleaning process, using the BetaFacts DJ 30JT solder paste and Fowler CBT-301 cleaner, has demonstrated a process capability comparable to or better than the Freon cleaning process. It can, therefore, be used as an alternative to the Freon-based process. However, strict process control methods (SIR, Omegameter) must be employed to ensure ongoing compliance with cleaning requirements.

1

2

1

The author states that the tests accomplished the objective of the study. She adds a note of caution but offers an evaluative summary and gives a go-ahead for the new product.

2

Engineers often place the outcomes at the end of a document in a summary. Depending on the practice that is adopted by a company, the summary could very well be used to open the report instead.

Model 13.B (1)

1

SPRINGS:
A LAB EXPERIMENT

Introduction

In their ability to expand and compress, to flex and stiffen, the spring is able to take on a variety of roles and importance. Used in a variety of ways, the spring has added significantly to our lives, from the retractable ballpoint pen to an automobile suspension system to tool balancers. Springs can be found almost everywhere, from aerospace technology to photocopy machines.

2

The following experiment explains the mechanisms of springs and how the mechanisms relate to one another.

Objectives

- To determine how the force of a spring depends on the spring's extension.
- To determine how the period of oscillation of an object attached to a spring depends on the mass of the object.
- To compare results with those predicted for an ideal system, first by comparing the product of the two graphs with the value $(4p^2)$* and then by comparing the x-intercept of the T^2 versus mass graphs with $(-m_s/3)$, where T is the period of oscillation, which is the time necessary for a complete repetition of the motion, and m_s is the spring mass.

3

Equipment

- Three springs of different masses
- Wooden ruler
- Stopwatch
- A box of different hanging weights
- Clamps
- Metal tower apparatus (to support spring, hanging weight, and ruler)

Lab Report 2: Mechanical Engineering

Model 13.B is an example of a typical lab report from a college environment. Labs are important tools in engineering education, and, as you know, there are lab manuals for a wide variety of procedures undertaken in college courses. A formal laboratory report of practices and findings can easily be constructed from many of the experiments included in college manuals, even if they use the popular fill-in-the-blank format. In truth, many corporations follow suit with laboratory "forms," but you need to be able to turn either a fact-filled form or raw data into a standard report format if it is requested.

Laboratory Reports

START DATE__________
COMPLETION DATE__________

TEST RESULTS

SUBJECT OF TEST__________

PURPOSE OF TEST__________

PROCEDURE__________

RESULT SUMMARY

SUMMARY OF RESULTS__________

CONCLUSION__________

ACTION TO BE TAKEN__________

1 Following the familiar formats already seen, this author easily constructed a professional version of one of his physics labs on the principles of springs.

2 The laboratory report begins with the customary background. The author presents the basic mechanism—the spring—and invites the reader to follow as he conducts and analyzes an experiment to demonstrate the principle that is the purpose of the lab.

3 The next section turns to the specific objectives. Like the first author, this writer cannot shift immediately to procedures. He must first establish the *specific* objectives for the demonstration of the *general* principle. The first author discussed solder pastes. This writer approaches springs through three specific points of analysis (the third analysis is identified here although it is omitted from the sample).

Model 13.B (2)

2

Procedure

Three springs of different masses were selected (large, medium, and small—henceforth, spring 1, spring 2, and spring 3, respectively). The three springs were weighed, and their respective masses were noted. The apparatus was then set up (see Figure 1) as follows:

1

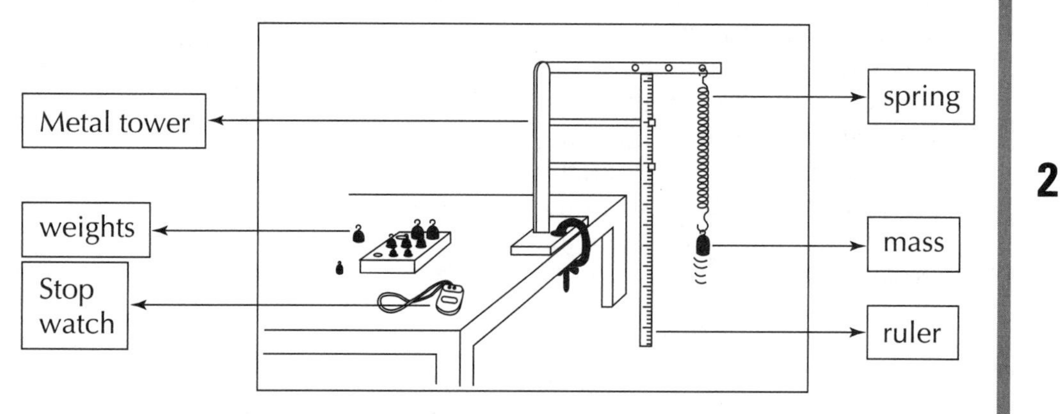

2

Figure 1. Apparatus

- The tower was clamped securely to the table.
- The ruler was placed exactly parallel to the spring.

3

The spring was hooked onto the metal tower. With no mass hanging from the spring, the initial point of extension X_o (see analysis) of the stretched spring was measured. Weights were attached to the spring one at a time, and increased incrementally. For the large and medium springs, weights were increased in increments of 20 grams. For the small spring, weights were increased in increments of 5 grams.

The following data were recorded for each spring:

- Eight values of different masses versus spring extension distance X (see analysis).
- Eight values of periods of oscillation (twice per mass) using the stopwatch.

4

(For one of the masses, six values of T were collected to produce an error bar). The procedure was repeated for each of the three springs.

1 Although this section describes a simple procedure using simple equipment, it contains the same elements as a professional laboratory report.

2 The author includes a graphic to illustrate the setup for the experiment. A pictorial diagram of a procedure is a very common element in a lab report.

3 The procedure is identified and explained, as is standard, in any lab report.

4 Note that procedures are not clear unless the instrumentation is illustrated, and that the instrumentation is unclear unless procedures define the tasks involved. The two features—illustration and description—are important to the clarity of the lab report.

Model 13.B (3)

Data

Table 1. Values Obtained for Spring 1

Spring #1: **mass = 151.366 g** **X_O = 0.295 m** **$X = (X_{measured} - X_O)$(m)**

M (kg)	$X_{measured}$ (m)	F_s (N)	*X* (m)	*T* (s)	*T* (s)	*T**2(s*2)	Om(s*2)*20
0.000	0.295	0.000	0	0.533 0.546	0.540	0.291	0
0.020	0.313	0.196	0.018	0.607 0.608	0.608	0.369	0
0.040	0.336	0.392	0.041	0.669 0.665	0.667	0.445	0
0.060	0.355	0.589	0.060	0.728 0.733	0.731	0.534	0
0.080	0.378	0.785	0.083	0.776 0.777 0.780 0.777 0.774 0.781	0.778	0.605	0.0211
0.100	0.400	0.981	0.105	0.828 0.827	0.828	0.685	0
0.120	0.423	1.177	0.128	0.870 0.880	0.875	0.766	0
0.140	0.444	1.373	0.149	0.933 0.917	0.925	0.856	0

1

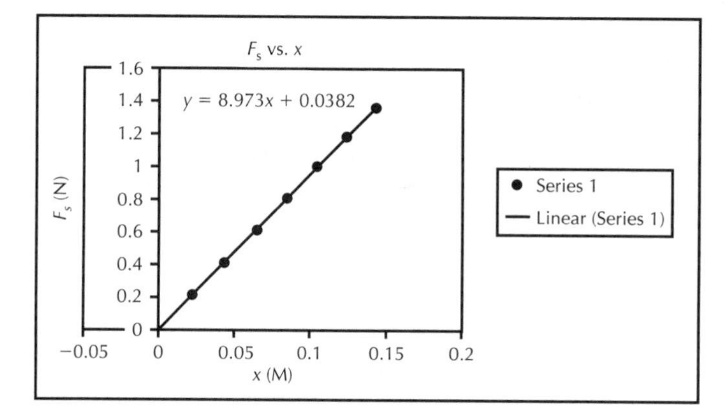

2

Figure 3. Graph of F_s vs. *x* for Spring 1

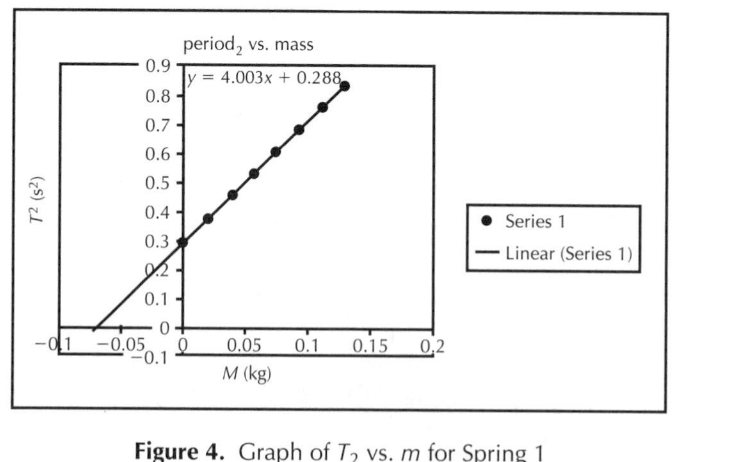

3

Figure 4. Graph of T_2 vs. *m* for Spring 1

1

Lab manuals for college courses provide the procedure but not the outcomes. This author completed the tables of data with precision. As in the Eldec calibration test form, the activities are defined, but the investigator still must determine the results.

2

Just as tables are ideal for plotting comparisons, charts are ideal for plotting factors that change over a period of time. The author used three tables and six graphs to summarize the experiment.

The author developed all the tables, graphs, and illustrations. The quality was exactly as you see it here (there were other graphics I had to omit). He had a keen sense of the appropriateness of graphic features that help develop his project.

Model 13.B (4)

(Additional data were omitted here.)

Analysis

Free body diagram of hanging mass

1

The mass in the diagram is in an equilibrium state. From Newton's third law of motion, the sum of the forces (ΣF_x) is
$\Sigma F_x = -F_s$(spring force) $+ Mg$ (M is hanging mass and g is gravity) $= 0$.
$\therefore F_s = Mg$.

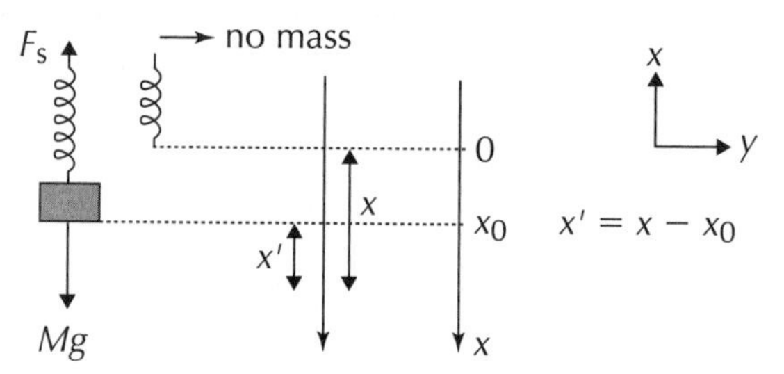

2

Figure 2. Free Body Diagram of the Mass Hanging from the Spring

Note: x' in data is called x, and x is called $x_{measured}$.

Conclusion

Response to Objectives

By analyzing the graph of F_s versus x for all springs, one can see that the best fit is represented by linear equations (see Figures 3, 5, and 7). This indicates that the force of the spring is linearly proportional to the extension of the spring.

$$F_s \propto x \qquad (1)$$

3

However, if x_0 had been chosen to be the nonstrech point of the spring, then the graph of F_s versus x would have passed through the origin.

An analysis of the graph representing T^2 versus M for all springs revealed that as the mass of the object increased, the period of oscillation squared (T^2) increased. Thus, T^2 is linearly proportional to the mass of the object (see Figures 4, 6, and 8).

4

$$T^2 \propto M \qquad (2)$$

1

The text shifts to the calculations that demonstrate the principles of the spring. Using his lab data the author completed well known equations.

2

The diagram on this page is a variation of the figure on page 2 of the report, Figure 1 was drawn by hand, and Figure 2 was generated on a computer. The first is an illustration of the apparatus, and the second is an illustration of the principles demonstrated in the experiment.

3

Calculations appear often in laboratory reports. The reports are seldom addressed to anyone outside of a professional audience. The findings can, therefore, be kept in their most technical form, expressed mathematically or in tables and graphs.

4

Notice that the equations are numbered near the right margin of the report. If you make frequent use of equations, it is convenient to number them, particularly if you want to refer to any of them in the body of the text.

Summary

- Follow established procedures when developing a company laboratory document.
- Compose a custom report using an outline style or a paragraph style or both.
- If you design a special document, follow the general format of a lab report, which is quite generic: purpose, procedure, results, conclusions, actions.
- If you prefer the outline style, use paragraphs for the opening section to state the purpose.
- Use graphics, math, and tables as appropriate.
- Determine whether the report is to be descriptive or evaluative. Evaluative reports should include recommendations.
- If the report will be distributed, use a peer review procedure to double-check technical details and math (see discussion on p. 137).

Activities Chapter 13

Use the chapter summary as your guide.

- *Use a laboratory report from one of your other courses as the subject matter for this exercise. Use either a tear-out lab from a textbook manual or longhand original lab from a course that did not use a manual.*
- *Develop a formal typed version of the lab. Assume that the instructor who originally requested the lab will be the reader. Assume that he or she requested the formal typed version for a presentation.*
- *Identify appropriate sections and develop the text. You might need the following sections:*

Purpose

Procedure

Results

Conclusions

Action

- *Be sure that the purpose is clear and thorough. Add the graphics that are needed (you may photocopy graphics from the lab book also). Assign each graphic a number—as in "Figure 1"—and refer to each figure as you write the report: (see Figure 1).*

- *Attach the original lab to the back of the modified laboratory report in an additional section with a cover page titled:*

Appendix

Original Lab Report

- *If you base your project on a tear-out lab report, be sure to restate everything in your own words.*
- *If you are in a drafting design program or a similar program that involves site reports, develop a formal typed version of one of the reports in which you examined a facility, a construction site, or a similar setting. For this project use the following basic headings for guidelines:*

Introduction

Background

Status (observations)

Advisory Discussion (or recommendations if provided)

When the project is returned to you:

At your instructor's request, resubmit *the project with the corrections that are indicated on the original.* Boldface *all corrections so that the instructor can see the revisions. Use the* Writer's Handbook *as your reference.*

Share a Project: First Option

Work with three other members of the class to discuss and edit the set of laboratory reports. You may use the same groups that were used previously, or you may develop new groups.

- *Consult Appendix A. Before, holding the meeting, review the sixth editing checklist for formal laboratory reports.*
- *On the day the projects are due, have each member of the editing group explain his or her project in terms of objectives and project development.*
- *Hand the papers around for a critical reading and editing.*
- *Have each member edit the texts for both writing errors and technical errors.*
- *Have each member write a one-paragraph critique at the end of each project.*

Share a Project: Second Option

A special task group of volunteers could undertake the following interesting project.

It would be instructive if class members who are currently working in a technical environment where they document laboratory procedures or use calibration tests or similar activities could present various laboratory reports or test forms from the workplace to the other members of the class. Perhaps someone in the class could also gather site reports. Four or five such members could form a group to discuss document design. Each member would present and explain his or her lab forms or reports.

In order for the class to have an opportunity to hear the discussion and see the documents, the special task group could present a panel discussion in front of the larger group, with the samples passed among the audience.

Work in Progress

Happy Endings

I was very lucky to have to write all these regulations in one language, English. Pacific Aero Tech is a company that could easily hire individuals who are bilingual and for whom English is their second language. OSHA requires that safety be understood in the language the individual employee knows best. In the future there might be a need to make these programs available in different languages!

Since I had seven other topics to compile and write, the final project was far from over. Tenacious time management was the sole encouragement to get them done. It was an overwhelming need to be finished that drove me to get the job done. The following is the list of those parts that I still had to write!

- *Accident, Incident Reporting*
- *Bloodborne Pathogens*
- *Employee Fire Loss Prevention*
- *Fire Extinguishers*
- *Flammable Materials*
- *Personal Protective Equipment*
- *Records and Reporting*

My method of writing is straightforward. I write first and edit for grammar and content at a later date. My editing consists of several trips to the computer for various reasons—first I look for obvious errors such as spelling and typographical errors. They are usually easy to find. Then I put the project away for awhile and return again to check for grammar and clarity. As a last resort I ask friends to read portions for anything that I may have missed. I was careful not to ask too much of them since there was so much to edit. I normally did my own editing and took several days to complete a subject. I believe that taking the time to do quality work is imperative.

After proofreading the entire project several times and making last minutes corrections, I placed my final draft in a three-ring binder similar to that of my original proposal. I separated the individual sections with dividers and labels, and inserted sheets of colored paper in between the sub-topics; this added a little splash of flash. I submitted another cover letter in the front as before, and added reference material to the back.

All in all it took me roughly six months to complete the safety regulation recommendations. It actually took less time for the president to implement some of my suggested practices. Since I discussed the status of each subject in small meetings with Karen and Matthew, they were aware that times were about to change and they were prepared to do so. By the time of completion, the employees were already applying Karen's mandatory hearing protection rule and all those who required the use of respirators were trained properly and were working with precautions. With the two largest safety programs underway the rest were sure to follow.

I have to admit that this assignment was a challenge to complete. Fulfilling the needs of a company that requires such an extensive overhaul in the safety department, was definitely energy draining and time consuming, but the result was the finest work that I ever produced. I am very proud of the whole project and relieved to have it complete—at last. *L.C.L.*

Offers for Services

"Make me an offer."

"Have a proposal on my desk by next week and we will see what we can do."

"These problems are beyond the committee's scope. Let's hire a consultant to analyze the problem and suggest recommendations."

Bids, estimates, and proposals are important parts of the engineering and technological world. The connections between an engineering service and a client are the documents that transact services for fees. These critical documents analyze problems and needs and offer solutions and outcomes—usually for a price. In the course of the mutual arrangements, the bid or estimate or proposal establishes a binding contract. It is for this reason that the three documents are the basic transactions of business agreements.

Engineers and engineering technicians offer services, and they often use some method of presenting bids or proposals. If engineers work within a company environment, these services are usually in-house, and proposals are often used for project development. If the engineers are self-employed or part of joint ventures, the services are directly offered to the public through a bidding process.

As the three opening quotations suggest, there are many uses for bids and proposals. They spell profit and loss for agencies and industries of every kind. Although the three documents perform similar functions, and although people speak of the three terms interchangeably, each one is usually used for a distinct purpose.

- A **bid** is a direct offer of services.
- An **estimate** is a tentative offer of services.
- A **proposal** recommends services. (The suggested services address a condition or solve a problem, and the author of the proposal often cannot act as the service provider.)

Observe that in all three documents, *money* is almost always involved and a *change* of conditions is the inevitable outcome of any of the three. Because expenditures and changes are often challenged, these documents are often scrutinized by contending interests who either do not want the expense or the change or both. In this respect bids and proposals are often political, although the person or business that offers the bid may never hear or see the heated discussions that can go on among the potential clients.

There are two basic types of contracts: *service bids* and *development proposals*. Most bids that offer a service for a fee are documents that provide information for technical and financial decision-making purposes. In this case bids that are designed to offer a client services will offer predictable activities:

An offer to install

An offer to treat

An offer to update

An offer to correct

An offer to design

An offer to produce

An offer to perform

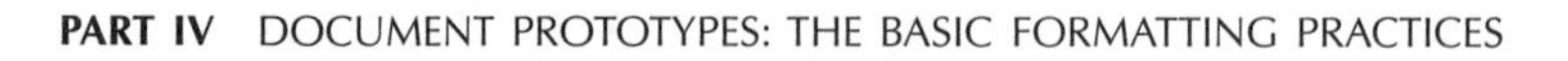

An offer to process

An offer to build

Proposals that are intended to offer developmental suggestions often have a different focus:

A proposal to monitor

A proposal to promote

A proposal to rate or compare

A proposal to coordinate

A proposal to expedite

A proposal to research

A proposal to solve

A proposal to study

Contract Strategy

The key concern you will address with a bid or proposal is "closure"; you want the offer to be accepted. The success of the document will depend on several factors, the first of which is the "need" of the client. The proposal must be on the mark. If you do not focus strictly on the needs of the client, the offer will be rejected, and the financial analysis will be inaccurate. If you are working with "bid requests," you will have the guidelines that precisely identify client needs. Use the bid request as a line-by-line plan for organizing your proposal. If there is no bid request, preliminary discussions or correspondence will be useful.

If, for example, you offer to do a six-terminal network installation for a small real estate office, there probably will be no bid guidelines. Public agencies and large corporations commonly use bid requests. In the case of the real estate office you must interview the owners and conduct a site tour with them. Carry a notebook and record all their requests. Ask for the opportunity to discuss the plan with the staff, particularly any other terminal users. Discuss the feasibility of completing the installation without interfering with work in the offices. With this information in hand, address a letter to the real estate agents who own the agency and list the needs they identified and those you perceived. Request corrections. This letter, once it is approved, serves as the basis for your bid.

Because the bid or proposal is intended to persuade the potential clients to accept the terms of the agreement, it is extremely important to show your professionalism by fully defining and understanding their needs. The preliminary dialog demonstrates your commitment while it gathers the precise information you need to analyze the potential of the contract. More importantly, if the needs are then precisely met in your bid, you create a winning image and a winning argument that will persuade your clients that you offer the *best* plan of action.

Notice that the bid strategy I have outlined avoids low-bid tactics. Instead, a "best service" approach is the preferred technique. If the bid happens also to be reasonable, then success is likely. Lowest-bid agreements are not particularly desirable, even though state and federal agencies often use bottom-line logic because they are distributors of third-party moneys—yours and mine. In truth, the parties to low-bid contracts are often frustrated by their attempts to use bottom-line logic, and the work that results may be inferior.

In addition to the "best service" approach to calculating a bid, there are other considerations that can be effective. Three options are based on the spirit of competition:

OFFER MORE
OFFER BETTER
OFFER EXPERTISE

Sample 14.A

-14-

GLOSSARY

The Catalyst 5505 is a high-performance, 5-slot chassis for the evolving Catalyst 5500 Series. It combines the size of the original Catalyst 5000 with the performance boost and added features of the Catalyst 5500. The Catalyst 5505 is ideal for high-performance wiring closets and data applications.

Cut Sheet A diagram that show all of the cabling and how and where it is run in a computer network. This greatly simplifies troubleshooting.

Fiber Optic Cabling is capable of carrying data at rates of 1000 Mbps and higher. Note: all cable chosen for this installation is plenum grade and complies with local and federal fire codes.

Fiber-to-copper media coverters are used for converting fiber connections to standard Ethernet cabling ports.

IT Information Technology

LAN (Local Area Network) Usually limited to a single room or building.

Netgear's 8 port dual speed hub enables you to mix and match 10 Mbps and 100 Mbps network devices on the same network with ease. It transmits information at 10 or 100 Mbps, according to the speed of the network devices attached. This hub is ideal for small networks in transition from 10 Mbps to full 100 Mbps.

WAN (Wide Area Networks) A network that has no limitations on distance.

Agencies that seek lowest-bid proposals are not likely to want more or better outcomes than are expected, but if the client is negotiating his or her options, then the competitive edge goes to the superior offer. Offering the *best understanding of the client's needs* assures the best outcomes and the most accurate cost analysis, but offering *more services* or *better services* may be to your advantage. If the discussion with the real estate office staff led to grumbling because there is not a seventh terminal station, perhaps you could offer one more station. If there is a superior business software application about which the real estate agents are unaware, you can offer a better product and perhaps roll it into the existing costs. And, by doing so, you demonstrate your *expertise* and make a unique, superior offer.

Good communication is critical, so a useful feature that might prove to be valuable is a glossary for your proposal. Most contract writing is handled with extensive appendices but there is little by way of clarification for readers who are not aware of the meanings of technical jargon. Since clients may be unfamiliar with your industry—or at least its more technical ramifications—a list of terms might enhance the proposal. Place the list before the appendices and explain its presence in the text of the proposal. The terms might include products as well as terminology as you will note in the list of terms on the left (Sample 14.A). This list was originally illustrated. Images of products add to a client's understanding. Since no other proposal is likely to have a list of terms, you are again showing additional expertise and interest.

Your concern for the client's interests and an attractive bid presentation should build a favorable impression of your professional standards, but there are other ways to assist the client in the decision-making process. Your professional qualifications may not be fully reflected in the bid proposal. The resume is probably a seldom-used tool in the world of contracted agreements, but similar documents can be quite useful. You might develop an employment history that explains your work experience in some detail. This document could include a brief list of major work-related accomplishments, namely, a short itemized presentation of your major contracts to date. For example, I maintain a list of articles that I have published, and I will occasionally show the list to a potential client (a publisher), because the articles are my work-related accomplishments as a writer. Similarly, note Teresa's list of engineering projects in Chapter 11 (Sample 11.L) in the *Writer's Handbook*.

Another helpful handout that you could prepare would be a List of Referrals. The list would contain the names of a few businesses, with the addresses and phone numbers, and the owners' names. If you are licensed or bonded or belong to professional associations, carry copies of verification that you can share with a potential client. All these documents can be part of your standard portfolio. I had the good luck to have a photo of me, taken at work, appear in a national magazine. I used the magazine as part of my presentations during contract discussions when I ran a contracting business. The photo was just another way to suggest expertise. The goal is to project know-how and integrity.

Model 14.A (1)

HENDERSON CONTRACTING SERVICES

Commercial Interiors and Exteriors

4011 East Evanston
Seattle WA 90000

July 10, 200X

TO: Dr. James Kent
Northwest Clinic
9800 35th Ave. NW
Seattle WA 90000-1234

PROPERTY: Commericial/Clinic

JOB DESCRIPTION: This is a bid for a paintout. The entire interior of the building is to be painted in accordance with the requests of both Dr. Kent and Dr. Savary. Full description appears below. Conditions attached.

1. Estimate inclusive of labor and materials for various interior painting.

   A. For painting, one-coat, all painted surfaces throughout the clinic, excepting attendant quarters and those areas listed below. Rooms include upper walls and all clerestories of hall, eastside entry, lobby, reception area, receiving, examination rooms (3), laboratory, X-ray room, kennel, and rear storage area. $XXX.00

   Suggested coating: premium eggshell latex "enamel." Uniform color throughout.

   B. For painting, one-coat, staff office, including clerestory. Suggested coating: premium eggshell latex "enamel." Color to be determined. $XXX.00

   C. For painting, one-coat, doctors' office and clerestory including bookcase wall. Owners to disassemble bookcase modules for access. Suggested coating: premium eggshell latex "enamel." Color: uniform with general color used throughout. $XXX.00

   D. For painting, one-coat, surgery room, surgery preparation room, and bath area. Suggested coating: premium eggshell oil-base enamel. Color: uniform with general scheme used throughout. $XXX.00

   Includes minor repair and priming of skylight area in surgery preroom.

Subtotal	$XXX.00
Plus tax	________

Bids and Estimates

Bids and estimates and proposals offer somewhat different conditions of agreement. A simple bid for services is quite uncomplicated. For a simple bid there are five considerations that you would examine.

What is the client's problem or need?

What are your solutions?

How much will it cost?

How will it be done?

How will it be paid for?

Contracting agents who offer bids of this type often use a statement tablet from a stationery store and they do not bother to deal with the first question on the list. They simply offer line-itemed services (solutions) at a cost for labor and materials, and they may include a suggested dateline.

If you want to develop a bid format that is effective, begin by constructing the document on a computer. This will make you look professional and will indirectly help eliminate competition that is less attentive to bid submissions (note Model 14.A). In addition, although bids, estimates and proposals look like formidable documents to create, once you have constructed one, it can serve as a prototype for future uses. In fact, the simpler versions can be used repeatedly if you make a template out of the bid (see Appendix B).

To Make a Template

1. Go to **File** on the menu bar.
2. Click **Save As**.
3. In the dialog box, click **Save as type** (it is near the bottom).
4. Select a location among the files.
5. Select **Document Template** (it is near the bottom).
6. File the template in any folder in the dialog box window or make a folder. Move the cursor to the selected file and double-click.
7. Important: Change the file name. Simply add a new date, for example (do *not* use the slash).
8. Click on **Save**.

This estimate spells out all the details of a contractor's offer. The costs are broken down by work areas, but notice that labor and material are figured as one cost in each case.

Model 14.A (2)

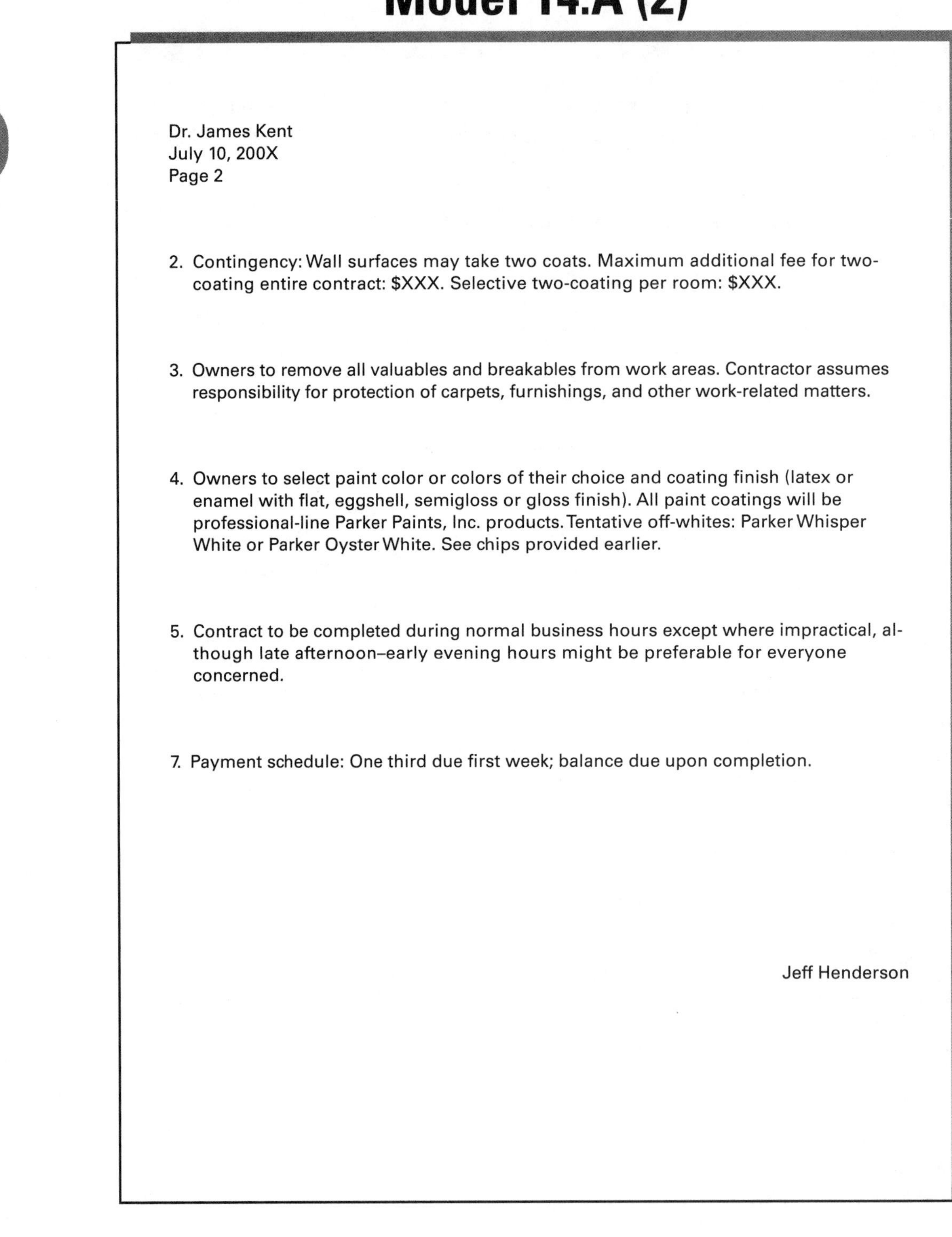

Dr. James Kent
July 10, 200X
Page 2

2. Contingency: Wall surfaces may take two coats. Maximum additional fee for two-coating entire contract: $XXX. Selective two-coating per room: $XXX.

3. Owners to remove all valuables and breakables from work areas. Contractor assumes responsibility for protection of carpets, furnishings, and other work-related matters.

4. Owners to select paint color or colors of their choice and coating finish (latex or enamel with flat, eggshell, semigloss or gloss finish). All paint coatings will be professional-line Parker Paints, Inc. products. Tentative off-whites: Parker Whisper White or Parker Oyster White. See chips provided earlier.

5. Contract to be completed during normal business hours except where impractical, although late afternoon–early evening hours might be preferable for everyone concerned.

7. Payment schedule: One third due first week; balance due upon completion.

Jeff Henderson

In the case of a simple bid, the problem and the solution can often be combined in one or two sentences of intent:

> Contractor will grade driveway, provide a 2″ bed of 5/8″ gravel, and install an asphalt surface as per the specifications below.
>
> T & D Incorporated offers to upgrade the existing four-terminal system to the desired level of performance (see below). The new installation involves a seven-terminal network, server, and software upgrade.

Each section of the bid can be given a heading. Either of the two preceding statements could be placed under the simple heading "Proposal." The sections that follow would involve costs, timetables, and special considerations such as the problem of how and when to allow for system downtime during hours of operation.

The cost analysis, in particular, should be thoroughly detailed. Divide the costs into *equipment, supplies, labor, subcontracts,* and *overhead costs.* In the network bid (see Model 14.B), for example, every computer cable must be accounted for even if an omission is only a $15 SCSI that can be bought at an electronics store if there is an oversight. You need to be thorough so that you do not lose money. The worst-case scenario is that of a contractor who submits a fixed bid, underestimates, and pleads for extra money halfway through a contract. No business is better than bad business in this case, and accurate cost estimating is the key.

One way to propose a bid is to offer, services for the cost of time and materials (TM). Another option is to offer a contract using the system known as *cost plus;* then you total the labor and materials and add a percentage of the total. In this case you are working at cost plus a percentage. The percentage can be identified as "overhead."

Small businesses frequently operate on narrow profit margins, so the bid should conclude with very specific guidelines for payment (see Model 14.A). If, for example, you will depend on the client for money to buy materials, this must be made clear:

> One third of the gross cost for materials and supplies is due in the first week.

Many contractors, large and small, use such a provision to avoid losses in a default. They can suffer the labor loss more easily than they can recover an outlay of their own capital. Other considerations can also be defined:

> Labor costs are to be handled as a weekly expense.

The failure to explain or identify work conditions or other agreements is a common oversight, In every contract, Henderson included comments of the sort that you see here.

Model 14.B (1)

EZ Network Solutions
42901 Hillshire Blvd., Suite 301
Bellevue, WA 90000
January 27, 200X

Mr. Scott Pearl, Director
Contract Services
Small Business Insurance Group
Smith Building
Fourth and Lincoln
Seattle WA 90000

Dear Mr. Pearl:

Thank you for allowing me this opportunity to present my solution to the networking needs of the Branch Claims Offices, known as BCO's at Small Business Insurance Group.

Our understanding of the needs for this system are as follows. First, you would like to install a new network between all the Field Claims Representatives and each BCO and its corresponding Business Center so that everyone in this network can access the new regional databases.

I believe that we can accomplish the above goals cost efficiently and with a minimum of business interruption. Attached to this letter is a full proposal to install and implement a complete information technology system at an individual BCO. This proposed system would be able to be replicated at all 38 of Small Business Insurance Group's BCOs with little variation from the attached system architecture. The advantage to this is that, with little variation in the systems attributes, the systems administration can be greatly simplified, which makes it much more cost effective and enables your production staff to maintain much of the system without outside consulting.

I will call you in two weeks time to talk to you about details of the attached proposal. In the meantime, if you have any questions please do not hesitate to call me at (206) 523-0000.

Cordially,

Aaron Noble
President, EZ Net

And, finally, the payoff on the contract should be defined so that no ill will develops after thirty days.

> All payments are due thirty days after the completion of the contract.

The simple bid can be presented as an *estimate* of probable cost, but the provisions of the estimate must be spelled out. The conditions are usually legally binding and you want to be quite specific in stating that your offer is *not* a fixed bid. You can define provisions for cost overruns created by clients:

> All additional emendations or requests for changes are expenses that are subsequent to this contract.

You can allow for large openings in cost modifications:

> All costs are subject to change.

You can allow for reasonable limits:

> This estimate may be exceeded by as much as 30% of the initial figures provided.

Commonly, contracts that are too risky to negotiate as a fixed bid are managed by proposing a "time and materials" offer or by using the cost-plus method. If the working relationship between the contractor and the client is well established, these two approaches can be mutually convenient, since the client has the right to stop the contract at any point if necessary.

To provide a polite touch to these legalities, add a cover letter to each bid or proposal. Depending on the nature of your client relationship, the letter can be quite formal (see Sample 14.B) or rather casual. Open by thanking the client for the opportunity to present your proposal. End by promising to call in a week or so to see if the client has any questions. The body of the letter can briefly highlight the proposal and it may also be used to handle additional details that aren't mentioned in the bid or proposal.

Model 14.B (2)

SYSTEM UPGRADE PROPOSAL FOR
HOLLAND LEGAL COPY SERVICE
from
plusNetworks

Proposal

Our recommendation to Holland Legal Copy Service is to avoid proprietary systems (both hardware and software) which may restrict both horizontal and vertical integration possibilities. Since your intention is to upgrade your existing computer facilities, the expense of the proposed basic computer system is one that will be incurred regardless of the type chosen. Basically, you should get the best system you can for your investment. OTS software guarantees compatibility with client software (it is identical), which eliminates the complications and the expenses of proprietary systems.

The most important point of our proposal is that it allows Holland Legal Copy Service to enter the field of optical archival systems without a sizable financial liability but with a solid hardware foundation that allows you to utilize any proprietary system (including Vmicro) in the future, should you choose to do so.

Background

Holland Legal Copy Services is having difficulty with the archival capacity and the retrieval capability of the existing computer systems. The company needs both hardware and software support for suitable high-speed scanning and a CD-ROM archival and retrieval-on-demand system with capabilities comparable to and compatible with the existing system proposed by Vmicro—but without the sizable investment required.

Benefits of Upgrading

Utilization of OTS generic hardware allows complete upgrade and expansion possibilities of all integral components without the limitations posed by proprietary systems. Since all hardware conforms to hardware specifications certified by Vmicro, switching back to their proprietary software at any later date poses no hardware limitations. Also, by utilizing a local systems design firm (plusNetworks), technical support is literally a phone call away, which eliminates complications that arise when companies are based in other states.

Utilization of OTS software also allows complete upgrade and expansion possibilities, as well as general compatibility with other hardware.

Proposals

Bids allow one company to offer services to another. Proposals offer the same option, but the documents are usually larger, more complex versions of a bid. Proposals also function as in-house management tools to facilitate changes within a company.

Proposals are attempts to solve problems. The point of a proposal is to offer *solutions,* but to do so it must clearly explain the *problem.* Because either expenditure or change (or both) will result from any modification brought about by a proposal, the problem must be identified in order to overcome the inherent tendency of the status quo to resist "more costs" or "more charges." The document attempts to *prove* that the proposal is workable and valuable.

Since the proposal seeks approval, it is persuasive in intent. In the case of a *service proposal,* you must convince the potential client that, among the contenders, you are making the best offer. In a *development proposal* submitted to your supervisor you would be similarly persuasive, perhaps in arguing the need for an additional secretary or new lab equipment.

Consultant firms develop a wide variety of proposals, and these proposals combine the concepts of a service proposal and a development proposal. The consultant is paid to offer the proposal rather than to execute it. In this case the proposal is *itself* the service rendered for a fee, and the client has paid for consultation. This proposal will involve professional recommendations that are being sought by a client. For example, if a company is downsizing, the management may try to avoid the appearance of disregarding employee interests. The management team may hire a consulting firm to analyze the problem and suggest downsizing solutions. The third-party implementation plan is a neutral and objective point of view that all parties might find tolerable in an unpleasant situation.

The following is a summary of the three varieties of proposals, which are the most likely proposals that you will encounter:

Proposal for Services (and Products) The proposal for services is a "bid" on a large scale. It is likely to include extensive explanations of needs and recommendations. You are most likely to draft this type of proposal if you are self-employed or negotiate contract awards between companies.

Proposals are generally more complicated than bids or estimates. They follow a problem-and-solution structure that involves a discussion of the existing conditions (see Background) and proposed improvements (see Benefits of Upgrading).

Model 14.B (3)

plusNetworks

2

There are additional benefits:

—Software will be owned by Holland Legal Copy Service, eliminating licensing and royalty fees and per-scan charges, which will result in lower overhead costs and lower costs to potential clients.

—OTS software generally maintains compatibility with other software, and this feature may provide application possibilities not considered in proprietary formats.

—By utilizing software already in use by the client, no training investment is required by the client or Holland Legal Copy Service. This also eliminates the need for an in-house training staff or the need to train existing sales staff to use a software system.

Risks

We judge your risk level to be very low. By not utilizing proprietary software, restricted compatibility with clients who already use proprietary wares may be encountered. However, if clients are utilizing such software, chances are very high that they are already doing business with another service bureau. Also, restricted compatibility depends on claim of full compatibility with existing legal software. If the claim is valid, compatibility problems should not be an issue.

The only other complication that could be of note is that of obtaining the OTS software that most adequately emulates the majority of tasks performed by your existing software. However, it is currently in stock.

Procedure

Hardware recommendations will be based on the requirements specified in our proposal (see attached) in order to maintain hardware compatibility.

Implementation will take four weeks of lead time and one week for service installation. Front offices will be scheduled by mutual agreement. It is in the best interest of both parties to seek low traffic hours for the period of the installation. Back-office upgrades will occur during normal business hours.

Susan Dunn
plusNetworks

In-house Development Proposals Development proposals are used inside corporate environments as the path of modification and change within the corporate system. Companies are usually conservative in their attitudes: if it isn't broken, don't fix it. The proposal is a tool that provides a plan of action and suggestions for change, so it *must* demonstrate need and overcome skepticism about its recommendations. You are likely to develop this type of proposal if you work within a company and deal with departmental needs.

Consultative Proposals Consultants offer expertise that is often of value to a company. They are contracted for their services, which include proposed recommendations. They are also selected because they represent an objective viewpoint independent of company persuasions. This type of proposal is only of concern to you if you work for an engineering consulting firm.

In most in-house development and consultative proposals, a logical argument has to be developed to convince readers to implement the suggestions in the proposals. Costs are usually rationalized as near-term investments for long-term gains, such as cost reductions or profits. Other changes are encouraged on the grounds that they will improve the quality or conditions of the company.

Procedure

The needs or problems that have to be met constitute the opening phase of the proposal. The proposal is frequently *requested,* in which case there is likely to be an understanding of need, often in writing. If there is documentation that defines the need, adapt the material to your opening comments. If the needs were properly prioritized, you can itemize them and briefly discuss each major consideration. Place the original documents in an appendix to the proposal.

If *you* initiate a proposal, the challenge is to identify your own needs or the needs you think should be addressed. If lawn crews are making an enormous disturbance outside company windows each Friday morning, you can send a brief proposal to your supervisor if you see a way to resolve the problem. At this level of importance, an e-mail memo might suffice. Explain the problem and offer recommendations and you have a development proposal!

The addition of a "risk" analysis is not common to all proposals. It is an acknowledged component of network proposal writing methods that are taught in certification programs.

Model 14.B (4)

plusNetworks

3

Hardware Recommendations and Estimated Costs

Personal Computer

—Acer PCI motherboard with Intel Triton chipset, AMI Plug & Play compliant BIOS w/FLASH EPROM support	
—133 MHz 80586 processor	$XX
—2 GB SCSI hard drive and interface (internal)	$XX
—16 MB RAM	$XX
—512 KB cache	$XX
—3.5≤ (1.44 M) and 5.25≤ (1.2 M) floppy disk drives	$XX
—2 High-speed serial ports (16550 UART), 2 parallel ports	$XX
—4X CD-ROM w/SoundBlaster	$XX
—28.8K Fax/Modem	$XX
—Light Pen and Card	$XX
—MPEG2 Video Board	$XX
—SVGA monitor (one of the following)	
—DEC 17 Multisync	$XX
—Amdek 17≤ .28, 1024 ¥ 768	$XX
—ViewSonic 17≤, 28, 1280 ¥ 1024 (n.i.)/1024 ¥ 768	$XX
—Sun Systems 21≤ SVGA	$XX
External Hard Drive	-N/A
Adapter Card for External CD-ROM Equipment	
—Buslogic BT-542B SCSI adapter	$XX
—SCSI cable	$XX
—External SCSI terminator	$XX
Scanner (one of the following)	
—Fujitsu 3097 high speed scanner	$XX
—Kodak 500 & 900/990	$XX
—VisionShape VS-1250	$XX
—TDS 2610	$XX
—Bell & Howell 6338A	$XX
Scanner Controller Card	
—Kofax 9250-1201	$XX
—Kofax 9250-1206	$XX
—Kofax 9250-1208	$XX
—Kofax 9275 (dual side scanning)	$XX
Estimated Total Cost (parts and labor)	$XXXX

Schedule of Activities

As you requested, we can begin Monday, September 8, and will conclude the upgrade by Friday, September 12. Evan will arrive at 9:00 a.m. on the following Monday, September 15, to walk through the changes with the staff. We are on call for the remainder of the week as needed.

Once you are ready to write a more involved proposal you must decide on a structure, because these documents are longer and more complex than the simple bid or brief memo proposal. In their least complicated form they can be developed with only three headings:

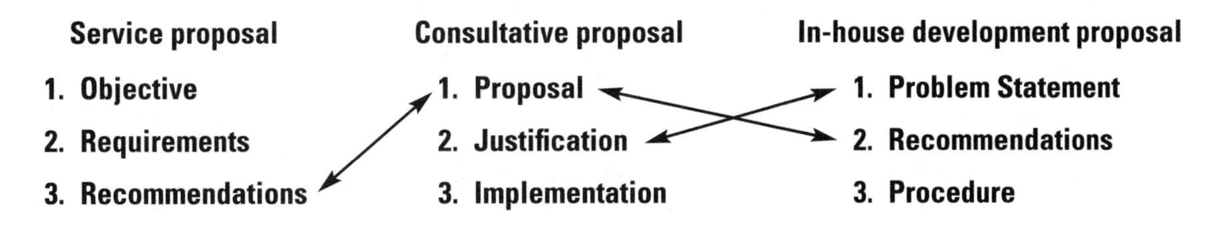

As the arrows indicate, each format differs somewhat from the others, and there are a great many variations on them. If more categories are needed for various considerations within the proposal, a somewhat larger structural model may be useful:

1) Proposal

2) Statement of the problem
(Reverse 1 and 2 if there is an opening summary.)

3) Procedure

4) Costs

5) Schedule of Activities

Note the similarity of this outline to the lab report formats you examined in Chapter 13. Headings and precise organization are a convenience in both environments.

At the most ambitious level, these documents can be elaborate affairs that may contain hundreds of pages:

1) Executive Summary

2) Description of Problems

3) Background

4) Purpose

5) Proposed Solutions

6) Benefits

7) Risk Assessment

8) Implementation

A detailed cost accounting is expected in any bid, estimate, or proposal. It is extremely important to account for the anticipated costs, which helps convince clients that their expense is justified.

Model 14.C (1)

WESTERN ELECTRONICS

EXPANSION PROPOSALS FOR THE DEVELOPMENT OF A MICROPROCESSOR PRODUCTION FACILITY IN PHOENIX, ARIZONA

Prepared
for Donald Jamison, President
and the
Board of Directors of
Western Electronics,
Portland, Oregon

A Production of Market Research Associates
812 Pike Street
Seattle, Washington

Report Supervision
by
Mary Westlin, Chief Analyst
Ronald J. Spangler, Market Analyst Director

Abstract

This report analyzes the production needs of Western Electronics for the period 2000–2015. The report recommends the development of a new plant facility in Arizona. The analysis examines the expected growth in Western Electronics and outlines the growing demand for IC and more sophisticated chip manufacturing and assembly. Geographical considerations are emphasized with reference to transportation and manufacturing of support needs, skilled employee pools, and cost considerations.

November 30, 200X

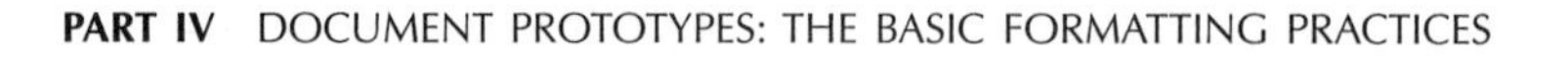

9) **Timetables**

10) **Personnel or Contracts (Needed Agents)**

11) **Facilities (Needed Space)**

12) **Equipment**

13) **Budget**

14) **Appendices**

The cover page and the table of contents from the Western Electronics proposal (Model 14.C) suggest the scale of a larger proposal. For your purposes a proposal that is developed in three to six subsections will be adequate, but observe the options available to you in the longer list. When any of these categories are needed, you can add subsections depending on your needs, or remove them when they are not appropriate.

The sections of a proposal can be developed in paragraphs, outlines, or lists, and there is probably a tendency to use all three format options as the proposal progresses. Here is a sample layout and suggestions:

1) **Proposal.** State your recommendations. Use paragraphs if this is the opening of the proposal. If you begin with the statement of the problems and place the proposal second, then you can itemize the recommendations in lists.

2) **Statement of Problem.** Discuss the problem. If you are seeking a contract, the problem should be explained in terms provided by the potential client or the supervisor.

3) **Procedure.** The procedure explains how you plan to get from the problem to the solution. Think of this as a proposed "strategy."

4) **Costs.** Be as precise as possible in this section. It is time consuming to research the cost, but it matters to your business success. Your client will appreciate a neat and orderly analysis.

5) **Timetables.** Be cautious and avoid unrealistic promises. From my experience, I suggest a win-win strategy. If you want to start on the tenth, tell the client it will be the fifteenth. If it will take two weeks, ask for three. If you then offer to arrive *early* and also finish *before* the deadline, you will be admired. If you start late, however, you leave the most common poor impression an otherwise highly respected contractor is prone to creating.

If you compare this formal cover with the student sample on page 174, you will realize that they are essentially the same in structure and detail. Once you master the conventions of writing, they serve many purposes.

Model 14.C (2)

TABLE OF CONTENTS

page

As I noted earlier, proposals should have a cover letter (see Sample 14.B), and many businesses make it a practice to include a cover letter with simple service bids as well. The cover letter is the contact point between the two business parties. Cover letters are cordial and add a slightly personal touch to the affairs of business. Chapter 12 includes the cover letters for the bid and one of the proposals that accompany this chapter. Page 324 presents the cover letter for Henderson's bid. Page 326 presents the cover letter for the large thirty-page Western Electronics proposal, two pages of which appear here. Follow-up letters are also appropriate, particularly if a bid or proposal has been accepted. There are always additional wrinkles to address in follow-up letters, but nevertheless, a brief letter of appreciation and acknowledgment is always in order to seal a contracted agreement (see p. 312.)

Illustrations and Supplements

It is often said that proposals must be realistic to succeed. The suggestion concerns costs, of course, but there is more to consider. Be as vivid as possible in the proposal, and simplify the complexity of the presentation where possible so that readers will not get bogged down in the details. The successful proposal writer knows when to relegate endless details to appendices. Moreover, appendices allow you to include vivid materials such as product descriptions that should be viewed as important considerations because the photographs, drawings, and technical specifics bring your proposal to life. Include vivid specification sheets and brochures and similar materials and refer the reader to them in the text. A vivid proposal encourages a "can do" attitude among clients. They can begin to see the results before the project begins. Even the Henderson bid included technical specifications and product descriptions for his paints, and he included sample color cards as well.

Another reason to be as vivid as possible is that there will likely be a variety of readers. Often the CEO, who will sign off on the technical contract, is someone who does not understand the technicalities. Architects will show blueprints to a client, but unless the client is an engineer, the blueprints do not visualize the product. At times, products such as blueprints clearly create a valuable image of expertise but still leave the clients unconvinced that they should accept a contract. As every architect knows, a simple floor plan and the realistic illustrations of the finished product help make the sale. As another example, the new computer network installation businesses that are appearing in every metropolitan area have quickly adopted innovative software that allows them to create floor plans that illustrate network layouts of computer installations for clients.

Proposals can involve hundreds of pages of text. As the document grows, the need for organization becomes very important. A table of contents is one obvious tool that will help a reader handle larger documents.

Model 14.D (1)

NETWORK DESIGN FOR THE AATP LAB

Introduction

This is a proposal to redesign the AATP Lab and is to be read by all those directly involved in the program. As a result of the current demand for certified IT people, namely, **MCSE** and **A+ Certified** (see glossary for all **boldface** terms), the AATP Lab has been quite successful since its birth nine months ago. The large enrollment of students and broad scope of classes offered (geared toward training students for certification—and ultimately jobs) has required that the lab be open seven days a week, and on some days as many as fifteen hours per day. Because of the overwhelming response, and the current lab configuration, it has been determined that the lab must be redesigned to accommodate more students and allow for growth.

1

Existing Conditions

- There is no standardization of hardware (i.e., monitors, keyboards, and mice.)
- The main **fiber optic** line into and out of the classroom (see Figure 2.a) is connected to a bridge that is sitting out on a table in the classroom, which is accessible to everyone.
- The servers for the class, including the Web server that has the AATP's Web page and other Web pages on it, are sitting out on two desks in the server/storage room. (See Figure 2.b.)
- In addition, there is no separate remote area equipped with computers for the lab technicians to perform daily tasks, such as testing, repairs, and upgrades.

2

3

A Proposal: Network Systems

Proposals can be constructed in a variety of styles. The plan suggested in the accompanying diagram is one of many possibilities, as is Model 14.D. This model is an in-house development proposal in which one agency of an institution is offering a proposal to meet the needs of another agency. The cover letter has been omitted.

Proposals

Introduction: ______

Statement of Conditions (needs): ______

Proposal (recommendations): ______

Procedures: ______

Costs: ______

Timetables: ______

1

It is appropriate to provide background for a proposal, although such an introduction is often omitted. The value of the background introduction is that if the letter of transmittal is removed, the proposal still has an appropriate introduction. In addition, in interagency situations, the client addressed by the service provider is likely to be a committee, so opening with a background discussion may be helpful to members who are unfamiliar with the proposal.

This author then moves directly to a description of existing conditions. The focus here is on problem areas that need attention. The various needs of the client must be identified so that the recommendations can address each concern in turn.

3

The existing situation can be examined in a number of ways. A list is quite convenient. The list can be ranked by importance, by cost, by location, or by other criteria. The service provider might structure the list in an order that corresponds to the way the proposal will deal with them.

Model 14.D (2)

Objective

This proposal identifies plans to redesign the AATP Lab to accommodate more students, in a more comfortable setting that promotes learning. It addresses the need of lab technicians to have a work area away from the classroom, and it gives the program coordinators more flexibility for scheduling classes and open lab hours.

Proposal

- The first step in the redesigning of the classroom will be to divide it in half. This will be accomplished by utilizing a drop-down beam in the center of the room and installing a soundproof wall, with a door to pass from one room to the other. (See Figure 1.)
- For standardization purposes we suggest purchasing new monitors, keyboards, and mice.
- The exposed fiber optic line will be rerouted underground and the bridge will be moved. This bridge and fiber optic line will then be reinstalled in a lockable rack in the server room. (See Figure 1.d.) This room will be locked and kept off limits to nonfaculty or staff members for stabilization and security reasons.
- Most of the existing cable (i.e., Category 5) can be used for the redesign of the physical network, but because of the growth of the classes, more cable will need to be purchased. All the existing cable will be reorganized and routed as needed along with the new cable, and a **cut sheet** will be made to keep track of how the

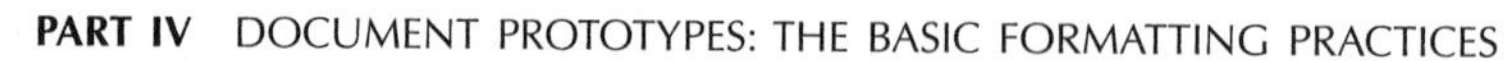

1

The author develops a brief subsection to provide an overview of the intended outcome of the proposal. Like the illustration on page 8, the objective is intended to be a brief overview of the planned outcome of the process.

2

The proposal is detailed but stated in clear language that the nonprofessional will understand. The proposal is designed to address the problems outlined in the section concerning existing conditions. (Several pages are omitted.)

3

Notice that figures are referenced when needed (although most are omitted from the sample). When figures are used in a document, it is important to tell the reader when to look at them so they see the connection between the text and the illustrations.

4

The use of bullets or numbers adds clarity in a proposal. Paragraphs are not a convenient way to list recommendations unless discussion is necessary.

Model 14.D (3)

cables are run, and to help out in troubleshooting the network—should problems arise.

1

- A separate space away from the classroom (see Figure 1.e) should be in place for the lab technicians to perform their job functions without getting in the way of the classes.

1

- The following hardware needs to be purchased to separate the rooms into two networks: A router (Cisco 2514), two switches (Cisco 2924), two port patch panels, two UPSs (uninterruptible power supplies), and two lockable racks to hold and organize all the hardware. (See Figures 3.)

2

Cost

- The estimated cost for the new wall and doors . . . . $xxxx
- The estimated cost for the new carpet and paint. . . . . $xxxx
- The estimated cost for the new tables and chairs . . . . $xxxx
- The estimated cost for of the hardware . . . . $xxxx
    - One Cisco 2514 router . . . . $xxxx
    - Two Cisco 2924 switches . . . . $xxxx
    - Two 24 port patch panels. . . . . $xxxx
    - Two UPSs . . . . $xxxx
    - Two server racks. . . . . $xxxx
- The estimated labor cost . . . . $xxxx

3

4

(For Complete Cost Analysis See Appendix)

1

Notice that the recommendations are not simply listed. Rather, the author includes the rationale behind several of the modifications.

2

Since it is awkward to address committees, the author uses a mixture of general terms and precise technical identifications of products. For some readers a "switch" is clear enough. For others, the type of switch is relevant.

3

Costs are the bottom-line consideration, but, by convention, the cost analysis is placed next to the end of a proposal. Perhaps the logic of this strategy is to convince the client of the needs and outcomes first so that there is less sticker shock concerning the price.

4

This author had a two-page cost analysis that was attached at the end of the proposal. What you see here is an abbreviated overview of the pricing that is adequate for many readers. The others are referred to the appendix.

Model 14.D (4)

4

Conclusion

This is a large project but a worthwhile one that will pay for itself in the months to come. When the lab is finished it will accommodate more students, which means more revenue. The lab will be much more stable, so troubleshooting time will decrease dramatically.

Schedule

The best time to start rebuilding the AATP Lab would be during a break in the school year so that class time is not interrupted. Because we sense that you are anxious to get going on the project and have it ready for the 200X school year, we should start the planning as soon as possible, beginning with measurements of the room, and bids for the new wall and doors. In addition, we will need bids on the furniture, carpet, paint, and hardware. Once these considerations are taken care of, we do not anticipate any problems with starting the rebuilding process on the first day of winter break.

Phase I

On the first day of winter break, the room is gutted and the redesign of the AATP Lab begins.

Phase II

Primary construction and cable installation occur between the third and tenth days of the break period.

Phase III

We complete the installation and move on to testing the network to make sure it is working properly. The lab will be up and running before the first day of winter quarter.

1

2

3

1

A proposal may or may not have a conclusion. This document is subtly designed to sell, and so the conclusion is yet another effort to build an image of a finished product—one that is worth the price. The proposal could have been constructed without the objective section, the floor-plan diagrams, and this conclusion section, but these features help the client see the finished product and help sell the proposal.

2

Timetables are important, and they should be honored. They deserve the same precision that is given to the cost analysis. Materials and labor (costs and timelines) deserve equal attention.

3

The timetable should be worked out by mutual agreement. Notice that there is often a problem with facilities or services that are generally open to the public. Timing is critical if a facility cannot close during construction.

This author used his computer skills to construct an attractive proposal complete with floor plans (in full color) and illustrations from manuals and Web sites. Following is a list of all the hardware and software used to complete this project:

Hardware

- AMD-K6 350 MHz (nonproprietary computer)
- 64 MB of RAM
- Western Digital 6.4 GB hard drive
- 17 in. Amptron monitor
- Hewlett Packard Deskjet 695C color ink-jet printer
- Black/white and color laser printers at photocopy store
- 100 MB Zip drive for backups

Software

- Microsoft Windows 98
- Microsoft Word 2000
- All diagrams built on Visio Professional 5.0

Model 14.D (5)

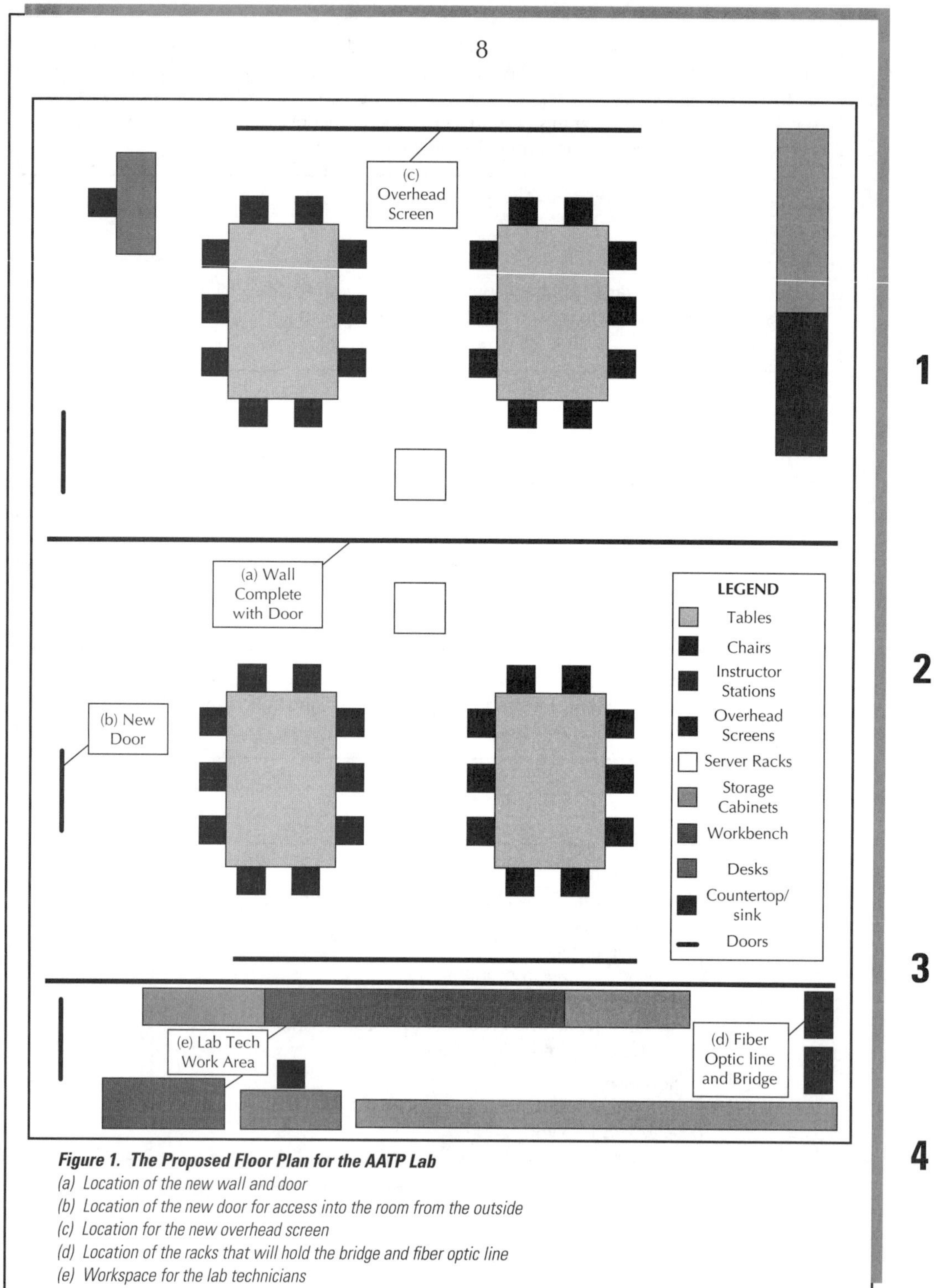

Figure 1. The Proposed Floor Plan for the AATP Lab

(a) Location of the new wall and door
(b) Location of the new door for access into the room from the outside
(c) Location for the new overhead screen
(d) Location of the racks that will hold the bridge and fiber optic line
(e) Workspace for the lab technicians

1

Because a substantial investment is usually involved in contract agreements, the service provider should be sure that there is a clear mutual understanding of every consideration involved in the agreement. Use of space is an important consideration, and drawings are critical.

2

Visual concepts are not clearly understood or shared in written agreements. Words do not render space design in an adequate fashion, so it is important to illustrate key concepts.

3

This author was quite thorough in developing a visualization of the installation. In other illustrations he used representational drawings that illustrated the servers and the terminals.

4

Drawings should be clear to the client. In this instance there is a legend (2) as well as a key that explains the main changes that are explained in the proposal.

Summary

- Bids are used to offer services.
- Estimates are used to offer services under certain conditions regarding cost.
- Proposals recommend services and are of several types
  - Elaborate bids
  - In-house proposals for changes within a company
  - Consultation services
- A bid should be thorough but as short as possible.
- A bid should address all the needs of the potential client.
- A bid should clearly explain the services (and products) to be provided.
- A bid should analyze the cost thoroughly.
- Longer proposals discuss the needs (or problems) of the client at length and provide clearly stated recommendations.
- Proposals usually involve a cost analysis and timetables for completion, and other sections as needed.
- It is a good practice always to include a cover letter with a bid, estimate, or proposal.

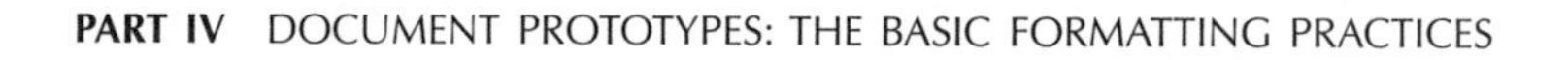

Activities Chapter 14

Use the chapter summary as your guide.

Service Bids *If you are self-employed or working for a small firm, develop a fixed bid for a contract and attach a cover letter. Feel free to use an existing bid from your files, but upgrade the presentation in light of the discussion in this chapter.*

Development Proposals *Create a proposal for a work-related problem you see at your place of employment. Examine the needs and suggest recommendations. Consider costs and other contingencies. Include a cover letter.*

Subjects for development proposals might be inspired by unwelcome changes in your workload, changes that are needed for improving the office environment, upgrades in computer equipment, improvements in safety practices, and so on.

If you are completing a college program with the intention of opening your own business, this is an opportunity to develop prototypes of important documents that you will use again and again. Develop (1) a simple bid, (2) a longer proposal, and (3) a cover letter for contracts that you anticipate in the near future.

Share a Project

Work with three other members of the class to discuss and edit the completed proposals. You may use the same groups that were used previously, or you may develop new groups.

- *Consult Appendix A. Before holding the meeting, review the seventh editing checklist for proposals.*

- *On the day the projects are due, have each member of the editing group explain his or her project in terms of objectives and project development.*
- *Hand the papers around for a critical reading and editing.*
- *Have each member edit the texts for both writing errors and technical errors.*
- *Have each member write a one-paragraph critique at the end of the project.*

Share the Preparation of a Document: First Option

Collaborate with three other members of the class to develop a proposal with a full text. This project will be developed over a period of several weeks, and you will have opportunities to meet to discuss design, research, production, and progress.

- *Select a familiar technical situation that involves contract proposals.*
- *Develop a text that will include a cover letter, a site analysis or other preliminary investigation, a proposal, and a cost analysis or estimate.*
- *Delegate parts of the task to group members.*
- *Discuss the client's needs, design considerations, research, and your progress at each meeting.*
- *Submit the final project to your instructor.*

Share the Preparation of a Document: Second Option

If you are not prepared to develop a technical proposal in your area of study, consider the following possibilities.

Collaborate with three other members of the class to develop a proposal with a full text. This project will be developed over a period of several weeks, and you will have opportunities to meet to discuss design, research, production and progress.

Select a familiar situation that involves one of the following topics or one of a similar nature:

- *A proposal to improve safety in the working environment in the technical labs at the college.*
- *A proposal to make computer upgrades in the campus computer labs.*
- *A proposal to create a network installation in the home of a committee member with children.*
- *A proposal to add a new two-car garage in the home of another team member.*
- *A proposal to make the home of one committee member more energy efficient.*
- *A proposal to protect the home and its contents of another team member by proposing a security system and other security measures.*
- *A proposal to community leaders concerning a specific local issue such as traffic congestion, air pollution, noise pollution, ordinances for signs and billboards or a similar matter.*

Share the Preparation of a Document: Third Option

Collaborate with the same three members who worked with you on the team comparison project or the team proposal projects. The final team submission is a project management report in which you explain how you developed the other projects.

- *Account for your time and effort in the team projects.*
- *Briefly describe the procedures that were used to tackle each project.*
- *Describe the roles of the individuals and their responsibilities.*
- *Describe the outcomes and explain any difficulties that emerged.*
- *Construct a flowchart to graphically represent the schedule you followed.*
- *Submit this project to your instructor as a memo (of 500 to 1000 words) compiled by the group.*

APPENDIX

A

Guidelines For Editing Projects

Peer Reviews

Editing is the process of suggesting or making text revisions in a document. The experience of editing another writer's work is probably new to you, so I would like to establish guidelines for your editing activities. When you write at work, your particular need to assist or be assisted in a writing project is a work-related matter: it is a peer review of a document that is in progress. There are two concerns: the technology and the writing. For now you should focus on the writing.

The easiest way to develop editing skills is to divide the tasks into areas of interest. You know from your own writing experience that editing is usually inadequately handled if it is rushed. The problem is that there are many details that draw attention. If there are three errors in a sentence, it is easy to focus on one and pass over the others simply because a reader tends to choose a point of interest—perhaps spelling errors. You can see this preferential attention in the corrections you receive from your engineering technology instructors. They focus on content errors of a technical nature.

A list of troublesome problem areas is an effective tool that can help you cover all aspects of editing. It is easy to overlook inconspicuous problems, but if you have reminders that tell you what to look for, the editing process can be a tightly controlled activity. In addition, if the problem areas are identified in clusters of editing considerations, the editing process is much more orderly and manageable. For example, your concerns here are quite specific:

word errors	**(spelling and usage)**
sentence errors	**(grammar and punctuation)**
paragraph errors	**(continuity of thought)**
document errors	**(introductions, transitions, and so on)**

In practice, I edit my own work in about four separate readings that focus on each level of development: words, sentences, paragraphs, overall document. The book you are reading has been edited many times, but true to form, I reread the publishing house copy—called a masterproof—three final times and made three different types of editorial corrections in three different colors (blue, red, and green). The blue notations concerned design work; the red notations concerned the text content; and the green notations explained my personal suggestions to the editor.

When I edit student projects, I work on all four levels in one reading because of time constraints. I ask you to do the same. Be patient with your own work and check the project carefully by using two, three, or more readings. There are usually time limitations when editing peer work, so you will have to manage on a single edit. However, you want to keep the idea of "editing clusters" in mind because the groups of editing considerations help organize the activity.

Use two types of editing practices. First amend the text as you read. Circle any problem you see and correct the problem above the circle. Do *not* alter the original text in any way because your instructor still has to evaluate the project. You can also draw arrows to a problem and explain your perception in the margin. In other words, you can make a bit of a mess in the interest of creating a better project for your coworker, but *add* corrections and thoughts; *do not remove* any material or make the original impossible to read.

Once you are done with the reading, stop and review your overall impression. Add a paragraph (or more) of your ideas at the end to explain what you think. This summary is your second task. Do not confuse the role of editing with "criticism." Tell the reader what was well done! The best way to suggest corrections is to match strengths and weaknesses. As you probably know from experience, strong criticism is seldom an effective learning tool. Peer evaluation means *helping*. The idea is not to decide what is wrong but to strive, as a team, to make a document better. Sum up with honest, but helpful and supportive, comments.

I have provided several editing checklists. Use them in order, but use only one for a given set of readings. When new projects are due, move on to the next checklist. Each list suggests edits. The lists are cumulative and expand as you study more of the text. If the projects are written in the approximate order in which they are presented in the text, there is an editing checklist for one or more of the document prototypes you are probably evaluating:

First Readings	**A Descriptive Analysis**
Second Readings	**An Evaluative Analysis**
Third Readings	**A Comparison**
Fourth Readings	**Memoranda**
Fifth Readings	**Business Letters**
Sixth Readings	**Lab Reports**
Seventh Readings	**Bids and Proposals**

If your instructor decides not to use an in-class editing session for these readings, use the checklists as handy editing reminders for your own projects.

First Readings

Editing the Descriptive Analysis

- Note spelling errors.
- Note punctuation errors.
- Note typographical errors.
- Look for subject-verb agreement errors.
- Look for pronoun agreement errors.
- Look for comma splices and sentence fragments.*

Prototype: Single-Topic Descriptive Analysis

- Does the introduction identify the issue in question?
- Is the description divided into logical components?
- Is the description handled objectively?
- Is the description handled without an evaluation?

* *For questions concerning any of the above problem areas consult the* Writer's Handbook.

Second Readings

Editing the Evaluative Analysis

- Note spelling errors.
- Note punctuation errors.
- Note typographical errors.
- Look for subject-verb agreement errors.
- Look for pronoun agreement errors.
- Look for comma splices and sentence fragments.*
- Is the cover page well designed?
- Is the text double-spaced with 11/4-inch margins?
- Is the font attractive and readable (12 point)?

Prototype: Single-Topic Evaluative Analysis

- Does the introduction identify the issue under discussion?
- Does the introduction take an evaluative position or indicate that there will be one?
- Is the text handled objectively?
- Is the evaluation clearly explained?

* *For questions concerning any of the above problem areas, consult the* Writer's Handbook.

Third Readings

Editing Comparisons

- Note spelling errors.
- Note punctuation errors.
- Note typographical errors.
- Look for subject-verb agreement errors.
- Look for pronoun agreement errors.
- Look for comma splices and sentence fragments.
- Is the cover page well designed?
- Is the text double-spaced with 11/4-inch margins?
- Is the font attractive and readable (12 point)?
- Does the introduction clearly state the point of the project?
- Does the introduction use the "outline" concept in order to state key points of interest?
- Does each paragraph have a topic sentence that relates to any plan that is outlined in the introduction?
- Does the author use transitions effectively?

Prototype: A Comparison

- Is the comparison handled by subject or by criteria?
- Is the text clearly constructed to reflect the subject organization or organization by criteria?
- Are the criteria identified and explained?
- If the criteria are of equal value, does the text devote equal time to each?
- If the criteria are ranked, does the text devote greater attention to high priority interests?
- Is the text evaluative or descriptive?
- If descriptive, is the description handled with clarity and without bias?

Fourth Readings

Editing Memoranda

- Note spelling errors.
- Note punctuation errors.
- Note typographical errors.
- Look for subject-verb agreement errors.
- Look for pronoun agreement errors.
- Look for comma splices and sentence fragments.*
- Is the text spaced with appropriate margins?
- Is the font attractive and readable (12 point)?

Prototype: The Memo

- Is the format correctly handled?
- Does the memo *inform, regulate,* or *document* events? Is it orderly?
- Does the opening paragraph use a historical link to relate the memo to another memo?
- Does the memo get to the point in the first several paragraphs?
- Is the content properly prioritized?

* *For questions concerning any of the problem areas above, consult the* Writer's Handbook.

Fifth Readings

Editing Business Letters

- Note spelling, punctuation, and typographical errors.
- Look for subject-verb agreement errors.
- Look for pronoun agreement errors.
- Look for comma splices and sentence fragments.*
- Is the text spaced with appropriate margins?
- Is the font attractive and readable?

Prototype: The Business Letter

- Are the parts of the letter correctly used?
- Is the tone polite?
- Does the first paragraph explain the intent of the letter?
- Is the text prioritized or otherwise structured in an orderly way?
- Are the paragraphs too short, just right, or too long?
- Does the concluding paragraph end with a polite comment?

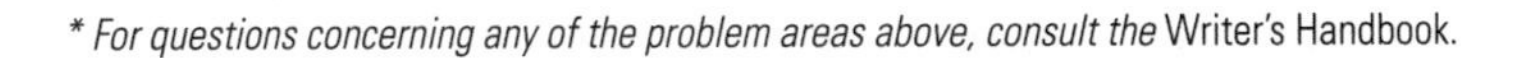
* *For questions concerning any of the problem areas above, consult the* Writer's Handbook.

Sixth Readings
Editing Laboratory Reports

- Note spelling, punctuation, and typographical errors.
- Look for subject-verb agreement errors.
- Look for pronoun agreement errors.
- Look for comma splices and sentence fragments.*
- Is the text spaced with appropriate margins?
- Is the font attractive and readable?
- Are appropriate graphics used?
- Does the text use figure numbers for the illustrations and reference the figures in the text body?

Prototype: A Laboratory Report

- Are the sections of the report clearly identified with titles?
- Is the purpose clearly indicated?
- Are all procedures thoroughly explained?
- Are conclusions stated with clarity?
- If action is an outcome, is a recommended or completed action indicated?
- Are there any technical errors?

* *For questions concerning any of the problem areas above, consult the* Writer's Handbook.

Seventh Readings

Editing Bids and Proposals

- Note spelling, punctuation, and typographical errors.
- Look for subject-verb agreement errors.
- Look for pronoun agreement errors.
- Look for comma splices and sentence fragments.*
- Is the text spaced with appropriate margins?
- Is the font attractive and readable?
- Are appropriate graphics used?
- Does the text refer to illustrations in the appendices?

Prototype: A Bid or a Proposal

- Are the sections of the bid or proposal clearly identified with titles?
- Is the need or problem clearly indicated?
- Are all procedures thoroughly explained or itemized by cost?
- Are recommendations stated with clarity?
- Is the cover letter clear? Is it properly structured? Is the tone cordial?
- Are there any technical errors?
- Are there any cost accounting errors?

* *For questions concerning any of the problem areas above, consult the* Writer's Handbook.

APPENDIX B

Templates and Tips

Document Templates

In all your writing activities it is extremely helpful to conserve your writing energy by using templates to construct your documents. A template is simply a mold in which you cast your essay or your business letter or memo. Since the layouts, particularly of the business letters, are time consuming, it is helpful to be able to reach for a document template and fill in the blanks, so to speak.

The software corporations were slow to respond to this word processing need, but beginning with *Word* 6 for MACs and *Word* 7 for Windows, the template files began to take shape.

The templates are of several types. The "no-text" type of template marks the location for the parts of a document, of a business letter, for example. The "full text" varieties provide a completed document that can be altered as desired. A third variety of template serves as a model to demonstrate mock-ups of such productions as brochures. Most of the templates that will be of value to you are of the first two types, as well as templates of your own making. You will want to construct your own templates so that you can have an *outline* template and a *lab report* template.

The following explanations describe what is available in

Word 6 for Macs

Word 98 or Office 98 for Macs

Word 7, for Windows

Office 97 for Windows

Office 2000 for Windows.

I selected these applications because they are evidently the most frequently installed word processing programs on most campuses. All the applications are similar but different. The instructions will also appear to be similar but there are important changes from program to program.

Note: The number of templates installed in your home computer or your computer lab systems can vary from installation to installation.

Word 6 for Mac

I. Available Templates

- Access

    Double-click on the **hard disk** icon.

    Double-click on the **Microsoft Office folder.**

    Double-click on the **Microsoft Word 6 folder.**

    Double-click on **Templates.**

- **Essay** templates are not available.
- **Memo icons:** There are three memo templates. There is *no* message content, and the templates are used for layouts.
- **Business Letter icons:** There are three business letter templates. The samples reflect the standard layouts and contain *no* message content.
- **Resume icons:** There are four resume samples that can be used as is or altered to suit your needs. These samples are designed for layout, but they contain clear content indicators.
- **Letters folder:** The letters folder contains documents *with* texts that contain underlined information that you should alter. Adapt but the entire document to your needs. Two of the full-text templates may interest you:
    - ✔ Resume Cover Letter
    - ✔ Application for Work (under "Curriculum Vitae Cover Letter")

There are additional form letters that are useful to company employees:

- ✔ Cover Letter for Enclosure (Explains items being forwarded)
- ✔ Cancel Notice
- ✔ Price Request
- ✔ Thank You Letter
- ✔ An Apology for Delay
- ✔ Response to a Complaint

II. To Save a Modified Template

You can alter the template to suit your needs or to conform to the suggestions of your instructors, your supervisor, or the recommendations in *Basic Composition Skills*. These modifications can be saved as a *new* template.

1) Go to **File** on the menu bar.
2) Click on **Save As.**
3) In the dialog box, click on **Save File as Type** (it is near the bottom).
4) Select **Document Template** and click.

Then, to place the template file in the folder, use steps 5 to 9:

5) Click on **Desktop** in the same dialog box.
6) Click on the **Hard Drive** in the dialog box.
7) Scroll down to the **Microsoft Office** folder, and double-click.
8) Scroll to **Microsoft Word 6** and double-click.
9) Scroll to **Templates** folder and double-click.

This places the template in the template folder!*

10) Important: Change the file name. Simply add a new date, for example (do *not* use the slash).
11) Click on **Save**.

III. For Help

1) Click on the **question mark** above and to the right of the menu bar.
2) Select **Microsoft Word Help.**
3) Select an icon or click on **Search.**

** Steps 5 through 9 are clumsy, and they were simplified in all subsequent applications. If this seems inconvenient, do not use templates. Instead, copy an existing project you have on file and type over it.*

Word 98 or Office 98 for Macs

I. To Find Templates of Essay Components

These templates are of the no-text variety.

1) Go to **File** on the menu bar.
2) Click on **New.**
3) In the dialog box select **Other Documents.**
4) Click on the **Thesis** icon.

   You will find:

Title Page
Table of Contents
List of Illustrations
Text Page
Glossary
Bibliography

For the projects in *Basic Composition Skills* you need only the two templates denoted in boldface.

II. To Find Business Writing Templates

1) Go to **File** on the menu bar.
2) Click on **New.**
3) Click on the **Memo** tab.

Memo icons: There are three templates. There is *no* message content, and the templates are used for layouts. For help, click on the **Wizard** icon.

4) Click on the **Letters and Faxes** tab.

Business Letter icons: There are three business letter templates. The samples reflect the standard layouts and have *no* message content. For help, click on the **Wizard** icon.

5) Click on **Other Document** tab.

Resume icons: There are resume samples that can be used as is or altered to suit your needs. These samples are designed for layouts, but they contain clear content indicators. For help, click on the **Wizard** icon.

III. To Save a Modified Template

You can alter the template to suit your needs or to conform to the suggestions of your instructors, your supervisor, or the recommendations in *Basic Composition Skills*. These modifications can be saved as a *new* template.

1) Go to **File** on the menu bar.
2) Click on **Save As.**
3) In the dialog box, click on **Save File As Type** (it is near the bottom).
4) Select **Document Template.** This function automatically places the template in a template file.
5) Double-click on the **Other Documents** folder to place a template in the folder.
6) Important: Change the file name. Simply add a new date, for example.
7) Click on **Save.**

IV. To Save an Original Template

If you build your own title page or any other new page layout, you can use it again and again. You can simply make a copy and modify it, or you can make a template and modify it. *Be sure to save the first outline you develop and the first lab report. Save them as templates.*

1) Go to **File** on the menu bar.
2) Click on **Save As.**
3) In the dialog box, click on **Save File as Type** (it is near the bottom).
4) Select a location among the files.
5) Select **Document Template** (it is near the bottom).
6) File the template in any folder in the dialog box window or make a folder (see part V.). Move the cursor to the selected file and *double-click.*
7) Important: Change the file name. Simply add a new date, for example.
8) Click on **Save.**

V. To Open a Personal Template Folder for Your Templates

1) Go to **File** on the menu bar.

2) Click on **Save As.**

3) Click on the **New** icon on the menu bar in the Save As dialog box.

4) Type a name for the folder on the **New Folder** screen.

5) Click on **Create.**

6) The new folder is created and opened automatically.

7) Important: Change the file name. Simply add a new date, for example.

8) Click on **Save.**

VI. For Help at Any Point

1) Click on **Help** on the menu bar.

2) Go to **Show Balloons** or **Microsoft Word Help.**

3) Activate the help program with a click.

4) Type in a question where indicated and click on **Search.**

Word 7 for Windows *(Office 95)*

I. To Find Templates of Essay Components

These template are of the no-text variety.

1) Go to **File** on the menu bar.
2) Click on **New.**
3) In the dialog box select **Publications.**
4) Click on the **Thesis** icon.

You will find:

Title Page
Table of Contents
List of Illustrations
Text Page
Glossary
Bibliography

For the projects in *Basic Composition Skills,* you need only the two templates denoted in boldface.

II. To Find Business Writing Templates

1) Go to **File** on the menu bar.
2) Click on **New.**
3) Click on **Memo** tab.

Memo icons: There are three memo templates. There is *no* message content, and the templates are used for layouts. For help, click on the **Wizard** icon.

4) Click on the **Letters and Faxes** tab.

Business Letter icons: There are six business letter templates. The samples reflect the standard layouts and contain *no* message content. For help, click on the **Wizard** icon.

5) Click on the **Other Document** tab.

Resume icons: There are three resume samples that can be used as is or altered to suit your needs. These samples are designed for layouts, but they contain clear content indicators. For help, click on the **Wizard** icon.

III. To Save a Modified Template

You can alter the template to suit your needs or to conform to the suggestions of your instructors, your supervisor, or the recommendations in *Basic Composition Skills*. These modifications can be saved as a *new* template.

1) Go to **File** on the menu bar.
2) Click on **Save As.**
3) In the dialog box, click on **Save File as Type** (it is near the bottom).
4) Select **Document Template.** This function automatically places the template in a template file.
5) Double-click on the **Publications** folder to place a template in the folder.
6) Important: Change the file name. Simply add a new date, for example (do *not* use the slash).
7) Click on **Save.**

IV. To Save an Original Template

If you build your own title page or any other page layout, you can use it again and again. You can simply make a copy and modify it, or you can make a template and modify it. *Be sure to save the first outline you develop and your first lab report. Save them as templates.*

1) Go to **File** on the menu bar.
2) Click on **Save As.**
3) In the dialog box, click **Save File as Type** (it is near the bottom).
4) Select a location among the files.
5) Select **Document Template** (it is near the bottom).
6) File the template in any folder in the dialog box window or make a folder (see Part V). Move the cursor to the selected file and *double-click.*
7) Important: Change the file name. Simply add a new date, for example (do *not* use the slash).
8) Click on **Save.**

V. To Open a Personal Template Folder for Your Templates

1) Go to **File** on the menu bar.
2) Click on **Save As.**
3) Click on the **New** icon on the menu bar in the Save As dialog box. (To locate the proper icon, rest the cursor on each icon until the label pops up.)
4) Type a name for the folder on the **New Folder** screen.
5) Click on **Create.**
6) The new folder is created and opened automatically.
7) Important: Change the file name. Simply add a new date, for example (do *not* use the slash).
8) Click on **Save.**

VI. For Help at any Point

1) Click on **Help** in the menu bar.
2) Go to **Answer Wizard.**
3) Activate the help program with a click.
4) Type in a question where indicated and click on **Search.**

Office 97 for Windows

I. To Find Templates of Essay Components

On one Office 97 version, you will be referred to a Web site for the templates. The templates have been placed on the Microsoft Web site, but you cannot download them unless you provide an e-mail address (which the company might use for other purposes. Your best option is to move on to section III and save your own templates. *Do not download on the computers in campus computer labs.*

Other versions of Office 97 may or may not have the essay component's templates. At least one of the versions removed any mention of essay templates.

II. To Find Business Writing Templates

1) Go to **File** on the menu bar.

2) Click on **New.** The dialog window contains seven folders.

3) Click on the **Memo** tab.

Memo icons: There are three memo templates. There is *no* message content, and the templates are used for layouts. For help, click on the **Wizard** icon.

4) Click on the **Letters and Faxes** tab.

Business Letter icons: There are three business letter templates. (There are three fax templates also.) The samples reflect the standard layouts and contain *no* message content. For help, click on the **Wizard** icon.

5) Click on the **Other Document** tab.

Resume icons: There are three resume samples that can be used as is or altered to suit your needs. These samples are designed for layout, but they contain clear content indicators. For help, click on the **Wizard** icon.

III. To Save a Modified Template

You can alter the template to suit your needs or to conform to the suggestions of your instructors, your supervisor, or the recommendations in *Basic Composition Skills.* These modifications can be saved as a *new* template.

1) Go to **File** on the menu bar.

2) Click on **Save As.**

3) In the Save-as-Type box, click on **Document Template.** This function automatically places the template in a template file.

4) Double-click on the **Publications** folder to place a template in the folder.

5) Important: Change the file name. Simply add a new date, for example (do *not* use the slash).

6) Click on **Save.**

IV. To Save an Original Template

If you build your own title page or any other page layout, you can use it again and again. You can simply make a copy and modify it or you can make a template and modify it. *Be sure to save the first outline you develop and your first lab report. Save them as templates.*

1) Go to **File** on the menu bar.

2) Click on **Save As.**

3) In the Save-as-Type box, click on **Document Template,** which will take you to a template folder (it is near the bottom).

4) File the template in any folder in the dialog box window or make a folder (see part V.). Move the cursor to the selected file and *double-click.*

5) Important: Change the file name. Simply add a new date, for example (do *not* use the slash).

6) Click on **Save.**

V. To Open a Personal Template Folder for Your Templates

1) Go to **File** on the menu bar.

2) Click on **Save As.**

3) In the Save-as-Type box, click on **Document Template,** which will take you to a template folder.

4) Click on the **Create New Folder** icon on the menu bar in the Save As dialog box. (To locate the proper icon, rest the cursor on each icon until the label pops up.)

5) Type a name for the folder on the **New Folder** screen.

6) Click on **OK.**

7) Double-click on the folder you created to open it.

8) Important: Change the file name. Simply add a new date, for example (do *not* use the slash).

9) Click on **Save.**

VI. For Help at Any Point

1) Click on **Help** in the menu bar.
2) Go to **Microsoft Word Help.**
3) Activate the help program with a click.
4) Type in a question where indicated and click on **Search.**
5) Select from the options or use the Search button. Click on the **Option** or the **Search** button and a window will open.

Office 2000 for Windows

I. To Find Templates Of Essay Components

1) Go to **File** on the menu bar.

2) Click on **New.**

3) In the dialog box select **Publications.**

4) Click on the **Thesis** icon.

You will find:

Title Page

Table of Contents

List of Figures

Glossary

Text Page

Bibliography

II. To Find Business Writing Templates

1) Go to **File** on the menu bar.

2) Click on **New.**

3) Click on the **Memos** tab.

Memo icons: There are three memo templates. There is *no* message content and the templates are used for layouts. For help, click on the **Wizard** icon.

4) Click on the **Letters and Faxes** tab.

Business Letter icons: There are three business letter templates. (There are three fax templates also.) The samples reflect the standard layouts, and contain *no* message content. For help, click on the **Wizard** icon.

5) Click on the **Other Document** tab.

Resume icons: There are three resume samples that can be used as is or altered to suit your needs. These samples are designed for layout but they contain clear content indicators. For help click on the **Wizard** icon.

III. To Save a Modified Template

The template can be altered to suit your needs or to conform to the suggestions of your instructors, your supervisor, or the recommendations in *Basic Composition Skills.* These modifications can be saved as a *new* template.

1) Go to **File** on the menu bar.
2) Click on **Save As.**
3) In the Save-as-Type box, click on **Document Template,** which will take you to a template folder.
4) Important: Change the file name. Simply add a new date, for example (do *not* use the slash).
5) Click on **Save.**
6) Note the location where the template is being saved. It may save to different locations.

IV. To Save an Original Template

If you build your own title page or any other page layout, it can be used again and again. You can simply make a copy and modify it or you can make a *template* and modify it. *Be sure to save the first outline you develop and your first lab report. Save them as templates.*

1) Go to **File** on the menu bar.
2) Click on **Save As.**
3) In the Save-as-Type box, click on **Document Template,** which will take you to a template folder.
4) Important: Give the file a name.
5) Click on **Save.**
6) Note the location where the template is being saved. It may save to different locations.

V. To Open a Personal Template Folder for Your Templates

1) Go to **File** on the menu bar.
2) Click on **Save As.**
3) In the Save-as-Type box, click on **Document Template,** which will take you to a template folder.
4) Click on the **Create New Folder** icon on the menu bar in the Save As dialog box. (To locate the proper icon, rest the cursor on each icon until the label pops up.)

5) Type a name for the folder on the New Folder screen.

6) Click on **OK** and the folder will automatically open.

7) Important: Change the file name. Simply add a new date, for example (do *not* use the slash).

8) Click on **Save.**

VI. For Help at any Point

1) Click on **Help** in the menu bar.

2) Go to **Microsoft Word Help.**

3) Activate the help program with a click.

4) Type in a question where indicated and click on **Search.**

5) Select from the options above or use the Search button. Click on the **Option** or the **Search** button and a window will open.

Acknowledgments

There are four texts and three instructor's supplements in the *Wordworks*™ series. In order to properly acknowledge the many people and organizations who have contributed to this project, I have chosen to first extend my thanks to those who assisted the endeavor in a larger context. Many more contributors helped shape the separate titles.

The Wordworks™ Series

The *Wordworks*™ project would never have come about without the patience and generosity of my colleague and assistant Patricia Britz. There were 4000 pages of manuscript, endless keyboarding tasks, and elaborate page spreads. The project would have been impossible without Pat.

My colleague, Dr. Rita Smilkstein (published by Harcourt Brace), deserves very special thanks. Dr. Smilkstein has read every manuscript page of the project and contributed endless ideas and support.

To Stephen Helba, executive editor at Prentice Hall, I owe a special thanks. Some years ago I came home one day and found twenty-five pages of contracts spread all over the floor under an old fax machine. From the beginning Stephen took a personal interest in the *Wordworks*™ project and has been a source of encouragement for the many years it has taken to develop.

To Dr. David Mitchell, President of South Seattle Community College, I owe thanks for a special favor. When he was the former dean of my campus, I sought his help in finding a space where I could create a dedicated classroom for my program. His response was immediate. We surveyed a prospect and agreed to the experiment. He budgeted a fully provisioned room of tables and chairs and blackboards and other features I requested. That dedicated teaching and learning environment played a major role in developing the concepts embodied in the *Wordworks*™ series.

I would like to thank Marc Vassallo, Lee Anne White, Jeff Kolle, and other editors of the Taunton Press in Newtown, Connecticut. It was their faith in my ability to produce cover stories, feature articles, and shorter pieces for the nationally known Taunton magazines that give me the gumption to try to write *Wordworks*™. It was a great boost to realize that each publication had a circulation of several hundred thousand readers and that, all told,

I had reached several million readers thanks to the Taunton staff. I also discovered the excitement of working with publishing teams and writing copy that included photographic compositions and line art concepts.

I would like to thank a number of my colleagues for their technical advice: Steve Anderson (physics), Lynn Arnold (network technology), Dale Cook (HVAC), Fred Edelman (CAD), Tom Griffith (chemistry), John Hagans (computer technology), Ralph Jenne (mathematics), Chris Sandars and Dennis Schaffer (electronics). Many of the diagrams in the text were first constructed with CAD by Stan Nelson, who generously donated his time to the project.

It helps to have the support of a close friend when facing the misgivings involved in an enormous project. Rob Vinnedge, one of the Northwest's finest professional photographers, assisted me as I developed articles for the Taunton Press even when he was trying to meet his own book and magazine deadlines that were taking him as far away as Hong Kong. I regret that there was no time in our busy lives to share any shoots for *Wordworks*™, but the deadlines were too tight. What matters the most is that Rob said, again, again, and again, "You can do it."

Basic Composition Skills for Engineering Technicians and Technologists

I am indebted to many former students for their enthusiasm and their willingness to appear in the *Wordworks*™ series. Many contributors developed projects that in some way appear in the *Wordworks*™ textbooks or the *Wordworks*™ supplements. A number of multi-page models appear in *Basic Composition Skills* and *Technical Document Basics*. Other models are located in the instructor's supplements for the two texts. The following contributors provided the selections that appear as multi-page models.

Richard Bigham
Janine Boyer Richards
Andrew Cameron
David Campbell
Corin Carper
Sean Caughlan
Jane Chateaubraind
Kathryn Chumbley
Ed Condon
Jennifer Dillard
David Hilderbrand
Mike Kang
Allan Kellner
Ricky Keokitvon
Ansar Khalil
Eric Long
Dornie MacKenzie
Mike Meagher
Ben Minson
Janine Michelsons
Scott Schaper
Steve Schattenbild
Patricia Renderos
Troy Sewell
Dawn Shephard
David Stinson
Linda Strout
Michael Summers
Don Thornton
John Touliatos

Mark Eskridge
Edward Ferraro
Kenneth Tim Forman
David Hale
Damon Harrell
Paul Hefty
Joe Mitchell
Susan Mutuc McCants
Greg Novlan
Michael Pitcher
Carrie Pratt
Salim Rabaa
Lucy Underwood
Anthony VanNorman
Mark Vansteenkiste
Jerald Yun

In order to properly develop two of the texts, I needed a wide variety of single-page or half-page samples that appear throughout *Basic Composition Skills* and *Technical Document Basics* and the instructor's supplements.

As the books took shape, I found more and more material that was appropriate, and the list of contributors grew. Fortunately I had considerable success in contacting everyone involved, even though some of the people identified below are now in Texas, Arizona, California, and Oregon. To all of you, a special thanks.

James Affeld
Kayla Agan
Mitch Agan
Ben Andrews
Roman Ariri
Lynn Arnold
Erinn Barnett
Tyler Beam
Bob Bergstrom
John Bienick
Thomas Booze
Janine Boyer-Richards
Douglas Bradley
Dewi Cahyantari
David Campbell
Sean Caughlin
David Clemmons
Nadine Hamby
Brian Hanners
Art Hedley
Eric Hesselgesser
Mike Huckaby
Wayne Jarvimaki
Ernie Jean
Jeremiah Jester
Glen Johnson
Leonard Kannapell
Aaron King
Lawrence Kitchen
Steven Knobbs
Anna Komissarchik
Brian Kraft
Jay Voravong Kruise
Matt Laundroche
Peter Mikolajczyk
Kristy Moody
Jim Murray
Paul Nguyen
KimAnh Pham
Nick Pierce
Marianne Pinyuh
Michael Pitcher
Randy Rosen
Zijad Saric
Martin Saxer
Hassan Shirdavani
Robert Simpson
Mark Skullerud
Robert Smith
Greg Speers
Tim Stacey

Carol Collins
Wes Dang
Kevin Donnelly
Michael Elliston
Diana Eng
Eugene Escarez
M.D. Esclavon
Randy Farlow
Mari Fortes
Chris Fraser
Bruce Fugere
Nazim Haji
Ron Lewis
Clarence Lim
LeAnne Livingston
Eric Long
Paul Luke
David MacDonald
Dornie MacKenzie
Sherry Sloan Manning
Blair Marshall
Ted Marier
Christine McCurdy
Janine Michelson
Sandra McKay Stimson
Glenn Sudduth
Jamison Surquy
Earl Thompson
Tuan Tong
Elaine Tritt
Vu Quoc
Jeremy Watts
Chris Wiederhold
Karen Woodruff
Zuo Yan

Apart from the contributions of the many former students identified above, I owe a special note of thanks to Laura Lippai for documenting an important project she developed for Pacific AeroTech Inc. The case study was a challenge to write and Laura handled the chore with enthusiasm. I also thank Karen Borgnes, President of Pacific AeroTech.

Two shorter descriptions of writing tasks were developed by Todd Rendahl and Kayla Agan. My thanks to you both. Special thanks to Zetron Inc. for permission to use the documents that accompany Todd's discussion.

I would like to thank Andreas Brockhaus for acting as a resource on computer-related matters. Andreas developed the computer tips for the text and worked with me to make the appendix of templates as accurate as possible.

The following organizations allowed me to use photographs from their archives:

Dover Publications

The Liaison Agency, New York, NY

Northrop Grumman Corporation

Roger-Viollet, Paris, France

Heritage House, University of Wyoming

In addition, I thank the Eldec Corporation for permission to use a page of testing procedures from their files.

Several sample documents that are used as illustrations are adapted from documents that I located at North Seattle Community College. I would like to thank Dr. Kathleen Noble, former President, and Vince Offenback, Associate Dean, for permission to use the documents.

Prentice Hall/Pearson Education

Special thanks to the production staff at Prentice Hall in Columbus, Ohio, and the Clarinda Company. Two production teams saw to it that the *Wordworks*™ series would go to press. At Prentice Hall, I would like to acknowledge the following staff:

Editor in Chief: Stephen Helba

Associate Editor: Michelle Churma

Production Editor: Louise N. Sette

Design Coordinator: Robin G. Chukes

Production Manager: Brian Fox

Marketing Manager: Jimmy Stephens

At Clarinda I would like to thank additional personnel for seeing the *Wordworks*™ series through the production stages:

Manager of Publication Services: Cindy Miller

Account Manager: Jennifer Graham

For her patience and skills, a special thanks to Barbara Liguori for copyediting the *Wordworks*™ series.

I would also like to acknowledge the reviewers of this text:

Pamela S. Ecker, Cincinnati State Technical and Community College (OH)
Dr. Harold P. Erickson, Lake Superior College (MN)
Patricia Evenson, Northcentral Technical College (WI)
Anne Gervasi, North Lake College (TX)
Mary Francis Gibbons, Richland College (TX)
Charles F. Kemnitz, Penn College of Technology (PA)
Diane Minger, Cedar Valley College (TX)
Gerald Nix, San Juan College (NM)
M. Craig Sanders, Bellevue Community College (WA)
Laurie Shapiro, Miami-Dade Community College (FL)
Richard L. Steil, Southwest School of Electronics (TX)
David K. Vaughan, Air Force Institute of Technology (OH)

David W. Rigby

Index